The Environment and the Human Condition

*An interdisciplinary series edited by faculty
at the University of Illinois*

BOOKS IN THE SERIES

Wildlife and People:
The Human Dimensions of Wildlife Ecology
Gary G. Gray

Sustainable Agriculture in the American Midwest:
Lessons from the Past, Prospects for the Future
Edited by Gregory McIsaac and William R. Edwards

Green Nature/Human Nature:
The Meaning of Plants in Our Lives
Charles A. Lewis

Justice and the Earth:
Images for Our Planetary Survival
Eric T. Freyfogle

Agricultural Biotechnology and the Environment:
Science, Policy, and Social Issues
Sheldon Krimsky and Roger P. Wrubel

A Wider View of the Universe:
Henry Thoreau's Study of Nature
Robert Kuhn McGregor

Environmental Program Evaluation: A Primer
Edited by Gerrit J. Knaap and Tschangho John Kim

Environmental Program Evaluation

Environmental Program Evaluation
A PRIMER

Edited by Gerrit J. Knaap and Tschangho John Kim

Foreword by Ravinder K. Jain

UNIVERSITY OF ILLINOIS PRESS

URBANA AND CHICAGO

© 1998 by the Board of Trustees of the University of Illinois
Manufactured in the United States of America
C 5 4 3 2 1

This book is printed on acid-free paper.

Library of Congress Cataloging-in-Publication Data

Environmental program evaluation : a primer / edited by Gerrit J. Knaap
and Tschangho John Kim ; foreword by Ravinder K. Jain.
    p.   cm. — (The environment and the human condition)
Includes bibliographical references and index.
ISBN 0-252-02334-X (cloth : acid-free paper)
1. Environmental policy—United States—Evaluation.
2. Environmental indicators—United States.
3. United States Army—Environmental aspects.
I. Knaap, Gerrit J., 1956– .
II. Kim, Tschangho John.
III. Series.
GE180.E63   1998
363.7'05—dc21
97-4760
CIP

Contents

Foreword / RAVINDER K. JAIN   xi
Acknowledgments   xiii

INTRODUCTION: Environmental Program Evaluation: Framing the Subject, Identifying Issues / GERRIT J. KNAAP AND TSCHANGHO JOHN KIM   1
 Environmental Program Evaluation: Framing the Subject   3
 Methodological Issues in Environmental Program Evaluation   5
 Utilization of Evaluations   12
 The Chapters That Follow   14
 Concluding Comments   17

PART 1: ENVIRONMENTAL PROGRAM EVALUATION

1. Program Evaluation and Environmental Policy: The State of the Art / ROBERT F. RICH   23
 Historical Perspective   23
 Development of Evaluation Research as a Field   27
 The Scientific Approach as It Relates to Environmental Policy   31
 Conclusions   38

PART 2: INSTITUTIONS, POLITICS, AND PRACTICE

2. Evaluation of Environmental Programs: Limitations and Innovative Applications / LAWRENCE S. SOLOMON   45
 Evaluating the Environment: Why the Difficulty?   46
 The GAO's Approach to Environmental Evaluation   52
 Future Prospects for Environmental Program Evaluation   56

3. Why Institutions Matter in Program Evaluation: The Case of the EPA's Pollution Prevention Program / WALTER A. ROSENBAUM   61
 About the Program   62
 Evaluating the EPA's Program Implementation and Self-Assessment   63

The Brief, Fitful, but Significant History of Federal Pollution Prevention   68
The EPA's Program Implementation and Self-Assessment   72
Success by Acclamation: The EPA Evaluates Itself   79
Conclusions: Data Dilemmas, "Takeoffs," and Other Lessons to Be Learned   80

4. Environmental Program Evaluation in an Intergovernmental Context / WILLIAM R. MANGUN   86
EPA Oversight of State Environmental Programs   87
State Air Pollution Control Programs   90
State Water Pollution Control Programs   92
State Environmental Program Evaluation Initiatives   94
Conclusion   102
Appendix   104

PART 3: ENVIRONMENTAL SCIENCE AND ENVIRONMENTAL CHANGE

5. The Role of Chemical Indexes in Environmental Program Evaluation / ROGER A. MINEAR AND MARK A. NANNY   129
Evolution of Analytical Methods   130
Impact of Changing Analytical Capabilities on Environmental Programs   135
Role of Chemical Measurements   137
New Dimensions   140

6. Biological Integrity: A Long-Neglected Aspect of Environmental Program Evaluation / JAMES R. KARR   148
Why Has It Taken So Long?   150
How to Measure Biotic Integrity   158
The Future of Biological Monitoring   168

7. Ode to the Miners' Canary: The Search for Environmental Indicators / HALLETT J. HARRIS AND DENISE SCHEBERLE   176
The Evolution of Ecological Indicators in Science   177
Defining and Selecting Environmental Indicators   180
Environmental Indicators in the United States and Canada   182
Green Bay as a Case Study in Ecosystem Management   189
The Power of an Indicator   195

PART 4: ECONOMICS AND EVALUATION METHODOLOGY

8. Economic Approaches to Evaluating Environmental Programs /
JOHN B. BRADEN AND CHANG-GIL KIM   203
Economic Models for Environmental Program Evaluation   208
Economic Valuation of Program Impacts   215
Risk Analysis   225
Economic Incentives for Pollution Prevention   229
Conclusion   230

9. Evaluation of State Environmental Programs: Observations on Selected Cases / DAVID H. MOREAU   238
Environmental Values in Benefit-Cost Analysis: A Historical View   238
Sedimentation and Erosion Control Program in North Carolina   241
Water Supply Protection Program   245
Conclusions   251

10. Endogenous Risk and Environmental Program Evaluation /
THOMAS D. CROCKER AND JASON F. SHOGREN   255
Self-Protection and Environmental Risk   256
Risk Assessment and Risk Management   258
Valuation of Environmental Risk   260
Risk Abatement and Transferable Risks   263
Conclusion   266

11. Pollution Prevention Frontiers: A Data Envelopment Simulation / KINGSLEY E. HAYNES, SAMUEL RATICK, AND JAMES CUMMINGS-SAXTON   270
Approach   272
Production Functions and Efficiency Frontiers   274
Data Envelopment Analysis   277
Application   279
Feasibility Assessment   280
Conclusion   284
Appendix 1   286
Appendix 2   286

PART 5: ADMINISTRATION, OVERSIGHT, AND ASSESSMENT

12. Using Environmental Program Evaluation: Politics, Knowledge, and Policy Change / MICHAEL E. KRAFT   293
   The Logic and Limits of Environmental Program Evaluation   295
   The Purposes of Environmental Program Evaluations   303
   Utilization of Environmental Program Evaluation: Conceptual Issues   305
   Putting Environmental Program Evaluations to Work   308
   The Promise of Environmental Program Evaluation   315

13. Congressional Oversight and Program Evaluation: Substitutes or Synonyms? / GARY C. BRYNER   321
   Defining Oversight   323
   Oversight Processes and Tools   325
   Oversight of Environmental Regulation and Programs   328
   Criticisms of Congressional Oversight   330
   Improving Congressional Oversight of Regulatory Agencies   333
   Resources for Oversight   334
   Coordinating Oversight in Congress   335
   Congress and the Regulatory Review Process   337
   The Role of the Executive Branch   339
   The Prospects for Increased Program Evaluation in Congress   341

CONCLUSION: Environmental Program Evaluation: Promise and Prospects / GERRIT J. KNAAP AND TSCHANGHO JOHN KIM   347
   The State of the Art   348
   The Place of Practice   349
   The How Come of Outcomes   351
   Diminishing Marginal Productivity   353
   The Meaning of Use   356
   A Research Agenda   358

Contributors   361
Index   367

Foreword

RAVINDER K. JAIN

This book represents a collaboration among scholars in many disciplines and among policy analysts from academic and public agencies. In addition to the significant contributions made by the authors here it is important to acknowledge the influence of a symposium on environmental program evaluation sponsored by the Army Environmental Policy Institute in 1992. This symposium was the necessary springboard for the development of this book. Because the institute's mission is to assist the army secretariat in developing proactive policies and strategies to address environmental issues, sponsoring a scholarly exchange in this area was in the long-term interests of the army. Without this foundation, the road to implementation would be precarious, at best.

Every field has its own esoteric language, whether it is nascent, as in environmental program evaluation, or well established, as in unit processes related to treatment technologies. The intent of this book is to break through these language and conceptual barriers to provide the most relevant information needed to understand environmental program indicators and related issues from a practical viewpoint. It is expected that this research would provide a basis upon which agencies could begin to develop measures and programs to detect the extent and rate of environmental progress.

With the growth of environmental laws and regulations and their consequential enforcement, most public agencies, and the army in particular, have grappled with how to best answer the question, "How well are we doing?" This question is fundamental to the investment strategies, programming issues, and budgeting justifications used in the competition for limited agency resources. However, compliance cannot be the only evaluative aspect that agencies recognize. It has become apparent that the mere number of notices of violations is not a substantive measure of success or failure of an environmental program. Rather, it may reveal an enforcement agency's work force and budget resource priorities and environmental media specific priorities.

Some of the most knowledgeable people in the field have attempted to capture the salient points of environmental program evaluation in this well-organized and timely volume. On a personal note, I would like to add that, as former director of the Army Environmental Policy Institute, it was my pleasure and honor to work with and learn from many of the contributors to this book.

Acknowledgments

A collection of essays is by definition a collaborative work, the quality of which cannot far exceed the quality of individual essays. Our primary debt therefore goes to the contributing authors, who not only wrote and rewrote the chapters herein but also stimulated and reviewed each other's work. For their essays and perseverance we are truly grateful. Like most projects of this nature, the final product reflects the work of many whose names do not appear here. Such contributors include discussants at the San Francisco conference at which many of these essays, in earlier forms, were first presented and reviewers who provided critical and insightful comments. These include Marina Alberti, Stanley Auerbach, William Colglazier, T. R. Lakshmanan, Laura Langbein, Si Duk Lee, Dan Mazmanian, Robert Olshansky, Bruce Piasecki, Phillip Prisco, Daniel Schnieder, John Stoll, Ned Stowe, and three anonymous reviewers for the University of Illinois Press. Elizabeth Mericke, then a graduate student at the University of Illinois at Urbana-Champaign, performed the thankless tasks of organizing the conference and assuring that something useful came back to Champaign. Our gratitude also extends to many others in Champaign-Urbana who helped to turn conference papers into readable chapters. These include Linda Butarian, Sheri Britten, and Elizabeth Dennison, former staff members of the Department of Urban and Regional Planning, and Karen Hewitt of the University of Illinois Press. Finally, we want to express our special gratitude to John Fittipaldi, senior fellow, Army Environmental Policy Institute. John had the insight to identify environmental program evaluation as an area of critical importance to the army and other public and private organizations and the determination to leverage the expertise of leading scholars in closely related fields. Without John's foresight and intellectual partnership, this work could never have been completed.

Environmental Program Evaluation

INTRODUCTION

Environmental Program Evaluation: Framing the Subject, Identifying Issues

GERRIT J. KNAAP AND TSCHANGHO JOHN KIM

When the environmental movement first swept the United States in the 1960s and early 1970s, there was widespread optimism that the adoption of government programs to protect and enhance environmental quality would soon solve the nation's environmental problems. Some thirty years and billions of dollars later, most of that optimism is gone. There is little doubt that significant progress has been made; fish have returned to many rivers, lakes, and streams; ambient lead concentrations have been significantly reduced; and toxic wastes are now closely monitored and controlled. There is considerable doubt, however, that the environment has been protected in the best possible manner (Portney 1990; National Academy of Public Administration 1995). According to the *New York Times* (March 21–24, 1993), the vanguard of a new environmentalism emerged in the 1990s. Proponents of this new environmentalism view existing and proposed environmental programs with increasing skepticism and insist that these programs and policies be driven by sound environmental *and* economic decision making.

The influence of this so-called pocketbook environmentalism at the federal level has resulted in a growing demand for economic evaluations of environmental programs. In 1980, the Supreme Court ruled that proposed environmental health standards could be invalidated if the standards were not justified by a quantitative risk assessment. In 1981, Reagan Executive Order 12291 required all federal agencies to submit regulatory impact analyses to the Office of Management and Budget to assure that the benefits of regulatory actions outweigh their costs. President Bush retained EO 12292 and established the Council on Competitiveness, headed by Vice President Quayle. Under Quayle's leadership, the council's charge was to review federal regulations to minimize economic burdens on business. Environmental regulations were a favorite target. Soon after taking office, President Clinton terminated the Council on Competitiveness and issued Executive Order 12866 on regulatory planning and review. EO 12866 retained executive oversight of the regulatory process but invites public participation and returns considerable responsibility to the agencies (Kraft 1996). After the 1992 midterm elections, scrutiny of environmental programs and regulations shifted from the executive office to the Republican-led congress. In 1995, the House of Representatives passed the Risk Assessment and Cost Benefit Act, which requires that no environmental, health, or safety rule be adopted by a federal agency unless the incremental risk reduction or other benefits "will be likely to justify, and be reasonably related to, the incremental costs incurred." The act failed to pass the Senate, but similar language will likely be proposed as existing acts come up for renewal.

Although the thrust of this new environmentalism is neither pro-environment nor pro-industry in principle, the clamor for evaluation comes most vociferously from industry and local governments seeking regulatory relief. As a result, the practice of environmental program evaluation became increasingly ex ante (conducted before the implementation of a regulation or program), uniscaler (measured in dollars), and conducted for a single purpose (to determine whether environmental programs or regulations should or should not be adopted). Environmentalists have opposed such forms of evaluation because, they argue, costs are immediate and easily measured while benefits (such as risk reduction) are long-term and difficult to measure in dollar terms. Further, environmentalists argue, because environmental program evaluation is fraught with uncertainty regarding the behavior of individuals and institutions, the responsiveness of natural environments to anthropogenic influence, and the costs and benefits of environmental change, environmental programs cannot be evaluated, except in the most general terms.

This debate over economic approaches to environmental program evaluation, unfortunately, presents a Hobson's choice: suffer inefficient and perhaps ineffective environmental programs or risk forsaking thirty years of environmental improvement. There is, however, another choice. Current and proposed environmental programs can be critically evaluated not solely to identify program failure or success but to improve these programs. Improvement requires information on how programs are conducted, how they affect the natural environment, and how evaluations can be used to facilitate programmatic change. As stated by Robert Bartlett, "Clearly desirable are multiple evaluations, done with a keen appreciation of the strengths and limitations of each approach and a frank recognition of the advantages of others" (1994:183).

Despite this need for multiple forms of environmental program evaluations, there are no obvious places to find them. No existing journal provides a forum for exchange on the subject, and a comprehensive computer search of the literature using the key words *environmental program evaluation* yields almost no citations. Relevant information is available because environmental programs have been evaluated by economists, political scientists, public administrators, sociologists, natural scientists, and others working in the public, private, and not-for-profit sectors; but the information is not organized or referenced in a manner that is readily accessible, while the need to access the information continues to rise.

Our purpose in preparing this volume is to provide students, scholars, and practitioners with multidisciplinary views on environmental program evaluation. Our goals are both modest and ambitious. Our modest goal is to contribute to knowledge on environmental program evaluation. We do not claim to provide a comprehensive survey of relevant literature or to identify universally appropriate methods of evaluation. We claim only that the chapters in this book offer an initial assessment of the state of the art, identify unresolved issues, and suggest ideas for evaluating environmental programs. Our more ambitious goal is to help establish the field of environmental program evaluation. It is our hope that this volume will stimulate future efforts to analyze, critique, and advance the work presented here. Such efforts, we hope, will lead to more effective environmental program evaluations and, ultimately, to more effective environmental programs.

The time for thinking about environmental program evaluation could not be better. Public concern for the environment remains genuine and pervasive, but rising demands for tax relief and less government intervention all but preclude new environmental programs that involve major public expenditures or federal regulations. Further, the newest pressing issues in environmental policy—biodiversity, international cooperation, environmental justice, and environmental restoration—will be difficult to address with existing environmental programs. If we are to make progress on addressing environmental issues we must find better methods of assessing the strengths and weaknesses of existing environmental programs, identifying what new types of programs are most sorely needed, and conveying the information effectively to environmental policymakers. These are the challenges of environmental program evaluation.

Environmental Program Evaluation: Framing the Subject

According to William R. Shadish, Thomas D. Cook, and Laura C. Leviton (1991) program evaluation is like medicine or engineering: pragmatic, instrumental, and conducted for the purpose of programmatic improvement. Environmental programs "usually have titles or labels and readily identifiable boundaries; typically, they are formally created by executives or legislators, have defined goals and are administered by a single agency or a relatively small group of agencies" (Bartlett 1994:172). While little has been written on environmental program evaluation, program evaluation as a research protocol dates back to 1923, when the American Association of Social Workers established a subcommittee on evaluation (see Rossi and Freeman 1993; Shadish, Cook, and Leviton 1991; and Rich in this volume for more detailed historical and literature reviews). Program evaluation emerged as a field of research in the 1960s, following the adoption of Great Society programs. As a result, program evaluations are common in the fields of education, health care, welfare, and criminal justice and in local government (Rossi and Freeman 1993).

Even where program evaluation is common, however, it has no firm or consistently applied definition. Leonard Rutman and George Mowbray define program evaluation as "the use of scientific methods to measure the implementation and outcomes of programs, for decision making purposes" (1982:12). Program evaluation is inherently instrumental; it is not concerned with value formulation, the policy-making process, or institutional change. Program evaluations are performed for many instrumental reasons: to foster more efficient program management, to help decide whether to expand or discontinue a program, or to compare the efficacy of one program to another (Rossi and Freeman 1993). In general, evaluations are conducted to improve the performance of an existing program. Despite their widespread use in other policy domains, however, evaluations of environmental programs are relatively rare.

The lack of systematic evaluations of environmental programs has multiple explanations. One is timing. Most environmental programs are relatively new and have only recently reached the stage at which evaluation is appropriate. Another is the difficulty of articulating well-defined goals. Whereas improvements in reading ability, reductions in crime, increases in life expectancy, and lower rates of poverty are relatively easy to express in quantitative terms, improvements in environmental quality, species diversity, ecosystem health, and ecological sustainability are much more nebulous concepts and are difficult to articulate as clearly defined and measurable goals. A related reason is complexity. Although social systems are complex it is far more difficult to isolate the causes of global warming than the causes of adult illiteracy. Finally, one explanation lies in the disciplinary divisions between the natural and social sciences. This division exacerbates the difficulty of reaching consensus on appropriate evaluation methods.

Methodological Issues in Environmental Program Evaluation

Methods in program evaluation are perhaps as numerous as the contexts in which they are applied. Categories of evaluation methods include qualitative designs, statistical designs, benefit-cost designs, participatory designs, formative designs, and summative designs, to name just a few (Hellstern 1986). While all the above labels have some ordering utility, methods for evaluating environmental programs can be grouped into three categories of focus: process, impacts, and efficiency. Process evaluations address program implementation; impact evaluations address program outcomes; and efficiency evaluations address program benefits and costs.[1]

Evaluations That Focus on Process

Process evaluations limit their purview to how activities are carried out. As such, evaluations of this type address the activities of individuals and organizations involved in program implementation. Process analyses are critical for program improvement. Without attention to process it is perhaps possible to assess *whether* a program has had impact and perhaps *whether* the value of program impacts exceeds program costs, but it is impossible to identify *why* the program had or did not have impact and *why* benefits did or did not exceed costs.

Process evaluations are often described as analyses of program implementation. According to Jeffrey L. Pressman and Aaron Wildavsky, the pioneers of implementation research, implementation means "to carry out, accomplish, fulfill, produce, complete" (1973:xiii). In essence, implementation is what happens between program initiation and program results. Analyzing implementation requires the analyst to confront three issues: what activities should be analyzed, how to conduct the analysis, and how to assess implementation success.

Implementation success is often judged against the stated goals of the program under evaluation. According to Daniel Mazmanian and Paul Sabatier, "this is to be expected, given the historical preoccupation of much implementation research with assessing the extent of goal attainment and analyzing the reasons for the inability of major policy initiatives to actually attain their stated objectives" (1989:10). The stated goals of many programs, however, are often poorly defined, can potentially conflict, and are sometimes misleading. Other times, a narrow focus on stated goals can cause the evaluator to miss important effects. Bartlett (1994) notes that the Superfund program may have cleaned up few

Superfund sites, but in assigning responsibility for clean-up costs, the program may have fostered pollution prevention and slowed the contamination of new sites. Academic researchers often choose their own measures of success, such as the impact of the program on the poor (Mazmanian and Sabatier 1989); the legitimacy, responsiveness, or fairness of the implementation process (Bartlett 1994); or economic efficiency. Professional evaluators often consider other issues as well, but progress toward stated goals is one issue a professional evaluator should be prepared to address.

Early researchers studying implementation (subsequently described by Sabatier [1986] as taking a "top-down" approach) suggested that it depends on three categories of variables: the tractability of the problem, the ability of the program to structure the implementation process, and the sociopolitical environment in the which the program is placed (Mazmanian and Sabatier 1989). Reflecting a top-down approach, analysts of social programs, for example, often use qualitative sources of information, such as open-ended interviews, direct observation of program activities, written questionnaires, and program records. From these sources data can be quantified and compiled using content analysis (Patton 1987). These views were challenged, however, by European scholars who offered a bottom-up approach (Mazmanian and Sabatier 1989). From this approach, implementation depends on the values, strategies, and interactions of "street-level bureaucrats" and those affected by the program under evaluation (the program clientele). Researchers with this view gather more of their information from program clientele than from program administrators. Since the late 1980s a synthesis has developed that recognizes the potential of both top-down and bottom-up variables.

Unfortunately, this synthesis does little to establish an intersubjective methodology for implementation research or to establish whether variables that characterize program structure and administration are more influential than variables that characterize program targets or street-level bureaucrats in determining program success. Further, these are precisely the salient issues in U.S. environmental policy today. Is the Environmental Protection Agency a better environmental manager than the Department of Energy or the Interior? Are programs more effective when administered by state rather than federal agencies? Do economic incentives work better than regulations? These remain hotly debated issues that probably contain few universal answers (see Mann 1982 and Downing and Hanf 1983 for early work on these issues; see National Academy of Public Administration 1994 and 1995 for recent reviews). Precise answers based on more than ideological foundations can be found only through sound research on environmental program implementation.

Evaluations That Focus on Environmental Impacts

Examination of the impacts of humans on the natural environment is as old as the notion of science itself. Modern analytical methods that closely linked government programs with environmental impacts, however, came only after the passage of the National Environmental Policy Act in 1969, which required all federal agencies to assess programs and policies that have significant impacts on the environment. This act spawned a large body of literature on environmental impact assessment. Environmental impact assessments, however, are generally conducted for programs and activities not specifically directed toward environmental improvement or preservation—such as highway, hydroelectric dam, and government building projects. Furthermore, they are predictive in orientation; they are designed to project the impacts of proposed policies or programs, not the impacts of those already in place. In addition, environmental impact assessments generally focus on the negative effects of programs on the environment and typically do not consider whether the program might be able to affect the environment favorably or in a more cost-effective or desirable manner.

Environmental program evaluations that focus on impacts attempt to measure changes caused by a program on the natural environment. Like process analyses, impact analyses require the evaluator to address three issues: what environmental variables should be measured, how environmental variables should be measured, and how environmental changes should be judged.

The issue of what variables to measure must be based on a conceptual model of how the program might affect the environment. An analyst might expect, for example, that treatment of municipal waste (a procedural change) results in a lower level of dissolved oxygen (environmental impact) in a waste-receptor stream. Dissolved oxygen, then, is a candidate for use as an environmental indicator. Continuing the example further, an evaluator might expect that sludge deposition in a poorly designed solid waste facility results in groundwater contamination. Thus contaminants in groundwater might then also serve as environmental indicators. Because environmental programs can have multiple effects, as the above example illustrates, the choice of environmental indexes can be complex. This complexity has given rise to a growing body of literature on environmental indicators.

Environmental indicators are a key component of environmental program evaluation; they provide necessary data for measuring the effectiveness of environmental programs. For this reason, indicators are receiving increasing attention by government agencies interested in the state of the environment and the efficacy of environmental programs. For example, in 1991 the EPA began to use environmental indicators to evaluate the success of its programs (U.S. EPA 1991). The selection of appropriate indicators and how to interpret changes in such indicators, however, remain controversial issues. As the debates over acid rain and global warming illustrate, reputable scientists still disagree on how to measure and interpret environmental changes (Levin 1992; Russell 1992).

The choice of indicators for evaluating environmental programs will vary, of course, with the environmental medium the program is expected to influence. The EPA, for example, developed the following categories for its Environmental Monitoring and Assessment Program: near-coastal waters, inland surface waters, wetlands, forests, arid lands, and agroecosystems. The choice of environmental indicators can also be categorized by disciplinary orientation. For assessing environmental change in Great Lakes ecosystems, for example, the Council of Great Lakes Research Managers (1991) identified the following types of indicators: physicochemical indicators, biological indicators, and socioeconomic indicators.

Physicochemical indicators measure physical or chemical changes in the environment (see Minear and Nanny in this volume). Examples of physical indicators for use in aquatic ecosystems include sedimentation rates, water temperature, water levels, water turbidity, and habitat assessments (both aquatic and terrestrial). Chemical indicators of water quality include dissolved oxygen, phosphorus, nutrients, and biochemical oxygen demand. Unfortunately, it is rarely feasible to monitor all chemical constituents in a system given cost and technological constraints (Council of Great Lakes Research Managers 1991). Choices must therefore be made concerning what indicators are most appropriate for each environmental program (see Harris and Scheberle in this volume).

Biological indicators of ecosystem health (and a program's effectiveness) can be considered in two classes: measurements that characterize the state of individuals or populations and measurements that characterize the state of community or ecosystem structure and function (Council of Great Lakes Research Managers 1991:16). The rationale for measuring individual and population health is that they provide indications of environmental conditions that physicochemical indicators cannot capture (see Karr in this volume). Biological indicators may, for example, indicate exposure to a lethal or debilitating combination of physicochemical stresses. Biological indicators include growth rates, carcinogenesis, behavioral effects, morphological changes, production or yield (fauna or flora), natality, mortality, population size and age structures, and range of species populations (Council of Great Lakes Research Managers 1991). An example of a controversial biological indicator is the northern spotted owl, which serves as an indicator of ecosystem health in the northwestern forests and as a measure of the effectiveness of the Endangered Species Act.

Measurements of ecological community health focus on all species present in a given habitat, where community refers to those species of particular interest to the observer. Measures of community structure include the number of species, the relative abundance or dominance of particular species, biomass, and food web (trophic) structure. Ecosystem *functional* processes are important indicators of ecosystem stability and can be useful indicators of environmental program success. Levels of productivity or decomposition are examples of this parameter.

Although difficult to translate into measures of environmental program effectiveness, socioeconomic measures help establish linkages between human activity and ecosystem/human health. It is difficult to detect long-term effects of environmental degradation on human health, but according to the Council of Great Lakes Research Managers, "epidemiological studies of exposed human populations provide the most convincing evidence of human health effects" (1991:30). The skin cancer rate in humans resulting from ultraviolet radiation serves as one example. Other common socioeconomic indicators of environmental health include infant mortality and morbidity rates, human life expectancy, and total health care expenditures.

After determining which variables to measure, a program evaluator who focuses on environmental impacts must identify the extent to which the variables have been affected by the program under evaluation. The evaluation literature identifies several types of impact evaluation designs: experimental, quasi-experimental, nonexperimental, cross-section, and time series (Mohr 1992). The first three differ in their use of control groups and other statistical requirements for quantitative analysis. Time series designs are considered a quasi-experimental design since they employ comparison groups that are not always identical. Non-experimental designs employ statistical analysis of continuous and categorical data without the use of control groups (Langbein 1980). Each design can be used at various points along the "outcome" line to assess, for example, the impacts of regulation of effluent emissions, the effects of effluent emissions on biochemical oxygen demand (BOD) levels, and the effects of BOD levels on fish abundance.

Although quite simple in concept, assessing the impacts of environmental programs has proven quite complex in practice. No two ecosystems are alike. Any given ecosystem can be affected by numerous climatic, seasonal, anthropological, and unknown factors—influences that are difficult to isolate from program effects. Data availability is typically poor. In assessing the impacts of the Clean Water Act in the Delaware River watershed, for example, Ruth Patrick states:

> Unfortunately, when we examined the available estuary and riverine data we found many defects. For example, the only available chemical data over the study period had been amassed by STORET, from a variety of sources. Different lab-

oratories had used different methods and instrumentation, their limits of determination had been different, and the accuracy of the endpoints had been affected by the types of instrumentation used. Because of this variability, it was decided that the best way to handle the data was to obtain a mean for each year and then a mean for selected groups of years. (1992:115)

Faced with these data limitations, Patrick and her colleagues were able to determine that in the Delaware study area BOD loadings and ammonia concentrations had decreased, toxic metals in the water column remained a problem, and considerable heavy metals lay in the bottom sediments. They were not, however, able to attribute any of these changes to specific programs or activities using statistical methods.

The final issue in program evaluations that focus on impacts concerns how to judge measured impacts. Once again, there are two options: a positive approach and a normative approach. Evaluations that take a positive approach are limited to measuring impacts. An evaluator might, for example, determine that a municipal treatment program caused a 20 percent increase in dissolved oxygen at a particular point in a waste-receptor stream. The positive evaluator would offer no assessment as to whether the increase was good or bad. Alternatively, an evaluator using normative standards might assert that a 20 percent increase in dissolved oxygen is a desirable but inadequate change.

For some environmental indicators, normative standards—at least for evaluating the direction of programmatic impacts—are easy to determine and widely accepted. An increase in dissolved oxygen in a waste-receptor stream can generally be viewed as an environmental improvement. But given uncertainty about "natural" environmental processes, and the difficulty of conducting controlled experiments, it is much more difficult to judge a change in species diversity, agricultural productivity, or regional climate. This complicates the task of monitoring progress on issues such as biodiversity, sustainable development, and global warming. More often than not, judgment of changes in environmental indexes are strongly influenced by the social and political context in which such judgments are made (Alberti and Parker 1991).

Many questions remain, therefore, for impact analysts of environmental programs. Are there indicators that provide meaningful measures of environmental quality? Is it possible to identify program impacts using these indexes? Can the identification of such impacts lead to better environmental policies?

Evaluations That Focus on Efficiency

Whereas program evaluations that focus on process and on impacts can be either positive or normative, program evaluations that focus on efficiency are explicitly normative. The efficiency question is essentially this: Does the environmental program (in the way it facilitates procedural change to influence the natural environment) constitute the best use of scarce resources? Most often, judgments concerning which program is best are established by economic definitions of costs and benefits; other times judgments are made on relative risks (see Braden and Kim in this volume). Despite growing mandates for their use by the federal government, economic evaluations of environmental programs remain controversial.

Objections to the use of economic approaches to environmental program evaluation are both technical and ethical. Mark Sagoff (1988, 1993), for example, offers a four-part ethical critique of economic analyses of environmental issues. First, he argues, resource allocation based on benefit-cost analysis represents a substitution of centralized planning for free markets. Second, benefit-cost analyses constitute an attempt to reflect individual preferences, which have been shown to be inconsistent, reversible, and intransitive, and thus inestimable. Third, benefit-cost analysis represents an attempt to satisfy (poorly defined) preferences, when what should be satisfied is the content and quality of individual lives.[2] Finally, benefit-cost analysis is grounded in microeconomic principles, when macroeconomic issues such as poverty, unemployment, interest rates, and savings are more important considerations. Most environmentalists view the issue in much more simple terms; they feel that environmental resources and all life forms cannot be measured in dollar terms.

Economists respond to these criticisms by pointing out that environmental programs involve the use of scarce resources, and in evaluating the use of scarce resources it is only reasonable to compare the value of what the program provides with the value of what is forsaken. The technical difficulties in conducting such comparisons, however, are formidable. Economic analyses of environmental programs begin with all the measurement difficulties that arise when attempting to evaluate program procedures and impacts. The economic analyst must know how the program induced procedural changes and what effects such changes had (or will have) on the natural environment. Only then is it possible to evaluate if the natural environment was changed in the most cost-effective fashion (see Moreau in this volume). Further, a complete efficiency analysis must compare the costs and benefits of the program under evaluation with the costs and benefits of all reasonable alternative programs—including the alternative of no program at all. Only under such a comparison is it possible to ascertain if the program is truly the best possible.

Estimating the costs of administering and complying with environmental programs pose no unusual difficulties, except perhaps that compliance costs must often be estimated using information from the complying industry. Estimating benefits, however, presents a host of difficulties. Air for breathing, water for recreation, vistas for viewing, and ecological systems are not traded in markets; their value, therefore, can only be estimated using indirect methods (see Braden and Kim in this volume).

In light of the difficulty of conducting comprehensive efficiency analyses, evaluations that focus on efficiency often adopt a more limited approach. Such approaches typically hold certain elements of a program constant and consider the effects of altering other elements of the program. Cost-effectiveness analyses, for example, evaluate the costs of alternative means of achieving the same outcome. Productivity analyses evaluate the productivity of alternative means of using the same resources (See Haynes, Ratick, and Cummings-Saxton in this volume).

Despite continuing debate, benefit-cost, comparative risk, and other forms of economic assessments will likely continue to be used to evaluate environmental programs. According to a benefit-cost analysis of Reagan Executive Order 12291, Paul R. Portney estimates that "Regulatory Impact Analysis will cover its cost, if not return it several times over" (1984:238). And according to a 1987 internal assessment by the EPA, benefit-cost analyses were found to be "increasingly useful tools in helping to provide the balance required in complex regulatory decisions" (1987:7-1).[3] Further, although the Risk Assessment and Cost Benefit Act of 1995 did not pass the 104th Congress, the passage of such an act remains a high priority of the new congressional leadership.

Utilization of Evaluations

Although the measurement of impacts and the application of appropriate research methods are important, how the results of an evaluation are utilized to affect programs or policies can change. It is not enough simply to design and implement a sound evaluation; to be influential it must mesh with the policy-making process. The purpose of most environmental evaluations is to determine whether to initiate or perpetuate the program: a favorable evaluation and a program is initiated; an unfavorable review and the program is terminated. If environmental programs were large in number, easy to evaluate, and costless to establish and dismantle, then only the best and most productive environmental programs would survive such an evaluation process. Because environmental programs are none of these things, both bad and good programs survive, and the bad programs are unlikely to become good ones. Using program evaluations for something other than a Darwinian process, however, presents another difficult problem.

As mentioned earlier, program evaluations can be used for multiple purposes.

These purposes range from the political (to further particular interests) to the scientific (to pursue truth). All evaluations, however, even the most objectively designed and conducted, will not be viewed as neutral assessments by the stakeholders of the program. Political conflict comes with the territory.

Impediments to Constructive Utilization

There are several impediments to the constructive use of evaluation findings. The first is a combination of methodological detractions (study design, appropriateness, and so forth) and a lack of utility of results. Michael Quinn Patton suggests that cumbersome, methodologically rigorous evaluations are falling out of favor because of their cost and time-consuming nature and notes that "traditional social science notions of rigor can limit the capability of both internal and external evaluators of thinking in practical, cost-effective ways about what is really useful" (1988:86). As a result, he suggests incorporating evaluation and monitoring into program delivery, employing simple, quasi-experimental designs (see Bryner in this volume).

A second impediment is often introduced by the evaluator. Eleanor Chelimsky (1987) argues that the evaluator shoulders much of the responsibility for ensuring that evaluation findings are integrated with policy formation. Often this fails to occur because evaluators do not understand the political domain of agency operations and are therefore not fully capable of performing and promoting effective evaluations. Evaluators must therefore assume responsibility for understanding and coping with political realities in their work (Mowbray 1988; see also Kraft in this volume).

Evaluators may also impede utilization by failing to present results appropriately or by presenting results that are not relevant to the policy deliberation. In the first case, the results are presented in a visually confusing manner or the writing is too technical. In the second case, the results are not useful because the study is too abstract, not relevant to the policy questions, or not timely. A client- or decision-maker orientation is the responsibility of the evaluator and must be assumed prior to commencing assessments (Patton 1978; Chelimsky 1987; Smith 1988).

Evaluators may also be uncomfortable or unwilling to promote evaluation results, feeling that this is beyond the scope of their responsibility or that they are not familiar with promotional activities. Often, however, such promotion can mean the difference between use and nonuse of results, whether the evaluators are internal or external to an agency. William P. Johnston Jr. (1988) responds to the lack of promotion of evaluation findings in describing the high acceptance of General Accounting Office (GAO) recommendations. As part of the GAO model, promotional activities after evaluations are completed account for much of the acceptance rate of between 51 and 77 percent. Though controversial from the standpoint of serious policy or program problems, one reason for this success is that the GAO pushes for incremental changes, such as behavior of organization staff and organizational structure, rather than global changes in policy approach or agency values. Furthermore, the GAO actively pushes Congress and agency officials to accept its recommendations, even tracking the utilization of results long after evaluations are completed (Johnston 1988; see also Solomon in this volume).

A third category of impediments is organizational or systemic barriers (Moran 1987; Patton 1988; Smith 1988). The first impediment in this category is the separation of the evaluation function from the policy-making function in government (see Bryner in this volume). This problem becomes acute when evaluators and policymakers do not understand each other's roles. Patton (1988) responds to criticisms of lack of cost-effectiveness of evaluations by providing a framework for program staff to integrate evaluations directly into program delivery. T. Kenneth Moran (1987) offers an information-based strategy for coupling evaluation research and decision making. This model provides continuous information for monitoring of a community service program, thus making the evaluation ongoing and offering needed information for administrative decision making.

Inevitably, organizational politics will partly determine the extent to which evaluation results are utilized. Such politics can generate support or opposition to evaluation results. Personnel conflicts may diminish the importance of evaluation findings to policy questions. As a response, M. F. Smith (1988) suggests that evaluators must ensure that evaluations are applicable to all levels of management. This facilitates broad support for evaluations within organizations and reduces the likelihood that evaluation reports will go unused. Flexibility with and adaptation to politically influential officials are other keys to acceptance of evaluation recommendations.

The Chapters That Follow

This brief introduction to environmental program evaluation suggests that economic analysis is a common, though controversial, form of environmental program evaluation used primarily to determine whether to initiate a new environmental program. It also suggests that analyses of processes and outcomes, which have been used for programmatic improvement or redesign in many areas of social policy, are far less common in environmental policy. The chapters that follow, therefore, begin to bridge the gap between the established field of program evaluation and the burgeoning field of environmental policy.

The chapters were written by authors with various disciplinary backgrounds and were presented as papers in a conference held in San Francisco in November 1992. Participants of the conference were given an early version of this introduction and asked to address issues in program implementation, environmental impacts, economic efficiency, or use. The intent was to explore the potential of various forms of evaluation and to consider multiple opportunities of the use of such evaluations. The chapters are organized into four sections followed by closing comments by the editors.

In part 1 Robert Rich reviews the history of program evaluation and assesses the state of the art. In so doing, he illustrates how program evaluation as a field reflects the influence of social forces and describes the field as moving to a middle ground where scientific methods are tempered with the practical concerns of the program administrator. This middle ground, he suggests, is the appropriate foundation for evaluating programs that focus on the environment.

Part 2 contains three chapters that address program implementation. Lawrence Solomon describes difficulties encountered by the U.S. General Accounting Office when evaluating environmental programs. Solomon illustrates these problems with specific examples of GAO efforts to evaluate groundwater, hazardous waste, and toxic waste programs. In response to data limitations, the GAO focused its efforts on evaluating the quality of environmental data. The subsequent chapter by Walter Rosenbaum presents a less optimistic view of data solutions to problems in evaluating federal environmental programs. Taking an institutional perspective, Rosenbaum suggests that problems in program evaluation reflect organizational values and loyalties as well as limitations in data. In reviewing the EPA's self-evaluation of its pollution prevention program, Rosenbaum offers several specific examples of how institutional factors influenced the process of evaluating environmental programs. Building on the themes introduced by Rosenbaum, William Mangun examines state responses to federal environmental initiatives and mandates. In reviewing air and water quality programs in several states, Mangun demonstrates how states vary in how they respond to federal initiatives and in how they evaluate their own environmental programs.

Part 3 contains three chapters that adopt chemical, biological, and ecosystem approaches to assessing environmental impacts. In the first of these chapters, Roger A. Minear and Mark A. Nanny reflect on the use of chemical indexes for environmental monitoring and measurement. According to Minear and Nanny, many aspects of environmental programs reflect new developments in analytical chemistry. Minear and Nanny review several key historical developments and examine the role of chemical indicators in environmental program evaluation. In the next chapter James R. Karr introduces the index of biotic integrity, which he offers as a broad-based and flexible tool to measure changes in environmental quality. According to Karr, monitoring biological communities through the use of such indexes is a promising and essential approach to evaluating environmental programs and protecting water resources. Next Hallett J. Harris and Denise Scheberle describe how environmental indicators were used to assess environmental program successes in Green Bay in Lake Michigan. Harris and Scheberle illustrate how indicators in Green Bay combine chemical and biological variables and reflect the influence of natural science and social science factors.

Part 4 contains four chapters on economic approaches to environmental program evaluation. John Braden and Chang-Gil Kim begin with an overview of major developments in environmental economics. Braden and Kim present the foundations of benefit-cost and risk analysis and describe methods used by economists to estimate the value of environmental improvement. They suggest that economic analyses and economic incentives can play a major role in the assessment and implementation of environmental programs. In the following chapter David H. Moreau describes how the State of North Carolina attempted to evaluate its erosion and sedimentation control program and its water supply protection program. Using a benefit-cost approach, Moreau illustrates both the difficulties of evaluating state-mandated but locally implemented environmental programs and the limitations of benefit-cost analysis as a method of environmental program evaluation. Further limitations of economic analysis are demonstrated by Thomas D. Crocker and Jason F. Shogren. These economists critique the risk assessment and management approaches taken by the EPA and other federal agencies to prioritize environmental programs. Typical risk assessments, according to Crocker and Shogren, ignore the distinction between endogenous and exogenous risk—that is, the distinction between risks that individuals can control and those they cannot. The failure to account for this distinction results in inappropriate rankings of environmental priorities and misguided risk management. A more appropriate risk management strategy, they argue, requires a simultaneous physical, biological, and economic approach to risk assessment. In the final chapter of Part 4 Kingsley E. Haynes, Samuel Ratick, and James Cummings-Saxton return to issues raised by Solomon and Rosenbaum: how to evaluate the performance of the EPA's pollution prevention program. Haynes, Ratick, and Cummings-Saxton offer a new measure of pollution prevention based on data envelope analysis, a method that calculates the relative efficiency of a plant by comparing the input the plant uses to the output it produces. With a simulation of the chemical industry, Haynes, Ratick, and

Cummings-Saxton demonstrate how the technique can be used to identify leaders in pollution prevention and to establish appropriate pollution prevention standards.

Section 5 contains two chapters that address the use of environmental program evaluation. Michael E. Kraft explores the use of program evaluations by chief executives and agency administrators by identifying the conditions that limit or promote the use of program evaluations. Kraft then applies these general findings to environmental program evaluations. Based on the available evidence, Kraft offers suggestions for improved utilization. Gary C. Bryner then focuses on the use of the environmental program evaluations by policymakers—that is, by members of the U.S. and state legislatures. Bryner focuses primarily on the role of congressional oversight as a means of assessing and reauthorizing existing environmental laws and in enacting new laws. Specifically, Bryner examines the goals of oversight, the mechanisms involved, the criticism of oversight, and how Congress might, through oversight, use program evaluations for formulating environmental policy.

In the conclusion we reflect on the material in the previous chapters. Based on these reflections, we offer an agenda for advancing the study and practice of environmental program evaluation.

Concluding Comments

As we stated at the beginning, our purpose in preparing this collection was to stimulate thinking on the topic of environmental program evaluation. This purpose reflects our belief that environmental program evaluation can and should be used to improve existing programs and to design better new ones. That said, we do not endorse pocketbook environmentalism in its most narrow form. Environmental programs should yield benefits that exceed costs; and in today's political climate, any discussion of environmental program evaluation must carefully consider the contributions of economic science. But a meaningful discussion must also consider the vagaries of program implementation, the difficulties in measuring and interpreting environmental change, the limitations of economic analysis, and the multiple opportunities for utilization. Without considering these issues environmental program evaluation can be reduced to an ideological exchange poorly concealed in the language of costs and benefits. The prospects for environmental program evaluation, we feel, hold much greater promise.

According to Walter Rosenbaum (1995), the 1990s represent a new environmental era in which policymakers can move beyond reactive policy making, identify and absorb lessons from two decades of environmental program experimentation, and begin to establish new programs that address the environmental issues of the future: "Having addressed many of the most apparently dangerous symptoms of environmental derangement—air pollution, water pollution, solid waste, and the rest—and having initiated a vast new enterprise of environmental research, [policymakers] now confront growing evidence that a durable solution to environmental sustainability will require profound changes in fundamental domestic and international social institutions" (330).

The students, scholars, and practitioners of environmental program evaluation have an opportunity to play a major role in fostering such profound changes and in designing such new social institutions.

NOTES

1. Bartlett (1994) offers an alternative classification of environmental program evaluations: evaluations that focus on process, outcomes, and institutions. Bartlett's process and outcomes categories correspond with our process and impacts classifications, except that his outcomes category includes evaluations that focus on efficiency. According to Bartlett, institutional evaluations "assess how processes work and outcomes are produced within a larger institutional framework created in part by policies and within which policies are made and remade. In short what is evaluated is political architecture" (1994:179). We recognize that effects on institutional structure are important, but we believe that significant changes in institutional structure should at some point be manifest in procedural, behavioral, or environmental change. For more on these issues see Rich, Mangun, Rosenbaum, and Kraft in this volume.

2. This is similar to the point made long ago by Boulding (1966) in "The Economics of the Coming Spaceship Earth."

3. For a critical view see Cooper and West (1988).

REFERENCES

Alberti, M., and J. Parker. 1991. "Indices of Environmental Quality: The Search for Credible Measures." *Environmental Impact Assessment Review* 11 (2): 95–100.

Bartlett, R. 1994. "Evaluating Environmental Policy Success and Failure." In *Environmental Policy in the 1990s,* ed. N. J. Vig and M. E. Kraft. 2d ed. Washington, D.C.: Congressional Quarterly Press. 167–88.

Boulding, K. 1966. "The Economics of the Coming Spaceship Earth." In *Environmental Quality in a Growing Economy,* ed. H. Jarrett. Baltimore: Johns Hopkins University Press. 3–14.

Chelimsky, E. 1987. "The Politics of Program Evaluation." In *Evaluation Practice in Review,* ed. D. S. Cordray, H. S. Bloom, and R. J. Light. New Directions for Program Evaluation 34. San Francisco: Jossey-Bass. 5–21.

Cooper, J., and W. West. 1988. "Presidential Power and Republican Government: The

Theory and Practice of OMB Review of Agency Rules." *Journal of Politics* 50 (88): 864–95.

Council of Great Lakes Research Managers. 1991. *A Proposed Framework for Developing Indicators of Ecosystem Health for the Great Lakes Region.* Windsor, Ontario: International Joint Commission.

Downing, P., and K. Hanf. 1983. *International Comparisons in Implementing Pollution Laws.* Boston: Kluwer Nijhoff.

Hellstern, G.-M. 1986. "Assessing Evaluation Research." In *Guidance, Control, and Evaluation in the Public Sector,* ed. F.-X. Kaufmann, G. Majone, and V. Ostrom. New York: Walter de Gruyter. 279–312.

Johnston, W. P., Jr. 1988. "Increasing Evaluation Use: Some Observations Based on the Results at the U.S. GAO." In *Evaluation Utilization,* ed. J. A. McLaughlin, L. J. Weber, R. W. Covert, and R. B. Ingle. New Directions for Program Evaluation 39. San Francisco: Jossey-Bass. 75–84.

Kraft, M. 1996. *Environmental Policy and Politics: Toward the Twenty-First Century.* New York: HarperCollins.

Langbein, L. I. 1980. *Discovering Whether Programs Work: A Guide to Statistical Methods for Program Evaluation.* Santa Monica: Goodyear Publishing.

Levin, S. A. 1992. "Orchestrating Environmental Research and Assessment." *Ecological Applications* 2 (2): 103–6.

Mann, D. E. 1982. *Environmental Policy Implementation.* Lexington, Mass.: Lexington Books.

Mazmanian, D., and P. Sabatier. 1989. *Implementation and Public Policy: With a New Postscript.* New York: University Press of America.

Mohr, L. 1992. *Impact Analysis for Program Evaluation.* Newbury Park, Calif.: Sage Publications.

Moran, T. K. 1987. "Research and Managerial Strategies for Integrating Evaluation Research into Agency Decision Making." *Evaluation Review* 11 (5): 612–30.

Mowbray, C. T. 1988. "Getting the System to Respond to Evaluation Findings." In *Evaluation Utilization,* ed. J. A. McLaughlin, L. J. Weber, R. W. Covert, and R. B. Ingle. New Directions for Program Evaluation 39. San Francisco: Jossey-Bass. 47–58.

National Academy of Public Administration. 1994. *The Environment Goes to Market: The Implementation of Economic Incentives for Pollution Control.* Washington, D.C.: National Academy of Public Administration.

———. 1995. *Setting Priorities, Getting Results: A New Direction for EPA.* Washington, D.C.: National Academy of Public Administration.

Patrick, R. 1992. *Surface Water Quality: Have the Laws Been Successful?* Princeton: Princeton University Press.

Patton, M. Q. 1978. *Utilization-Focused Evaluation.* Beverly Hills, Calif.: Sage Publications.

———. 1987. *How to Use Qualitative Methods in Evaluation.* Newbury Park, Calif.: Sage Publications.

———. 1988. "Integrating Evaluation into a Program for Increased Utility and Cost-

Effectiveness." In *Evaluation Utilization,* ed. J. A. McLaughlin, L. J. Weber, R. W. Covert, and R. B. Ingle. New Directions for Program Evaluation 39. San Francisco: Jossey-Bass. 85–94.

Portney, P. R. 1984. "The Benefits and Costs of Regulatory Analysis." In *Environmental Policy under Reagan's Executive Order,* ed. V. Kerry Smith. Chapel Hill: University of North Carolina Press. 226–40.

———, ed. 1990. *Public Policies for Environmental Protection.* Washington, D.C.: Resources for the Future.

Pressman, J. L., and A. Wildavsky. 1973. *Implementation.* Berkeley: University of California Press.

Rosenbaum, W. 1995. *Environmental Politics and Policy.* 3d ed. Washington, D.C.: Congressional Quarterly Press.

Rossi, P. H., and H. E. Freeman. 1993. *Evaluation: A Systematic Approach.* Newbury Park, Calif.: Sage Publications.

Russell, M. 1992. "Lessons from NAPAP." *Ecological Applications* 2 (2): 107–10.

Rutman, L., and G. Mowbray. 1982. *Understanding Program Evaluation.* Newbury Park, Calif.: Sage Publications.

Sabatier, P. 1986. "Top-Down and Bottom-Up Approaches to Implementation Research: A Critical Analysis and Suggested Synthesis." *Journal of Public Policy* 6 (Jan.–Mar.): 21–48.

Sagoff, M. 1988. *The Economy of the Earth.* Cambridge: Cambridge University Press.

———. 1993. "Environmental Economics: An Epitaph." *Resources* 111:2–7.

Shadish, W. R., T. D. Cook, and L. C. Leviton. 1991. *Foundations of Program Evaluation: Theories of Practice.* Newbury Park, Calif.: Sage Publications.

Smith, M. F. 1988. "Evaluation Utilization Revisited." In *Evaluation Utilization,* ed. J. A. McLaughlin, L. J. Weber, R. W. Covert, and R. B. Ingle. New Directions for Program Evaluation 39. San Francisco: Jossey-Bass. 7–19.

U.S. Environmental Protection Agency. 1987. *EPA's Use of Benefit-Cost Analysis: 1981–1986.* Washington, D.C.: Environmental Protection Agency Office of Policy, Planning, and Evaluation. EPA-230-05-87-028.

———. 1991. Concept paper on use of indicators for effectiveness of environmental programs. Washington, D.C.: Environmental Protection Agency Office of Policy, Planning, and Evaluation.

PART I

ENVIRONMENTAL PROGRAM EVALUATION

1

Program Evaluation and Environmental Policy: The State of the Art

ROBERT F. RICH

As Michael Quinn Patton (1986) points out, the idea of scientifically evaluating government programs is a relatively new idea to government and public policymakers. In the area of human services, the General Accounting Office developed auditing models solely for record-keeping purposes: e.g., How many people are being served? How much did a unit of service cost? The auditing models did not even begin to ask questions about the success of programs or interventions/initiatives and the factors that might lead to replicability. As the amount spent on human services began to shrink in the 1970s and 1980s, scholars and practitioners began to ask difficult questions about the viability and effectiveness of programs as opposed to posing the old stand-bys: Are people happy? Is it usual and customary practice?

For environmental policy, program evaluation became particularly important in the 1990s when legislators were asking critical questions about legislation that had been implemented in the 1970s and 1980s: Superfund, the Endangered Species Act, the Clean Water Act, the Safe Drinking Water Act, the Nuclear Waste Policy Act, and others. "Particularly in the United States, where the dominant public ideology has always been pragmatism, demand for evidence of success has become increasingly insistent, impatient, and narrowly focused" (Bartlett 1994:167).

Historical Perspective

From a historical perspective program evaluation and assessment are not new concepts. Indeed, one might argue that the German sociologist Max Weber was one of the early founders of the field of evaluation research. In his work on objectivity in the social sciences, Weber argued: "The question of the appropriateness of the means for achieving a given end is undoubtedly accessible to scientific analysis. . . . We can also offer the person, who makes a choice, insight into the significance of the designed object. We can teach him to think in terms of the content and the meaning of the ends he desires and among which he chooses. . . . An empirical science cannot tell anyone what he should do—but rather what he can do—and under certain circumstances—what he ought to do" (1949:54).

Weber's argument can easily be extended to the area of program evaluation as we now understand it. The focus of evaluation research or program evaluation is to assess the significance of a particular program or policy initiative. Moreover, program evaluation is concerned with assessing the effectiveness of particular interventions and with documenting costs and benefits of particular initiatives.

Even though the idea of formal program evaluation was new to government in the 1960s and early 1970s when the demand to assess the effectiveness of new social program initiatives arose, it was not a new process in human thinking and problem solving. Evaluation is part of our daily lives; we regularly make judgments about what is good and bad (Perloff and Rich 1986). Program evaluation has, since the 1970s, become an integral part of how government determines whether programs have been successful and of determining whether a program has produced a positive return on the investment of public dollars. The field of environmental policy has, however, not made similar progress. It is only since the late 1980s that the well-accepted human service evaluations have been joined by program evaluations in the areas of national defense, risk, technology, transportation, agricultural practices, and the environment (see Solomon in this volume).

Each decision maker or policymaker has an implicit model for assessment (evaluation) that he/she has developed over time. As human beings we all on a day-to-day basis accept or reject ideas and strategies for change that are proposed to us. These "models" for assessment are often implicit; yet, they are extremely important because they serve as the basis for choice and decision making.

This mode of inquiry was steeped in scientific tradition and legitimated by well-known and respected academics. Evaluators have considered their field to be a branch of applied social research (see Berk and Rossi 1990). The aim was to provide a "mirror" in which program administrators and policymakers could view their agencies or organizations (Perloff and Rich 1986). These scientists argued that programs designed to have an impact on human problems should be carefully monitored, periodically evaluated, and based on systematically researched models. Evaluation research was seen as part of the process of "rational program development" (Twain 1975). By using evaluation research, politicians and administrators could reduce the uncertainty that so often accompanies the development of social programs. Moreover, future program designs could be informed by empirical results measuring the impact of past interventions. What has been the impact? What were the intended effects? What were the unintended effects?

The focus on evaluation research as a formal mode of inquiry is directly related to the development of the field of policy sciences, which dates back to Harold Laswell's work. In an essay published in 1951, Laswell points out: "In the realm of policy more attention has been given to planning, and to improving the information on which staff and operational decisions are based. The policy scientist is far more interested in evaluating and reconstructing the practice of society than in . . . the higher abstractions from which the values are derived. The choice carries with it the de-emphasizing of much of the traditional baggage of metaphysics and theology" (1951:3).

Clearly, over time, evaluation research and program evaluation became identified as a branch of policy research (Wholey 1989). In this context, much of the attraction resulted from the development of methodology and instrumentation. The rising influence of economists and psychometricians in particular seemed to indicate that the work of social scientists would be more widely accepted when their research methods more closely resembled those of physical scientists (Laswell 1951).

The Great Society years were the zenith of program evaluation and the development of evaluation research as a discipline or field of inquiry. As Henry J. Aaron points out in *Politics and the Professors* (1978), scholars played a large part in the development of social programs in the 1960s. They also played a large part in advocating the systematic and comprehensive assessment of these programs through the use and application of social theory and the most up-to-date and sophisticated statistical techniques. As applied social science and policy research began to focus increasingly on questions of consequences and impacts, the field of program evaluation became more prominent (see Rossi and Freeman 1989, for a general discussion).

During this period the most popular approach for program development and program evaluation was borrowed from psychology: the so-called experimental approach. The originator of this approach as it applies to evaluation is Donald T. Campbell. In his seminal article entitled "Reforms as Experiments," he argues:

The United States and other modern nations should be ready for an experimental approach to social reform, an approach in which we try out new programs designed to cure specific social problems, in which we learn whether or not these programs are effective, and in which we retain, imitate, modify, or discard them on the basis of apparent effectiveness on the multiple imperfect criteria available. Our readiness for this stage is indicated by the inclusion of specific provisions for program evaluation in the first wave of the "Great Society" legislation, and by the current congressional proposals for establishing "social indicators" and socially relevant data banks. (1969:409)

Prior to 1964, the amount of money spent annually on program evaluation was not more than a few hundred thousand dollars; by the end of the 1970s, millions of dollars were devoted to this activity. In addition, the activity was institutionalized through the establishment of offices of program, planning, and evaluation in many federal and state agencies. It is also worth pointing out that social legislation of the 1970s and early 1980s required administrators to evaluate their new programmatic initiatives; in many cases, the requirement specified the use of an external evaluator.

Not only did evaluation research become institutionalized in government but it also became a profession in many academic circles. The field has specific educational and training programs, journals, professional societies, and continuing education programs. Program evaluation as a discipline is seen as a subset of social science research and part of the growing field of policy research; many policymakers have come to believe that this systematic form of assessment can help them in their problem-solving activities.

Because it is an applied branch of social inquiry, evaluation research comes with the built-in expectation that its results will be used and have an impact on program administration and policy development. As Carol H. Weiss notes: "Evaluation research is designed for utility. Its purpose is to answer a practical question of decision makers who want to know whether to continue a program, extend it to other sites, modify it, or close it down. If the program is found to be only partly effective in achieving its goals, as is often the case, evaluation research is often expected to say something about the aspects that are going awry and the kinds of changes that are needed" (1972:1).

As one reflects on this historical development, it is not surprising to find that evaluation research and program evaluation are a complex enterprise that includes a broad set of questions, techniques, and expectations with respect to utilization of the results produced. Since the 1960s government officials have carried out program evaluations under the umbrella of such activities as auditing, monitoring, planning, policy analysis, inspection, development of indicators, program analysis, and research. Former Assistant Secretary for Policy, Planning, and Evaluation Joseph S. Wholey stipulates that "program evaluation or evaluation comprises definition of programs and of intended program performance, assessment of program performance and of factors that influence performance, and communication of evaluation findings to policy makers, managers, staff, and interested others" (1989:5).

Development of Evaluation Research as a Field

The function of research is, primarily, to produce understanding and knowledge. Understanding is so important because it requires those conducting research to ask the difficult "why" questions; this mode of inquiry leads one to causal analysis and the development of theory. Donald T. Campbell represents the school of thought that tries to extend the methodology of science as we know it in the laboratory to nonlaboratory settings: "My central concern over the past twenty-five years has been as a methodologist trying to extend the epistemology of the experimental method into non-laboratory social science. . . . Since 1969 at least, this concern has been focused on applied social science, on treating the ameliorative efforts of government as a field experiment" (1988:291). Campbell is identified as the leading social scientist who has tried to apply complex and sophisticated methods to real world problems.

Campbell, W. Edward Deming, and many other social scientists have argued that the best way to evaluate interventions and programs is through the experimental method—or at least as close as one can come to this method:

> We are inclined to emphasize the role of experimentally gathered information in the shaping of social policy because such information is most helpful in learning the causal relationships among program elements and outcomes. If an effect can be demonstrated in a group of units (persons, places, or institutions) chosen at random and subjected to a specified treatment while a similar group that is not treated does not show the effect, one can be reasonably confident that the treatment produced the effect. Such confidence cannot so readily be reposed in non-experimental evidence, even though sophisticated methods of analysis can be used to reduce the ambiguity of causal inference. (Riecken and Boruch 1974:5)

As Campbell has noted, the attempts to implement high quality program evaluation designs and rigorous analysis plans has resulted in high levels of frustration stemming from the realities of the political system (1988). At the same time, program evaluators find themselves in the position of proposing "novel procedures for political decision making" that are designed so that the impact of programs can be evaluated (1988:291).

This school of thought within the program evaluation field has developed in several interesting directions. New instrumentation, statistical techniques, and methodologies have been developed to evaluate carefully and systematically governmental interventions from the "experimental" perspective. As part of this effort, many articles have been written about threats to validity and the need to develop "quasi-experimental" designs when pure experimentation is not possible.

It became clear to members of this school of thought that real-world field settings—settings in which programs would be implemented—did not lend themselves to the limits of laboratory experimentation. In their book entitled *Quasi-Experimentation,* Thomas D. Cook and Donald T. Campbell note: "Even though most of the controls associated with the laboratory cannot—and should not—be created in field settings, classical designs based on random assignment can nevertheless sometimes be implemented in such settings" (1979:2).

Proponents of this school of thought were interested in developing methodologies that approximated valid scientific research (what Laswell pointed to as approximating the physical sciences as closely as possible) and they were interested in influencing policymakers to design programs and regulations that were open to experiments or quasi-experiments: those that were open to random assignment and clear causal analysis.

Campbell and his colleagues realized that this required a philosophical orientation on the part of administrators that included the importance and validity of this type of research:

> The political stance would become: this is a serious problem. We propose to initiate policy A on an experimental basis. If after five years there has been no significant improvement, we will shift to policy B. By making explicit that a given situation is only one of several that the administrator . . . could in good conscience advocate, the administrator can afford honest evaluation of outcomes. Negative results . . . do not jeopardize his job for his job is to keep after the problem until something is found that works. (Campbell 1969:410)

One does not simply want to move from policy A to policy B in a mechanical fashion—i.e., "A has not worked so let us move to B." The adoption of policy B should be made with an understanding of the reasons why A failed.

Over time, scholars realized that it may not be sufficient to emphasize findings and validate the "truth" as measured through experimental designs or rigorous causal analysis (which employs sophisticated statistical techniques such as regression analysis, path analysis, and log-linear models). In his 1988 article "The Experimenting Society," Campbell argues:

> Societies will continue to use preponderantly unscientific political processes to decide upon ameliorative program innovations. Whether it would be good to increase the role of social science in deciding on the content of programs tried out is not the issue here. The emphasis is, rather, on the more passive role for the social scientist as an aid in helping society decide whether or not its innovations have achieved desired goals without damaging side effects. The job of the methodologist for the experimenting society is not to say what is to be done, but rather to say what has been done. The aspect of social science that is being applied is primarily its research methodology rather than its descriptive theory, with the goal of learning more than we do now from the innovations decided upon by the political process. (297)

Hence, proponents of this school of thought continued to emphasize the importance of methodology and statistical applications. It was through the development of methodology that one could best serve as the "mirror" that government looked to the social sciences to provide. Even though these scholars developed some knowledge of the political process, they continued to believe that evaluation should hold, wherever possible, to the principles of experimentation, random assignment, causal analysis, and a careful examination of any conclusions based on causal inference.

Given this basic paradigm, members of this scientific school of thought laid out a set of rules for the "proper conduct" of program evaluations:

1. Evaluations should be conducted by someone or some group external to the agency; the evaluator cannot have any stake in the success or failure of a program.
2. Agencies and other sponsors should encourage, if not require, the replication of evaluation research results. It is only through replication and secondary analysis that we can become more comfortable with our findings. This aspect is particularly important the further we get away from experimental designs.
3. Evaluators should be encouraged to uncover unintended effects as well as look for the intended effects. Once again, through this kind of activity it is possible to increase organizational learning.
4. Evaluators must not be influenced by the views of the program administrators. It is the job of the administrator to advocate on behalf of his/her programs; similarly, it is the responsibility of the evaluator to provide a fully honest assessment of the impacts of the programs—both good and bad. It is imperative that these two roles remain separate.

This school of thought continues to be dominated by the academic disciplines, the journals in the field, and the American Evaluation Research Association, which continues to provide support for this group. However, it is important to note that the school of thought has, since the mid-1970s, come under increasing scrutiny and criticism. Some of the more common criticisms include:

1. These external evaluators are not sensitive to the political process; hence, they represent a real threat to the program administrators responsible for a particular program. As a consequence, program administrators resist evaluation.
2. The evaluators are preoccupied with the numbers. It is said that they have a methodological toolbox and search for a problem instead of having a carefully defined problem and looking for the tools/instruments best suited to working on that problem. Some evaluators seem to care more about the eloquence and sophistication of their tools than they do about solving real problems.
3. Many evaluators do not give sufficient attention to distinguishing between *formative* and *summative* evaluation. Formative evaluation is designed to provide feedback and monitoring as an initiative is being implemented. The feedback should, presumably, help in fine tuning and changing programs as they are evolving. The summative evaluation, on the other hand, is designed to look at overall impacts and consequences after an appropriate period of time has passed.
4. Insufficient attention is given to how the results of evaluation research are going to be used (Rich 1979). Evaluators need to pay more attention to questions such as, Why is the information being asked for? What use will program administrators make of the evaluation results? Are program administrators open to receiving the results of evaluations? In response, the more academically based evaluation researchers would argue that insufficient attention is being paid to the misutilization and intended nonutilization of evaluation research results (Cook, Levinson-Rose, and Pollard 1980).
5. The work of the academic community is too costly and does not provide useful information in a timely manner. Hence, there is a need for more internal studies and "quick and dirty" studies that provide some short-term answers on how programs are progressing.
6. Many government officials contend that the academic community is to blame for the growth in consulting firms and so-called beltway bandits who during the 1970s and early 1980s captured the majority of the evaluation funds that were being allocated by federal agencies. This indictment of academia is based on many of the criticisms listed above. Academics counter

that these government officials find the consultants convenient because they will "parrot" back what the clients want to hear, being dependent upon them for future paychecks. At a minimum, one realizes that there is a real tension between these views, which in turn has affected the way program evaluation and policy research is viewed in governmental and academic circles.

These critiques and others have led to the development of a more "practical approach" to evaluation research. The so-called practical approach is based on the need to make evaluation and assessment techniques available to government officials in an effective and efficient manner. Further, suggests Wholey (1989), the widespread availability of personal computers has brought many evaluation procedures within the reach of managers and staff at all government levels.

The Scientific Approach as It Relates to Environmental Policy

The "scientific methods" described above have been tested and applied primarily in the fields of human service delivery. It is less clear that this paradigm can be (or should be) applied to environmental problems (Bartlett 1994; Vig and Kraft 1994).

Since the early 1980s environmental policymakers seem to have been preoccupied with making political judgments about the "success" or "failure" of particular initiatives (e.g., the Clean Air Act and Superfund). Robert Bartlett argues quite convincingly that "our addiction to simple judgments will be no easier to shake than our compulsive affection for simple, complete answers to complex sets of policy problems" (1994:183). In addition to the danger in looking for simple answers to complex problems, those concerned with environmental policy should also be cognizant of several other factors: an investment in environmental quality is a long-term investment for which we may not be able to scientifically document results in the short run; outcome-related results may be very difficult to measure; and political pressures may outweigh the significance of any research results (see Landy, Roberts, and Thomas 1990, for a general discussion of environmental policy).

If objective, outcome-oriented evaluation is not particularly appropriate at this point in the development of environmental policy, what kinds of analysis are relevant? Some studies have focused on output as opposed to outcome measures, such as the number of hazardous waste sites inspected, the number of actions taken to protect the habitat, the number of species recovery plans that have been approved, the number of environmental impact statements that have been submitted, and the amount of pollution in the air or water (Rosenbaum 1991; Bartlett 1994). Each of these outputs tells us something important about the implementation of particular environmental policy initiatives (see Kraft in this volume for more discussion).

In addition to focusing on outputs, environmental evaluations may also be "process-oriented." In this context process has two key dimensions that relate directly to evaluation: an assessment of the process (i.e., the technology) for improving environmental quality and the decision-making process associated with improving the environment. Environmental policy making has emphasized the importance of citizen input and of "democratic participation." It is through the dialogue between citizens and policymakers that effective and workable policy is developed. From an evaluation standpoint, one must, therefore, be concerned with an assessment of the mechanisms for participation and not just outputs or outcomes from specific policy initiatives. Hence, the merits of the process used to determine certain environmental standards are as important to policymakers as the effect of these standards on the environment (Majone 1986). If the process is flawed then its application is also likely to be problematic.

Another dimension of process-based evaluation that applies to environmental policy is the development of rules. In theory, the implementation of well-designed rules should lead to desirable outcomes. Rules are separable from individual actors; the application of rules should not be dependent upon a single actor. In the environmental area, the existence of certain specified rules should lead to one desired outcome.

As already alluded to, the political context is critical to take into account in this regard. Public policy represents the codification of mainstream values. Current values emphasize the importance of using market-based mechanisms and incentives to reach environmental goals; these approaches are favored even if their effectiveness has not been documented. "Because our current policies have not yet solved our environmental problems, and no true solution is in sight, does not necessarily mean that doing nothing would be a better choice" (Bartlett 1994:183). Relying on the market to regulate behavior will not necessarily produce desirable results, just because the approaches used since the early 1970s have not yet been documented as "successful." Mainstream values also call for giving more discretionary authority and power to the states. Yet, do we have documentation that the states have the capacity to deal with these problems? How will the states fit into what is increasingly being recognized as a global problem? Throughout the history of environmental policy, environmentalists have faced the trade-off between investment in economic growth and investment in the environment. As long as this trade-off is resolved in favor of economic growth, it will not matter how effective (from a scientific perspective) environmental policy initiatives are.

Given this context for conducting program evaluation in the environmental arena, what are some practical alternatives to the more traditional approaches?

Approaches to Practical Evaluation

Evaluability Assessment To begin with, Wholey and his colleagues developed the innovative and creative concept of "evaluability assessment" (Schmidt, Scanlon, and Bell 1979; Rutman 1980; Nay and Kay 1982; Wholey 1983, 1987). The evaluability assessment approach stipulates that not every program or program element is ready to be evaluated at any given point in time. Indeed, some programs may not be open to a rigorous evaluation approach. The program administrator and evaluator share an interest in determining whether a program (or program element) is ready for assessment. The advocates of evaluability assessment contend that a great deal of time, frustration, and money could be saved by conducting a simple evaluability assessment.

A program administrator should commission an evaluability assessment to answer the following questions:

1. Do the administrators of the program agree on what the goals/objectives of the program are? (If the answer is no, then the program cannot be evaluated at that time.) This is a particularly critical question because so many evaluations suffer because the goals of the program cannot be specified. In these cases, the evaluator either specifies the goals or accepts one stakeholder group's version of the goals. This can (and often does) lead to a discrediting of the evaluation.
2. Do the administrators of the program agree on what outcome measures are appropriate? What results is the program trying to achieve, and for whom? How does one go about assessing "success?" Once again, it is crucial to have agreement among the key stakeholder groups; without this agreement, it is very easy to discredit the evaluation.
3. Do the key stakeholders share a model of how the program that they helped to design is supposed to work (Wholey 1987; Nay and Kay 1982)? For any program, the model should diagram how the program is designed to work: A should lead to B; in turn, B should lead to C, D, or E; C should lead to X, Y, and Z; D should lead to K, L, or M; E should lead to Q, R, or S, and so forth. If one has a real understanding of a program, it should be possible to draw such a diagram and explain it. If it is not possible to do this, then one should question the evaluability of the program.

4. Does the program have a clear (or fairly clear) management structure that shows that it can be implemented?
5. Do the key stakeholders have a clear plan for how the results of the evaluation will be disseminated and used? Once again, in the absence of such an understanding, it may be difficult to conduct an effective evaluation. (see Nay and Kay 1982 for a full description of the evaluability assessment process).

Evaluability assessment is an important and practical tool because it accepts the posture that evaluation cannot and should not be undertaken without an examination of the readiness of an agency and its administrators to conduct this type of costly inquiry.

Rapid Feedback Evaluation Rapid feedback evaluation, coined by Wholey (1983), augments the evaluability assessment instrument and is designed for agencies who need information "yesterday," do not have the resources to invest in collecting new information, and need to provide some information of an evaluative nature.

This type of study would horrify academics who see themselves as the originators of the field of program evaluation. Instead of specifying a model and following careful rules of causal analysis, this type of evaluation requires reliance upon data and information that are readily available within the agency. The purpose, here, using small samples and small amounts of information, is to "produce a preliminary evaluation of program performance along with an analysis of designs for any further, more conclusive evaluation work" (Wholey 1989:8).

Performance Monitoring In the mid-1980s and early 1990s, government officials began to focus on the concept of "performance." Are government agencies and public programs performing well or, at least, adequately? Some feel that performance should be thought of in terms of levels of productivity, cost-effectiveness, and cost efficiency. Others contend that by concentrating on performance, the analyst can examine dimensions of intraprogram performance as well as performance across programs. The emphasis here is often on comparing a unit's (school, clinic, agency) performance with the same unit's performance at another point in time. Hence, one can build a time series on performance of one unit or more over time.

It should be noted that the emphasis here is on monitoring and describing what is occurring. The performance monitoring method does not explain performance; this approach does not try to establish causal linkages or relationships nor does it try to account for or explain a particular level of performance or change in performance over time.

This type of program evaluation is already widely used in the environmental arena in the formative stages of evaluation. Monitoring is typical to document progress in air and water quality; it is also used for inspections and licensing (see Solomon and Rosenbaum in this volume for more documentation).

Qualitative Evaluation Some of the critiques of the experimental model of evaluation often contend that this approach requires or encourages one to focus on trivial issues or trends: only those dimensions of a problem that are operational and measurable. Michael Quinn Patton and others have responded by suggesting that an alternative or supplementary approach might involve qualitative evaluation. According to Wholey (1989), qualitative evaluation can help policymakers gain the perspectives of clients and direct service providers, communicate the meaning of quantitative findings, and interpret changes in program performance. As a proponent of this approach, Patton stipulates that he uses "the term evaluation quite broadly to include any effort to increase human effectiveness through systematic data-based inquiry" (1990:1).

Qualitative data represent some of the most important data that one could collect: "Qualitative methods consist of three kinds of data collection: (1) in-depth, open-ended interviews; (2) direct observation; and (3) written documents. . . . The data for qualitative analysis typically come from fieldwork. During fieldwork the researcher spends time in the setting under study—a program, an organization, a community or wherever situations of importance to the study can be observed" (Patton 1990:3).

Once again, the so-called practical methods employ modes of inquiry and methods of data collection that violate the assumptions of the traditional approach. For evaluability assessment, performance monitoring, rapid feedback evaluation, and qualitative evaluation, the practical approach has been used to try to respond to the critiques that program administrators, legislators, and other government officials have made to the program evaluation field.

Other Developments in the Evaluation Area

In addition to the development of these practical methods since the early 1980s, evaluators have created other tools that are associated with program evaluation. Many of these tools have received quite a bit of positive attention over time.

Benefit-Cost Analysis, Cost-Effectiveness Analysis, and Cost-Efficiency Studies With the downturn in the economy during much of the post-Vietnam era, government officials at all levels have emphasized the importance of cost savings, cost cutting, performance, and effectiveness. In other words, the demand has been for accountability—especially in terms of financial management.

Benefit-cost analysis is particularly attractive from the perspective of legislators and many other government officials. It provides a summary statistic that indicates whether "on balance" a program or initiative is worthwhile. Moreover, this form of analysis provides a basis for comparison because it is seen as using a standard metric (i.e., costs).

Critics of benefit-cost analysis maintain that even though dollars are being used as a standard metric, one is still comparing apples and oranges in many cases. For example, how is it possible to compare health, transportation, welfare, and infrastructure programs using benefit-cost analysis? Perhaps one could examine the same program over time, but the comparisons, the critics maintain, are quite problematic.

Moreover, while it may be relatively easy to measure costs, it is difficult, if not impossible, to measure intangible benefits (see Moreau in this volume). Even if one could attach monetary values to certain benefits, they would be rather meaningless. Labor economists have, for example, applied benefit-cost analysis to health policy questions (see Braden and Kim in this volume). In this case, the primary benefit has been seen as length of life. The benefit is monetized by measuring the projected earnings (in salary) of various individuals grouped by profession (e.g., doctors, lawyers, teachers). The analyst is, then, placed in the position of comparing the value of teachers versus clerical workers versus lawyers to society. Although this type of analysis may be simple and attractive, it is quite problematic, especially in the area of social benefits.

Cost-effectiveness analysis, on the other hand, represents a more useful form of analysis. In this case, the questions being addressed are, Is the program/initiative under study reaching the objectives for which it is designed? If so, what is the cost of achieving these objectives? The cost may not be a single number; it may, rather, be a range of costs for achieving a particular set of goals. Cost efficiency simply adds the dimension of cost to the achievement of the desired outcome.

Development of Performance Indicators In addition to cost and benefit measures, the demand for accountability has also encouraged analysts to develop performance indicators. Like performance monitoring, these indicators are designed to provide a perspective over time for how an agency, program, or initiative is progressing.

For a particular policy area, analysts and managers ask a set of questions: What outcomes are we trying to achieve? Are there quantitative measures that could capture one outcome or more? Can these measures be collected and easily displayed over time? There are many examples of this that are currently being used in government. The unemployment rate, for example, has become an important indicator for the performance of the economy. A clearance rate (i.e., the number of crimes solved over the number of crimes unsolved) is an indicator that is often used in the criminal justice system (Skolnick 1975).

As with cost studies, these indicators are attractive because they are simple, easy to interpret, and easy to collect. Yet, because care and attention need to be paid to interpreting these numbers, they are easily open to misuse. This is particularly common when administrators are looking to numbers to advocate for a program (Campbell 1988).

User Satisfaction Studies Along with demands for accountability in policy circles there is a growing concern for taking the views of the client/consumer into account. How, this view goes, can one deliver effective programs without taking the needs, preferences, and assessments of the client/consumer into account?

With these questions in mind, a whole new class of studies has developed, including needs assessments, user satisfaction studies, and opinion surveys. These studies focus on the views and attitudes of particular segments of the population. In some cases clients of a particular program are being compared with a "control group," whose members could be clients but are not at the time the study is being conducted. In other cases, the clients are being studied without any direct comparison. In all cases, program administrators and analysts assume they need to take these views seriously. It is almost like a Nielsen rating for public policy programs.

Information Management Techniques Within the field of policy research and in the area of program evaluation there is also a growing concern for problems of information management. There is so much information being produced in the United States and in other countries, how can one possibly remain informed and up-to-date, even in one's own specialty?

Program evaluators have answered this question by designing and employing two very useful techniques: meta-analysis and knowledge synthesis. Both are forms of information reduction. In the case of meta-analysis the idea is to look for similarities and themes across a group of studies. Which findings are held in common? What is unique? This helps in establishing levels of reliability and validity.

In the case of knowledge synthesis, many techniques have been developed to reduce and summarize large amounts of information. In some cases there is also an attempt to put the information into some framework. Social, economic, and performance indicators are examples of synthesis. Similarly, citation analysis is also a form of information reduction within a particular framework. The popular computer-based bibliographic search is also a form of information reduction and synthesis.

The potential problem with synthesis techniques is the same as that outlined above with respect to performance indicators and benefit-cost analysis: one can easily forget that these are summary numbers; they do not replace the need for judgment and for putting the numbers into a broader context.

Indeed, in the context of policy research, there is a growing demand for "management indicators." Can a set of measures or numbers be developed that help to monitor the performance of individual managers and of groups of managers? What kinds of performance are we most interested in monitoring? What kinds of summary measures does one want to examine over time?

Conclusions

Since the establishment of the U.S. Environmental Protection Agency, environmental issues have become increasingly important at all levels of government in both the civilian and military sectors. Environmental policy is still a relatively young field. Consequently, the role of evaluation may be particularly important as we try to address serious problems in a timely fashion.

From the standpoint of evaluability assessment, it is worth posing the question, To what extent are environmental programs ready for formative and summative evaluation? What evaluation designs should be employed? Can we usefully employ experimental or quasi-experimental designs? The answers are not clear. It is also not clear that users really want to receive the results of such studies.

Establishing a framework for evaluating environmental policies is a complex task. Often, it is problematic for policymakers to define the environmental issue that is to be addressed by public policy. The scientific and technical basis may be uncertain, leaving wide latitude for debate and interpretation. Competing political views and stakeholders contribute their different interpretations to the policy design. Each brings to the table differing perceptions of risks and costs concerning the given environmental problem or the proposed policy response. These competing values may combine to create a policy that is at odds with the best scientific evidence, engineering expertise, or policy analysis. For example, from an environmental science perspective, environmental problems may be systemic in origin, scope, and impact, but the policy response is often piecemeal and certainly asystemic.

It is clear that it will not be possible or desirable to apply one evaluation strategy uniformly. The practical evaluation techniques described above may be the most helpful at this point in time. The evaluability assessment is probably invaluable because it will help the diverse stakeholders focus on what is realistic to evaluate and what is not. It may then be possible and useful to develop performance indicators and cost-effectiveness studies out of this type of assessment.

The current emphasis on practical techniques is important and appropriate. One does, indeed, need to take the views of the consumers and the clients into account. They are not bystanders in the process of policy design and implementation. However, we also do not want to stop asking the "why" questions. There is a need to formulate new explanations. Validity, reliability, and replicability are not outdated concepts that are relevant only in the domain of academia. One cannot make a cost-effective or cost-efficient decision without addressing these issues.

However, one should not lose sight of the value of the traditional school of evaluation research. In the long run there is not a need for immediate results, thus quasi-experimental designs are to be encouraged. For example, additional studies are needed that compare the effectiveness of various policy instruments or combinations of policy instruments on the same environmental problem to see which policy instrument(s) will achieve the intended objectives at the least cost or disruption to society. Other studies might examine the impact of specific policies across environmental media (e.g., the impact of specific air pollution control policies on surface water quality). At this moment we do not have the answers that quasi-experimental design can provide us with—the answers to the "why" questions.

In the field of environmental policy, the issues are pressing, knowledge is limited, and the stakes are high. As a result, neither purely practical nor purely scientific approaches to evaluation will suffice. Indeed, environmental policy may be the area in which quasi-experimental designs and practical evaluation techniques can work together to produce sound policy interventions. To a significant extent, the practice of environmental program evaluation and the quality of the environment for this and future generations depends on it.

REFERENCES

Aaron, H. J. 1978. *Politics and the Professors.* Washington, D.C.: Brookings Institution.

Bartlett, R. 1994. "Evaluating Environmental Policy Success and Failure." In *Environmental Policy in the 1990s,* ed. N. J. Vig and M. E. Kraft. 2d ed. Washington, D.C.: Congressional Quarterly Press. 167–88.

Berk, R. A., and P. H. Rossi. 1990. *Thinking about Program Evaluation.* Newbury Park, Calif.: Sage Publications.

Campbell, D. T. 1969. "Reforms as Experiments." *American Psychologist* 24 (Apr.): 409–28.

———. 1988. "The Experimenting Society." In *Methodology and Epistemology for Social Science,* ed. D. T. Campbell. Chicago: University of Chicago Press. 290–314.

Cook, T. D., and D. T. Campbell. 1979. *Quasi-Experimentation.* Chicago: Rand McNally.

Cook, T. D., J. Levinson-Rose, and W. E. Pollard. 1980. "The Misutilization of Evaluation Research: Some Pitfalls of Definition." *Knowledge, Creation, Diffusion, Utilization* 1 (4): 477–98.

Landy, M. K., M. J. Roberts, and S. R. Thomas. 1990. *The Environmental Protection Agency: Asking the Wrong Questions.* New York: Oxford University Press.

Laswell, H. 1951. "The Policy Orientation." In *The Policy Sciences,* ed. D. Lerner and H. D. Laswell. Stanford: Stanford University Press. 1–15.

Majone, G. 1986. "Analyzing the Public Sector: Shortcomings of Current Approaches—Part A. Policy Science." In *Guidance, Control, and Evaluation in the Public Sector,* ed. F.-X. Kaufman, G. Majone, and V. Ostrom. New York: Walter de Gruyter. 61–70.

Nay, J. N., and P. Kay. 1982. *Government Oversight and Evaluability Assessment.* Lexington, Mass.: Heath.

Patton, M. Q. 1986. *Utilization-Focused Evaluation.* 2d ed. Beverly Hills, Calif.: Sage Publications.

———. 1990. *Qualitative Evaluation and Research Methods.* Newbury Park, Calif.: Sage Publications.

Perloff, R., and R. F. Rich. 1986. "The Teaching of Evaluation in Schools of Management." In *Teaching of Evaluation across the Disciplines,* ed. B. G. Davis. San Francisco: Jossey-Bass. 29–37.

Rich, R. F., ed. 1979. *Translating Evaluation into Policy.* Beverly Hills, Calif.: Sage Publications.

Riecken, H., and R. Boruch, eds. 1974. *Social Experimentation.* New York: Academic Press.

Rosenbaum, Walter A. 1991. "Curing Regulatory Incapacity at EPA: What Can Policy Studies Add to Reilly's Rosary?" Paper presented at the annual meeting of the American Political Science Association, Washington, D.C., Aug. 28.

Rossi, P. H., and H. E. Freeman. 1989. *Evaluation: A Systematic Approach.* 4th ed. Newbury Park, Calif.: Sage Publications.

Rutman, L. 1980. *Planning Useful Evaluations: Evaluability Assessment.* Beverly Hills, Calif.: Sage Publications.

Schmidt, R. E., J. W. Scanlon, and J. B. Bell. 1979. *Evaluability Assessment: Making Public Programs Work Better.* Rockville, Md.: U.S. Department of Health and Human Services, Project Share.

Skolnick, J. H. 1975. *Justice without Trial.* New York: John Wiley.

Twain, D. 1975. "Developing and Implementing a Research Strategy." In *Handbook of Evaluation Research,* ed. E. Struening and M. Guttentag. Beverly Hills, Calif.: Sage Publications. 27–52.

Vig, N. J., and M. E. Kraft, eds. 1994. *Environmental Policy in the 1990s.* 2d ed. Washington, D.C.: Congressional Quarterly Press.

Weber, M. 1949. *Methodology of the Social Sciences.* Trans. and ed. Edward S. Shils and Henry A. Fincy. New York: Free Press.

Weiss, C. H. 1972. "Utilization of Evaluation Results." In *Evaluation Research: Methods of Assessing Program Effectiveness,* ed. C. H. Weiss. Englewood Cliffs, N.J.: Prentice-Hall. 110–28.

Wholey, J. S. 1983. *Evaluation and Effective Public Management.* Boston: Little, Brown.

———. 1987. *Organizational Excellence: Stimulating Quality and Communicating Value.* Lexington, Mass.: Heath.

———. 1989. "How Evaluation Can Improve Performance." In *Improving Government Performance,* ed. J. S. Wholey, K. E. Newcomer, and Associates. San Francisco: Jossey-Bass. 1–12.

PART 2

INSTITUTIONS, POLITICS, AND PRACTICE

2

Evaluation of Environmental Programs: Limitations and Innovative Applications

LAWRENCE S. SOLOMON

Since the late 1960s program evaluation has been a major part of tracking the effectiveness of programs in the public sector. Such evaluations have traditionally focused on studies in education, income maintenance, criminal justice, housing, and health. These evaluations are generally conducted to uncover the extent of progress toward the eradication of poverty or other social problems (Shadish, Cook, and Leviton 1991). Beginning in the 1980s, however, such "human service" evaluations have been joined by studies that consider technology, national defense programs, transportation, risk, agricultural practices, and the environment. Of these areas, the environment has become particularly crucial as an object of program evaluation.

The growing presence of the community right-to-know legislation, solid and hazardous waste recycling programs, and pollution prevention legislation define a great need for environmental program evaluation (U.S. EPA 1991b). The resources dedicated to these programs argue for an understanding of how well the programs have worked. As noted by the U.S. Environmental Protection Agency (1988), several hundred billion dollars have been spent to mitigate environmental problems. As of 1990, the costs of all pollution control activities were estimated to be $115 billion per year, with this amount expected to increase to $170 billion annually by the year 2000 (U.S. EPA 1991a).

In light of the massive resources spent on environmental programs, evaluators are faced with some critical problems when attempting to assess these programs. In this chapter I delineate those concerns and, in response, present evaluation designs developed by the U.S. General Accounting Office (GAO).

The development process of environmental evaluation methodology is not a purely scientific exercise devoid of political concerns. Within a national policy context, the EPA is affected by interest groups, business and industry, and other stakeholders, including supporters and opponents of environmental regulations. Since its inception, the EPA has been soundly criticized by all of these groups. Pro-environmental forces are interested in how well the EPA is addressing environmental problems; groups with fewer environmental concerns favor the use of evaluation to demonstrate the lack of EPA progress, hence demonstrating what is perceived to be a waste of resources.

Both supporters and opponents of environmental regulation favor environmental program evaluation but may seek to use the evaluation results differently. If the quality of information used in environmental program evaluation is poor, stakeholders can selectively choose the study results that support their respective positions, thereby intensifying conflict over the mission of the EPA. While environmental program evaluation will never be an apolitical exercise, the improvement of information used in evaluation will, if only incrementally, decrease the subjectivity surrounding the effectiveness of EPA programs.

Evaluating the Environment: Why the Difficulty?

Environmental programs can be evaluated via impact, process, and efficiency designs (see the introduction). In 1991 the EPA began using an "environmental indicators" approach (U.S. EPA 1991b) to study program impact. This approach seeks a bottom-line response to the question, "Is the environment any cleaner as a result of this program?" This type of design, while end-focused, does not do justice to the *means* by which such goals are achieved. If, for example, an impact evaluation shows little or no improvement in environmental quality, a process evaluation may be required to illuminate major reasons. Finally, resource limitations may lead to the invocation of efficiency studies, in which economic dimensions are central.

Regardless of the evaluation design chosen, precision in the evaluation of environmental programs has not been established. Clifford S. Russell (1990) has extensively examined major problems endemic to environmental monitoring and enforcement. As he notes, the measurement of qualities and characteristics of pollutants differ from source to source; for example, nitrogen has various chemical states between a gas (appearing in waste gases) and a liquid (appearing in wastewater). In addition, time is crucial in monitoring and enforcement activities. The frequency with which monitoring is conducted has a direct effect upon the amounts and types of reported environmental discharges. Furthermore, different environments required different monitoring schedules; for example, the extent of water monitoring is insufficient in urban areas (Russell 1990) while air monitoring activities should be increased in rural areas (Portney 1990).

Based upon a survey conducted by Resources for the Future, Russell delineates specific flaws in the current system of environmental monitoring and enforcement, all of which impact the ability to obtain reliable and valid data. The current system has a heavy emphasis upon self-monitoring by industry, infrequent and nonrigorous auditing of this monitoring by regulators, and relatively lenient treatment of defining violations and enforcing violation sanctions. Roger C. Dower (1990) has demonstrated the implications of this system for the hazardous waste area; as he notes, significant disagreement exists among the EPA, the Congressional Budget Office, and the Chemical Manufacturers Association about the ultimate fate of hazardous wastes. In addition, the estimates of the numbers of abandoned hazardous waste sites run from 130,000 to 600,000, depending on the definition used.

The practitioner's viewpoint adds understanding to these points: the GAO's modis operandi is to evaluate environmental programs on the basis of federal legislative directives. But, practical concerns and resource availability may lead to design programs that deviate from these directives; in some cases such deviation is justifiable. In 1994 the GAO found a great deal of diversity in state pollution prevention program goals and activities (U.S. GAO 1994). Even though federal legislation emphasizes source reduction of waste, several state programs continue to emphasize waste recycling, treatment, and disposal.[1] Thus, guiding legislation may have little effect on program intentions and activities. Further, state programs have been required to adapt to unique political and economic climates. In some states, popular support for pollution prevention programs has been muted, particularly where economies are depressed. Pollution prevention can prove economically beneficial for companies over the long term, but many firms are concerned about short-term survival (see the discussion of "total cost assessment" in Tellus Institute 1991). In some of these cases, state program officials attempt to "hook" companies by offering help in recycling, treatment, and disposal. Once a bond of trust is developed the program officials begin to push source reduction techniques.

In addition to these reasons, measurement flaws, inability to determine causality, the "fear of evaluation," lack of usable standards, and resource constraints account for the difficulty in evaluating environmental programs.

Measurement

Although nationwide environmental initiatives have been in effect since the late 1960s, evaluators have been unable to systematically quantify the extent of progress made in addressing these initiatives. Instead, policymakers and the general public have been left with disparate types of anecdotal evidence. In a comprehensive review, the GAO (1995b) found that many of the data the EPA uses to characterize environmental quality are either incomplete, missing, or obsolete—problems that encompass all media areas. In particular, in the hazardous waste area, the chief reason for the lack of evidence is incomplete and flawed data. For example, as the EPA noted in *Pollution Prevention 1991: Progress in Reducing Industrial Pollutants* (1991b) available data do not cover the multitude of pollutants (e.g., air particulates, fugitive emissions, chemicals) or all the environmental media (air, land, and water). Environmental data that do exist tend to be incomplete and exhibit serious measurement flaws. For example, the GAO (1991f, 1992) found that EPA questionnaires used to gather hazardous waste information from industry include ambiguities in the definitions of key concepts, terms, and questions; governmental reporting requirements; and the relationship of production amounts and types to the volume of hazardous waste produced. In 1995 the GAO found several other examples of measurement problems in the hazardous waste area. A review of facility data for 45,000 waste handlers in one EPA region identified more than 15,000 reliability errors (U.S. GAO 1995c). The GAO (1995c) reviewed data elements in the groundwater area that found reliability error rates from 8 to 37 percent. These data problems result in concrete consequences; for example, as noted by the GAO (1995c), benefit assessments (e.g. in-depth analyses of the benefits and risks of pesticides) conducted by the EPA are often imprecise because they are likely to be based upon poor data. These ambiguities cause reliability and validity problems for waste data.

Inability to Determine Causality

Donald T. Campbell and Julian C. Stanley's seminal book of 1963 provided the basis for evaluation research on federal programs that were developed by Joseph Wholey and colleagues: "Essential conditions for successful evaluation of a federal program are the existence of the methodology and sound measurements that will make it possible to distinguish the program's effects, if any, from the effects of all the other forces working in a situation—to isolate what happened as a result of the program from what would probably have happened anyway" (1970:86).

Methodology used to evaluate the environment greatly falls short of this ideal. For example, as shown above, flawed hazardous waste data can partially be traced to weaknesses in measurement instruments. While EPA procedures for developing these instruments contribute to data problems,[2] the complex nature of the environment provides the most challenge. Consider the case of nonpoint source pollution, which is diffuse runoff of pollutants from farmlands, forests, city streets, construction sites, mines, and other areas.[3] This type of pollution is often mobilized during rainstorms and other types of inclement weather. By its nature, nonpoint source pollution is inherently difficult to isolate and trace. Unlike, for example, firms that discharge wastes via National Pollution Discharge Elimination System permits, such pollution cannot easily be traced to a source at which even rough estimates of volume and toxicity could be made. Therefore, even if a public program is instituted to mitigate the effects of nonpoint source pollution, the failure to measure its outcome in terms of impacts on environmental media makes a causal inference difficult to determine.

A similar problem has occurred in attempting to evaluate changes in surface water quality (U.S. GAO 1991g). By the year 2000, the federal government will be spending in excess of $16 billion annually to assist municipalities in constructing and upgrading wastewater treatment plants (U.S. EPA 1991a). In 1991 the GAO was asked to determine the condition of the nation's water quality, how this quality had changed over time, the pollution sources that degrade water quality, and the effects of the Construction Grant Program. Because of sparse time series data and other methodological problems, the GAO was not able to draw definitive and generalizable responses to these questions.

Environmental evaluation is also at the mercy of the lack of ability in performing trendline analysis on pollution reduction activities. For example, in the pollution prevention area, the EPA (1991b) and the GAO (1991f, 1992) agree that there is a lack of understanding about how long a pollution prevention action should take to show appreciable effects on the environment. In particular, process changes at the facility level vary greatly in the amount of time taken to show effects.

"Fear of Evaluation"

"Fear of evaluation" is certainly not exclusive to the environmental field. However, the resistance to evaluation exhibited by environmental program officials has roots endemic to this substantive area. The evaluation of the effectiveness of hazardous waste minimization provides an example. While the data problems in this area may be extensive (U.S. GAO 1992), these problems may not prohibit those in positions of power from using data to suggest that a program is ineffective, especially in a time of fiscal austerity.[4] While some evaluators and policymakers are sensitive to data quality issues, it is likely that bottom-line conclusions could be derived without considering the complexity of this problem.

Performance Standards: Nonexistent or Inappropriately Used

What defines an effective environmental program? Some environmental program evaluation areas have long suffered from a lack or inappropriate use of performance standards. For example, as noted by the National Roundtable of State Pollution Prevention Programs (1991), the effectiveness of state-level pollution prevention programs is difficult to determine because of the absence of indicators that demonstrate success in pollution reduction. While the EPA has *regulatory* standards, none exist for pollution prevention effectiveness (see Haynes, Ratick, and Cummings-Saxton in this volume).[5]

An inappropriate use of performance standards has also hampered efforts in evaluating program effects in the groundwater protection area. Groundwater has many other uses than for drinking, including irrigation of crops, watering livestock, and supporting aquatic life. Even so, the GAO found that states continue to use drinking water standards as groundwater standards, even though actual groundwater quality may, in some cases, surpass drinking water standards (U.S. GAO 1989). This is particularly important since, by implication, the use of drinking water standards would potentially allow increased degradation of groundwater.

Resource Constraints

Given the difficulties of evaluating the environment, it can be argued that environmental program officials may be unable to retain or justify resource procurement for such evaluation. As noted by William R. Shadish Jr., Thomas D. Cook, and Laura C. Leviton (1991), funds for evaluation declined markedly through the 1980s. Since environmental evaluation is a part of the less "tried and true" physical systems evaluation arena,[6] budget and other resource shortfalls can limit its practice. While one can argue that evaluation may justify the expenditure by identifying weak programs, the current legislative arena at the state and federal levels may complicate this assertion. For example, a majority of states have enabling legislation that expressly mandates pollution prevention or recycling programs (U.S. EPA 1991b). At the federal level, public pressure reflected in the Community Right-to-Know Act strongly suggests that the government is expected to address environmental concerns. This legislation is primary in the environmental arena because of its extremely visible roots, the disaster in Bhopal, India. This accident caused officials to examine closely the potential for chemical accidents in the United States and to give citizens access to information about hazards. As noted by Riley E.

Dunlap (1992), public awareness of environmental protection issues began to rise significantly during the early 1980s and reached unprecedented levels by the 1990 Earth Day celebration.[7] In particular, during this time period the public has shown increasing levels of concern over specific environmental problems, including air and water quality and the greenhouse effect. For these reasons, barring severe funding shortages, environmental programs are not likely to cease operations.[8] However, priorities may allow environmental evaluation to be conducted only marginally, especially since data problems are in evidence (U.S. GAO 1991f, 1992). For example, as noted by the GAO (1991c), the EPA has been unable to evaluate progress in achieving waste minimization partially because of a lack of staff and funding while faced with admittedly flawed or unavailable data (U.S. EPA 1991b). Furthermore, as the GAO found in the groundwater protection area, states rely on information from the federal government on drinking water standards (U.S. GAO 1991b). Resource constraints do not allow states to gather this information on their own. Generally, states report that they need more federal guidance in this area and more information than is provided by the federal government. In 1995 the GAO (1995a) studied a related issue under the requirements of the Clean Air Act Amendments of 1990. These requirements call for states to collect, analyze, and report a great deal of information on air quality to the EPA, and a concern was raised in Congress that these activities would cause a burden to the EPA and the states. The GAO found that these requirements were indeed in excess of EPA program needs. However, while the EPA noted that less data would satisfy program needs, other important data that could contribute to a better program would not be obtained. As a result, less evaluative data would be made available (see Mangun in this volume).

As demonstrated above, environmental programs are difficult to evaluate. While several reasons for this difficulty emerge, the issue of information quality and accessibility forms the basis for all of the aforementioned problems. The relationship of information to the determination of causality suggests that program effects cannot be parceled out so as to determine endogenous, exogenous, intervening, and irrelevant factors. To operationalize concepts, an evaluator must be able to draw upon standards of performance, and these standards do not often exist. Even the aspect of "fear of evaluation" shows a direct connection with information; a program may work quite well but faulty measurement muffles this "fact." And, without the ability to measure program performance, environmental program officials will likely hesitate to dedicate large resource efforts toward developing and implementing complex program evaluations.

The GAO's Approach to Environmental Evaluation

The above discussion indicates how the GAO has concentrated on methodological issues in environmental program evaluation. The GAO applies and utilizes methodological techniques in a quest to understand the quality of federal programs. Specifically, the GAO concentrates on the quality of environmental data, the adequacy and appropriateness of methodological techniques used to gather these data, and the integrity of analytic techniques. By so doing, the major methodological constraints that exist in developing fully operational evaluations are highlighted, whether they be oriented toward *program* process, efficiency, or (perhaps most importantly) outcome measures that demonstrate effects upon environmental media. By recommending methodological changes to the EPA and other federal agencies, evaluators can improve data collection procedures, measurement quality, information availability, and information accessibility sufficiently so as to allow valid determination of program quality. The following summarizes the GAO's findings regarding methodology in major environmental substantive areas, including hazardous waste, reproductive and developmental toxicants, and groundwater protection.

Hazardous Waste

Some of the most challenging questions for evaluators include the determination of how much hazardous waste is being generated, how much the volume generated has changed over time, and how to characterize the various types of hazardous waste. Problems in hazardous waste data quality are apparent at both the federal and state levels. This is an important consideration, since the EPA and state environmental agencies share responsibility for hazardous waste management and environmental reporting (U.S. GAO 1990). According to the GAO's findings, this federal-state partnership has proved problematic in ensuring the development of reliable and valid hazardous waste data (U.S. GAO 1990). The EPA has not required states to use a standardized form to report necessary hazardous waste data, which has resulted in inconsistent information. Furthermore, the EPA has not required states to submit data in a disaggregated form. If data were submitted in such a manner, the EPA would be better able to delve into data inconsistencies (U.S. GAO 1990). These problems have greatly hampered the federal-state program Superfund Amendments and Reorganization Act Capacity Assurance, an exercise whereby states demonstrate that they have adequate capacity to handle anticipated waste production (U.S. GAO 1991e). In 1991, the EPA began operating the Resource Conservation and Recovery Information System (RCRIS) to assist the federal-state program in managing hazardous waste information. The GAO (1995c) evaluated RCRIS in 1995 and found that it had not met some of its primary objectives. Although the system was developed to be flexible (to meet the needs of individual states), most users found data entry and retrieval difficult.

The GAO has also evaluated measurement instruments used to collect hazardous waste data. Under the Resource Conservation and Recovery Act reporting system, the EPA has conducted a biennial census of hazardous waste generators, a sample survey,[9] and special tracking of several hundred chemical releases, the Toxic Chemical Release Inventory (U.S. GAO 1990). These measurement instruments contain flaws that seriously compromise hazardous waste data quality (U.S. GAO 1991f, 1992). To demonstrate the extent of data reliability and validity, the GAO administered a retest of the National Survey of Hazardous Waste Generators to a subsample of respondents (U.S. GAO 1991c, 1991f). Several concepts, terms, and questions are not defined for respondents in the survey. For example, the term *waste minimization program* was not defined for respondents, allowing those who filled out the survey to justify nearly every action as evidence of a program to reduce waste output. Some of the officials told the GAO that what was claimed to be an overt waste minimization program was nothing more than an employee suggestion program (U.S. GAO 1991f, 1992).

Particular problems with hazardous waste data reliability exist. As shown in table 2.1, for nearly every facility surveyed vast differences exist between the amounts of waste reported to the EPA and the amounts reported to the GAO. This problem is a result of faulty data definition and reporting. The federal-state partnership has been the cause of some of these discrepancies. For example, some states' requirements about the inclusion of hazardous waste types are more stringent than the federal requirements.[10] These requirements have also caused confusion among respondents to the EPA Generator Survey (U.S. GAO 1991f, 1992). A further problem is that codes used to categorize waste types and the source of these wastes are neither hierarchical, mutually exclusive, nor exhaustive, resulting in responses that greatly rely upon the individual judgments made by survey respondents. Because of these methodological flaws, the GAO has recommended that several improvements be made in EPA data collection activities. These recommendations include the clarification of hazardous waste coding schemes, improving record-keeping systems for industry, and intensifying quality assurance and control procedures for hazardous waste data.

Currently, there is little agreement between the EPA and industry on how to document, measure, and report waste types and amounts. Even though industry officials must certify the accuracy of data that are reported, no system exists that could verify the quality of these data. The general animosity between

Table 2.1. Total Waste Stream Production Amounts (Tons) Reported to the EPA and the GAO

Facility	Reported to EPA	Reported to GAO	Difference
A	347,636	222,635	125,001
B	195,500	7,791	187,709
C	435,400	476,833	−41,432
D	273,576	430,917	−157,341
E	283,333	283,333	0
F	652,104	2,272	649,832
G	183,382	23,760	159,622
H	279,791	233,197	46,594
I	1,218,333	1,513	1,216,820
J	175,900	48,864	127,036
K	634,205	634,205	0
L	546,797	980	545,817
M	219,305	219,071	234
N	1,952,000	5,287,187	−3,335,187
O	255,908	113,349	142,559
P	417,041	127,873	289,168
Q	322,049	32,624	289,425
R	242,609	18,731	223,877
S	209,846	468,181	−258,334
T	963,714	119	963,594
U	186,381	186,381	0
V	217,281	246	217,034
W	2,566,665	2,566,665	0
X	278,756	280,902	−2,146
Y	805,437	805,437	0
Total	13,862,949	12,473,066	1,389,884
Total absolute difference			8,978,764
Mean absolute difference			359,151

Source: U.S. GAO 1992:21.

the EPA and industry is further attenuated by the laborious reporting requirements, some of which have little value to company operations.

Reproductive and Developmental Toxicants

As noted above, environmental program evaluation has been hampered by a lack of usable standards of performance. Although standards of performance must exist to assess the effectiveness of any program, they cannot be developed in the absence of clear designations of environmentally harmful substances. Such was the case in the GAO work on reproductive and developmental toxicants. The GAO was asked by the Senate Committee on Governmental Affairs to evaluate federal regulatory actions on environmental chemicals known to cause adverse reproductive and developmental outcomes (U.S. GAO 1991d). These regulatory actions could not be evaluated without an authoritative list of relevant toxicants, and such a list could not be located.[11] Therefore, the GAO developed a final list of thirty chemicals by conducting a literature synthesis, surveying experts in the field of toxicants, and analyzing data on each of the chemicals from two national databases. Quite assuredly, without this list, meaningful program evaluations could not be conducted. Even when data on reproductive and developmental toxicants are available, GAO findings suggest that federal agencies are often unlikely to examine these data when making regulatory decisions (U.S. GAO 1991d). The GAO analyzed the rates of examining reproductive and developmental data for the Consumer Product Safety Commission, the Food and Drug Administration, several offices within the EPA, and the Occupational Safety and Health Administration. Overall, these offices used such data only 44 percent of the time even while making 138 major regulatory decisions.[12] As a result, agencies may be unlikely to utilize data to conduct evaluations of programs geared toward mitigation of the effects of reproductive and developmental toxicants. On the basis of these findings, the GAO recommended that the heads of the four agencies vastly improve the availability of data on reproductive and developmental toxicants. Specifically, the agency heads were told to review regulations on chemicals to ensure that they provide sufficient protection against reproductive and developmental diseases (and to revise them if necessary), perform separate analyses for reproductive and developmental outcomes in risk assessments, and ask Congress for the power to demand reproductive and developmental toxicity test data from entities that manufacture, import, sell, emit, or discard critical chemicals.

Water Quality

The GAO has also done extensive methodological work in the area of groundwater protection; a 1991 study examined the vulnerability of groundwater to contamination by pesticides (U.S. GAO 1991a). The EPA's proposed strategy for management of groundwater protection is to allow states great freedom to create management plans based on individual geographic areas, since hydrogeologic vulnerability to contamination varies from area to area.[13] Within states, the county could be used as the basis for these plans, but only if such a unit makes methodological sense. The GAO found, in fact, that the variability of hydrogeologic vulnerability is not significantly smaller *within* counties when contrasted to the state or national level of analysis. As in other substantive areas, problems of measurement become a great obstacle in developing reliable and valid measures of groundwater vulnerability to pesticide contamination.

Therefore, the ability to conduct evaluations of EPA programs to protect groundwater from potential contamination is hampered. As a result of this finding, the GAO has recommended that the EPA provide explicit guidance to the states on how to determine the geographic scale at which vulnerability assessments should be conducted to assure adequate protection.[14]

In the area of water quality, the GAO (1986b) developed an evaluation method of the Construction Grants Program. This method grew from the knowledge gaps identified in the EPA's evaluations of the program (U.S. GAO 1986a). The GAO identified that past evaluations of the program merely identified the changes in sewage-treatment plant efficiency that resulted from upgrades; some of the evaluations did not show the connection between plant discharge and the changes in water-stream quality. The GAO developed a method in which four areas are included: the effects of a sewage-treatment plant upgrade upon discharge levels, the history of water quality in the receiving stream, how these two measures relate over time, and the consideration of alternative explanations for the quality of stream water.

Future Prospects for Environmental Program Evaluation

As suggested at the beginning of this chapter, several problem areas emerge for the environmental program evaluation field. These include information quality and accessibility, an inability to determine causality, the "fear of evaluation," lack of usable standards of performance, and resource constraints. Of these dimensions, information problems are likely to come first in the priority hierarchy, as they give impetus to other problem areas. For example, in a study of the quality of waste minimization data, the GAO concluded that one of the major problems of hazardous waste data is based upon faulty record-keeping systems of industry (U.S. GAO 1992, 1991f). However, without measurement categories that have better methodological precision, improved record-keeping will not aid in the problem.

This argument holds for problems of determining causality. More precise measurement is required to allow environmental evaluators to determine causal priority, the effects of intervening and confounding factors, and the extent and nature of measurement error. Perhaps the fear of evaluation would be somewhat tempered by a greater level of confidence in measurement abilities; if environmental program officials were more confident about evaluative data, they would be less likely to dread the evaluation process. Furthermore, standards of performance cannot be derived without better measurement. Finally, while resource constraints affect the practice of program evaluation, greater confidence in data quality may allow agency fiscal officers to see more use for such studies.

The mission of the GAO is to evaluate the success of federal programs as objectively as possible. When evaluating environmental programs, the GAO has been hampered by weakness in both the quantity and quality of data. As a result, the agency has developed and applied some innovative approaches to environmental program evaluation. Although limited in the ability to evaluate program outcome, the GAO is very adept at evaluating the quality of environmental information. Because of its unique status as a federal legislative agency, the GAO is routinely able to access information to conduct such evaluations. The EPA (and other federal agencies, departments, and offices) is legally required to cooperate with the GAO. As a matter of course, private organizations also cooperate with the GAO, even though they may not be required to. Therefore, the GAO is able to triangulate its evaluation results and can provide fair and balanced evaluations.

The GAO will continue in its mission to evaluate methodology used by federal agencies on environmental programs. The chief goal is to vastly improve, through agency recommendations and reports to congressional officials, the quality of federal environmental data. With this accomplished, the bottom-line results of environmental programs will become available for evaluation.

NOTES

The opinions expressed are those of the author and not necessarily those of the U.S. General Accounting Office.

1. See *The Pollution Prevention Act of 1990,* which states: "The Congress hereby declares it to be the national policy of the United States that pollution should be prevented or reduced at the source whenever feasible. Pollution that cannot be prevented should be recycled in an environmentally safe manner whenever feasible; pollution that cannot be prevented or recycled should be treated in an environmentally safe manner whenever feasible; and disposal or other releases into the environment should be employed only as a last resort and should be conducted in an environmentally safe manner."

2. As noted by the GAO, organizational difficulties and resource constraints at the EPA have contributed to low quality data on the environment (see U.S. GAO 1991c, 1991g).

3. Agricultural runoff accounts for the greatest portion of nonpoint source pollution (50 to 70 percent), mostly from soil erosion, pesticides, and fertilizer applications. Other sources include urban runoff (5 to 15 percent), hydromodification (the change in surface waters when modifications occur because of lake drainage, flood prevention, or reservoir construction) (5 to 15 percent), resource extraction (1 to 10 percent), forestry operations (1 to 5 percent), construction practices (1 to 5 percent), and land disposal (1 to 5 percent). By volume, sediment (sand, clay, silt) is the largest contributor to nonpoint source pollution.

4. However, see the argument below on resource constraints. While environmental programs may be seen as ineffective by officials in power, public pressure is unlikely to allow massive cuts of environmental programs. The "fear" may thus be somewhat exaggerated among some program officials.

5. However, the American Society for Testing and Materials has a pollution prevention work group that is striving for standard-setting (Warren 1992).

6. The assumption is that traditional human services evaluation, as a more established area, would be more likely to withstand fiscal constraints.

7. To reach these conclusions, Dunlap examined public opinion polls conducted by Roper, the National Opinion Research Center, Cambridge, and the *New York Times*/CBS.

8. Alternatively, programs may change organizational form by moving to other parts of the government. However, the immediacy of environmental degradation requires action of some sort.

9. This is the 1986 National Survey of Hazardous Waste Generators.

10. States are allowed to exceed federal requirements, but must not be any less stringent.

11. The earlier efforts of the Congressional Research Service (CRS) to locate such a list were also unsuccessful. The CRS requested eight agencies to identify reproductive toxicants. Six of these agencies either failed to respond or stated that they did not develop such lists.

12. The most frequent reasons for the nonuse of the data included "the office mandate did not require it," the data were "unavailable," and the "agency focus was not on reproduction and development."

13. The strategy was issued in final form in October 1991.

14. The type of guidance is the responsibility of the EPA. The GAO merely recommends that guidance should be provided; it is beyond the scope of the GAO's mandate to make this determination.

REFERENCES

Campbell, D. T., and J. C. Stanley. 1963. *Experimental and Quasi-Experimental Designs for Research.* Chicago: Rand McNally.

Dower, R. C. 1990. "Hazardous Wastes." In *Public Policies for Environmental Protection,* ed. P. R. Portney. Washington, D.C.: Resources for the Future. 151–93.

Dunlap, R. E. 1992. "Trends in Public Opinion toward Environmental Issues: 1965–1990." In *American Environmentalism: The U.S. Environmental Movement,* ed. R. E. Dunlap and A. G. Mertig. Philadelphia: Taylor and Francis. 89–116.

National Roundtable of State Pollution Prevention Programs. 1991. *Position Paper: Evaluating Pollution Prevention Program Effectiveness.* Minneapolis: Waste Reduction Institute for Training and Applications Research, Inc.

Portney, P. R. 1990. "Air Pollution Policy." In *Public Policies for Environmental Protection,* ed. P. R. Portney. Washington, D.C.: Resources for the Future. 27–96.

Russell, C. S. 1990. "Monitoring and Enforcement." In *Public Policies for Environmen-*

tal Protection, ed. P. R. Portney. Washington, D.C.: Resources for the Future. 243–74.

Shadish, W. R., Jr., T. D. Cook, and L. C. Leviton. 1991. *Foundations of Program Evaluation: Theories of Practice.* Newbury Park, Calif.: Sage Publications.

Tellus Institute. 1991. *Alternative Approaches to the Financial Evaluation of Industrial Pollution Prevention Investments.* Boston.

U.S. Environmental Protection Agency. 1988. *Environmental Progress and Challenges: EPA's Update.* Washington, D.C.: U.S. Environmental Protection Agency.

———. 1991a. *Environmental Investments: The Cost of a Clean Environment.* Washington, D.C.: Island Press.

———. 1991b. *Pollution Prevention 1991: Progress on Reducing Industrial Pollutants.* Washington, D.C.: Office of Solid Waste and Office of Toxic Substances. EPA 21P-3003.

U.S. General Accounting Office. 1986a. *Water Quality: An Evaluation Method for the Construction Grant Program—Case Studies.* Washington, D.C. GAO/PEMD-87-4B.

———. 1986b. *Water Quality: An Evaluation Method for the Construction Grant Program—Methodology.* Washington, D.C. GAO/PEMD-87-4A.

———. 1989. *Groundwater Protection: The Use of Drinking Water Standards by the States.* Washington, D.C. GAO/PEMD-89-1.

———. 1990. *Hazardous Waste: EPA's Generation and Management Data Need Further Improvement.* Washington, D.C. GAO/PEMD-90-3.

———. 1991a. *Groundwater Protection: Measurement of Relative Vulnerability to Pesticide Contamination.* Washington, D.C. GAO/PEMD-92-8.

———. 1991b. *Groundwater Quality: State Activities to Guard against Contaminants.* Washington, D.C. GAO/PEMD-88-5.

———. 1991c. *Hazardous Waste: Data Management Problems Delay EPA's Assessment of Minimization Efforts.* Washington, D.C. GAO/RCED-91-131.

———. 1991d. *Reproductive and Developmental Toxicants: Regulatory Actions Provide Uncertain Protection.* Washington, D.C. GAO/PEMD-92-3.

———. 1991e. *SARA Capacity Assurance: Data Problems Underlying the 1989 Assessments.* Washington, D.C. GAO/T-PEMD-91-4.

———. 1991f. *Waste Minimization: EPA Data Are Severely Flawed.* Washington, D.C. GAO/PEMD-91-21.

———. 1991g. *Water Pollution: Greater EPA Leadership Needed to Reduce Nonpoint Source Pollution.* Washington, D.C. GAO/RCED-91-10.

———. 1992. *Waste Minimization: Major Problems of Data Reliability and Validity Identified.* Washington, D.C. GAO/PEMD-92-16.

———. 1994. *Pollution Prevention: EPA Should Reexamine the Objectives and Sustainability of State Programs.* Washington, D.C. GAO/PEMD-94-8.

———. 1995a. *Air Pollution: EPA Data Gathering Efforts Would Have Imposed a Burden on States.* Washington, D.C. GAO/AIMD-95-160.

———. 1995b. *Environmental Protection: EPA's Problems with Collection and Management of Scientific Data and Its Effort to Address Them.* Washington, D.C. GAO/TRCED-95-174.

———. 1995c. *Hazardous Waste: Benefits of EPA's Information System Are Limited.* Washington, D.C. GAO/AIMD-95-167.

Warren, J. 1992. "Development of ASTM Pollution Prevention Standards." Paper presented at the spring conference of the National Roundtable of State Pollution Prevention Programs, Raleigh, Apr. 23.

Wholey, J., J. W. Scanlon, H. G. Duffy, J. S. Fukomoto, and L. M. Vogt. 1970. *Federal Evaluation Policy: Analyzing the Effects of Public Programs.* Washington, D.C.: Urban Institute.

3

Why Institutions Matter in Program Evaluation: The Case of the EPA's Pollution Prevention Program

WALTER A. ROSENBAUM

Few approaches to pollution management in the federal government have been more often extolled, and more habitually ignored, than pollution prevention. Beginning in 1976, congressional law and the U.S. Environmental Protection Agency (EPA) regulations repeatedly declared it public policy that pollution prevention should be not just a major pollution management strategy but the preferred approach. Despite persuasive economic and technological arguments in favor of pollution prevention programs, and some tentative EPA excursions into related activities during the 1980s, not until 1990 did Congress deliver to the EPA a forceful, credible, and irresistible legislative mandate for agency-wide pollution prevention initiatives. Pollution prevention has finally acquired political cachet—it was, for instance, among the EPA major budget initiatives for FY 1996 (U.S. Senate Committee on the Environment and Public Works 1995:37). It has become an important item on any agenda for environmental program evaluation at the EPA.

Since 1990 the EPA has repeatedly attempted to evaluate its pollution prevention programs. Two of the most elaborate efforts, in 1990 and 1992, are especially worth examination. They illustrate in specific agency terms the problems inherent in evaluating environmental programs identified by the General Accounting Office (GAO) and reported by Lawrence S. Solomon in chapter 2. They also betray problems that proved to be inherent in later program evaluations as well. There is ample evidence, as these cases illuminate, that the EPA's pollution prevention program assessments were flawed by inadequate and often inaccessible data, difficulties of causal inference, nonexistent performance standards, and, especially, "fear of evaluation" by both evaluator and evaluated. In another important perspective, the EPA's pollution prevention programs are also one of the earliest, most ambitious efforts to achieve an "integrated" or "cross-media" approach to federal pollution management so often advocated as an alternative to the media-based pollution strategies embodied in almost all federal legislation such as the Clean Air Act and the Clean Water Act (Bartlett 1990). The EPA's experience throws into sharp relief many practical evaluation problems distinctive to this regulatory approach.

But there is more. Fear of evaluation turns out, like the symptom of an organizational neurosis, to express a more complex, deeply imbedded set of institutional conflicts and concealed needs that have shaped the whole process of managing and evaluating pollution prevention programs at the EPA. In short, institutional interests and forces, already exerting a potent influence upon program evaluation within the agency and among its regulated firms, can be expected to have an equally significant impact upon the EPA's efforts to reform its evaluation procedures in accord with the recommendations made by the GAO, as discussed at the conclusion of Solomon's chapter. Thus, the EPA experience opens a broad window on the institutional forces likely to be continually operative in shaping program evaluation.

The fate of the EPA's evaluation of its pollution prevention program, at least so far, also imparts credibility to the cautionary observations made by Robert F. Rich about program evaluation strategies developed by academic experts (see his chapter in this volume). One principal obstacle likely to be encountered in efforts to promote better program evaluation at the EPA—and likely to appear in program venues far beyond the EPA's pollution prevention activities—is administrative stonewalling inspired by organizational values and loyalties (especially in the media offices). Another likely obstacle to better evaluation will be subversion by the agency's political process that strives more diligently to enhance the EPA's tenuous congressional popularity by demonstrating program results than to reveal program deficiencies. Additionally, political tensions inherent to the relationship between the agency and its regulated interests create potent disincentives for the affected firms to provide prompt, accurate information to the regulators. It is problematic, in fact, whether such practical approaches to program evaluation as those described by Rich would even be attempted today.

About the Program

The statutory setting from which this analysis begins is the Pollution Prevention Act of 1990 (PL 101-508), the most recent and comprehensive legislative mandate for a federal waste minimization program. The act, however, is virtually a reprise of many earlier, uncoordinated pollution strategies attempted, or contemplated, at the EPA. The act rests heavily on this inherited administrative history, which affects the response of all the program's major stakeholders—Congress, the EPA, regulated interests, private advocates, and critics—to its current implementation. Evaluating the institutional or behavioral impacts of current pollution prevention activities at the EPA makes sense only in the context of this developmental history, which, among other things, creates many of the criteria essential to the evaluation process.

Throughout this discussion, the term *pollution prevention* is used interchangeably with *waste minimization*. In current statutory and regulatory approaches, these terms refer specifically to "in-plant practices that reduce, avoid, or eliminate the generation of hazardous waste so as to reduce risks to health and environment" (U.S. Office of Technology Assessment 1986:13). It is important to emphasize that these terms do not refer to recycling or any other action that reduces or concentrates waste in volume or dilutes it after it is generated—a distinction often lost to waste producers and handlers. In the implementation process, as later analysis reveals, confusion in definitions has become a major impediment to program implementation.

A proper beginning to assessing the institutional forces involved in the EPA's program evaluation is the consideration of two related questions: How did the EPA in fact implement the program and with what consequences? How did the EPA evaluate its own implementation and its consequences? The latter question raises explicitly the matter of what constitutes a program evaluation. The meaning of any program or policy "evaluation," like other aspects of policy analysis, is not self-evident in the literature of policy analysis. Writers ambitious (or inexperienced) enough to attempt a comprehensive explanation of policy analysis typically warn the initiate, as does political scientist Randall Ripley, that "none of the phrases has a fixed and widely accepted meaning" (1985:3) or observe, as does policy analyst Dennis J. Palumbo, that "it is not possible to define evaluation . . . in a way that does not reflect a particular methodological or epistemological bias" (1987:15). So, to evaluate policy, environmental or otherwise, is to choose both policy and technique. Before turning to the EPA's experience with pollution prevention, it is necessary to discuss briefly the methodology for evaluating the EPA experience.

Evaluating the EPA's Program Implementation and Self-Assessment

The EPA's approach to its own program evaluation has been relatively straightforward, fairly simple, not very reliable, and easily described if not easily defended. A more demanding task for the analyst interested in understanding the EPA experience is to evaluate the EPA's own program evaluations. An independent protocol is needed that addresses a second set of criteria for program implementation and impact. The protocol would determine how one should evaluate the evaluator.

In policy evaluation analysts commonly become their own methodologists. Not surprisingly, the paradigms of policy evaluation, like those of policy analysis in general, usually represent a synthesis of different approaches chosen by each analyst according to particular needs, intellectual conviction, or research opportunities (Nagel 1990). The inevitability of methodological choice in program evaluation, and the consequent need for selection criteria, may not be apparent, however, if the evaluation task is defined very broadly, for then the implicit choices involving alternative assumptions about the nature of the policy process may be obscured.

The Inevitability of Methodological Choice

The issues involved in choosing an independent evaluation methodology for the EPA program implementation become readily apparent if one considers the general problem of identifying the institutional and behavioral impacts a program should have on an agency—that is, what one is supposed to evaluate. Consider, for instance, the matter of selecting how behaviors should be judged. Policy analysts commonly suggest, like William N. Dunn and Robert F. Rich, that a policy or program can be evaluated only if there are "clearly specified goals and/or consequences, and a set of explicit assumptions that link policy actions to goals and/or consequences" (Dunn 1981:349; Rich in this volume). The presumption is that behaviors should be judged, at least, by the standard of program goals or anticipated consequences. In fact, Dunn argues that a program cannot be evaluated unless such explicit performance criteria exist.

But what if goals or anticipated consequences are unclear, vague, or inconsistent? For instance, a reading of the preamble to the Hazardous and Solid Waste Amendments of 1984 (HSWA), the first legislative declaration of a federal waste minimization program, reveals a clear confusion of congressional intention that will confound the quest for clear policy goals as a standard for program evaluation. In this instance, Congress failed to clarify whether it intended that waste minimization should be considered before other forms of waste management in industrial handling of hazardous materials (U.S. Office of Technology Assessment 1986:13–14). Moreover, program or policy goals are often, perhaps usually, vague. "Given the . . . grandiosity and diffuseness of program goals," writes evaluation expert Carol H. Weiss, "there tends to be little agreement, even within the program, on which goals are real—real in the sense that effort is actually going into attaining them—and which are window-dressing" (1987:51). Additionally, implementation research has repeatedly documented that program goals or objectives often change in the process of program evaluation. In the case of the Pollution Prevention Act, for example, the EPA grants to the states initiated after 1993 to encourage state pollution prevention programs allowed the states to reward recycling and other after-production processes despite clear statutory language in the act that pollution prevention, not post-production recycling, was the most preferred approach (Blomquist 1995:427–28). In the case of the HSWA, one could make a persuasive case that goal confusion in the legislation itself was an important explanation for its implementation problems. In this instance, the evaluator must make a decision concerning whether confusion of purpose inherent in HSWA is itself an important variable in explaining program outcomes.

What to Measure: The Importance of Implementation

The literature on the policy process and policy analysis provides a great variety of behavioral variables that can be considered when examining the institutional and behavioral impacts of public policies (Schneider and Ingram 1990). From the analyst's viewpoint, the more difficult task is to reduce the number of variables commonly acknowledged to be important to a concise array known to be relevant to the program to be examined and about which information is obtainable.

The evaluation of the EPA's response to its mandated pollution prevention program will focus on program outputs and institutional impacts as expressions of program implementation. The explanation can be briefly stated. First, public policies involve both process and product. A vast body of theoretical and empirical literature demonstrates convincingly that it is impossible to understand or to explain various policy "outputs" or "outcomes" unless one examines the procedures by which outputs or outcomes are created (Nakamura and Smallwood 1980; Ripley 1985; Mazmanian and Sabatier 1983; Alkin 1990; Gumm and Wasby 1981)—to look at what is now commonly called implementation. The importance of the implementation process in evaluating policy impacts will become readily apparent, for example, when the EPA's efforts to implement pollution prevention policies is specifically examined. In particular, it will be apparent that the cross-media nature of these policies creates distinctive administrative implementation problems that, in turn, directly affect the quality of program outputs and impacts. In short, the "institutional" consequences of policy in this perspective include the impacts upon the institutional objectives of policy and the institutions implementing the policy.

Implementation, in its most general sense, involves the "process of carrying out, accomplishing, fulfilling, producing and completing policy" (Pressman and Wildavsky 1973:iii)—in short, almost all the varied activities usually subsumed under the idea of "administration." Moreover, there is a politics of implementation as well as of organization. The political values, actors, and influences inevitable in the administration of public policies such as pollution prevention are an expression of the institutional forces at work in an agency's evaluation of its own activities as well. Thus, they deserve recognition in understanding not only how an agency carries out a policy but also how it assesses its own behavior.

The Importance of Politics to Analysis

Examining the institutional forces acting on an agency's internal policy evaluations seldom endears an investigator to the agency's leadership or whets the leadership's appetite to discover how its evaluation process may have been amiss. There may well be positive benefits, however, to deliberately seeking out the political and institutional forces active in agency self-evaluations. Ignoring the reality of institutional or political influences is unlikely to teach policy evaluators much they need to learn about the utility of their theories in the real administrative world. Peter deLeon's admonition is sensible: "It is . . . incumbent upon policy scientists to turn their analytic light on their own discipline . . . and ask how new theories will operate in the cauldron of the political kitchen" (1988:303).

Talking the language of politics and institutional interest may also get the leadership's attention. It might be the only way to provoke useful change in agency behavior because this language is more persuasive and relevant to the leadership than a politically sanitized version. The attempt to quarantine evaluation theory from politics, write Gary D. Brewer and Peter deLeon, is one reason why the literature on policy evaluation so often reveals "that evaluation has either failed to have much impact on decisions and operations or that much more fundamental social science remains to be done" (1983:357). Carol H. Weiss is more blunt: "Perhaps one of the reasons that evaluations are so readily disregarded is that they address only official goals. If the evaluator also assessed a program's effectiveness in meeting political goals—such as showing that the administrator is 'doing something' or that the program is placating interest groups . . . he might learn more about the measures of success that decision makers value" (1987:54). Small wonder the current literature increasingly eschews policy evaluation divorced from the political setting in which policymaking ordinarily transpires. Nor is it surprising that Rich should list first among the deficiencies in academic program evaluation revealed through experience that "external evaluators are not sensitive to the political process" (see Rich in this volume).

Bringing explicitly "political" variables within the scope of this study is particularly useful because most agencies are loath to do it in their own evaluation processes. A striking illustration of calculated apoliticism can be found even in the evaluation methodologies of the highly respected GAO, whose reports to Congress are often cited for the quality of their professionalism and objectivity. William T. Gormley's study of 120 GAO reports between 1980 and 1991 revealed that the GAO consistently avoided an examination of institutional variables—for example, the professionalism of state legislatures or the resources and relationships among state agencies—when evaluating state or federal implementation of environmental regulation and related activities (1992:14). While the inclusion of institutional variables would often seem valuable, even advisable, in the GAO's analyses, it appears that the exclusion of institutional considerations is a deliberate strategy to avoid congressional criticism and controversy.

The Research Variables

The paradigm in table 3.1 identifies how the essential variables in this evaluation of the EPA's program implementation are conceptually organized.

Table 3.1. Significant Variables to Be Examined when Evaluating the Institutional and Behavioral Impacts of Federal Pollution Prevention Programs

Impacts	Implementation	Outputs
Resources	Mandated	Intended
Activities	Other	Unintended

Once what agency activities to examine is decided, appropriate techniques for obtaining information about the behavioral and institutional impacts of policy and standard(s) for evaluating the impacts must be selected. Implicit in these tasks is a decision concerning what specific programs or policies to study. It is helpful to briefly review the history of the EPA's pollution prevention programs to identify the policies relevant to evaluation and to clarify a few additional policy terms.

The Brief, Fitful, but Significant History of Federal Pollution Prevention

When it comes to evaluating the institutional interests in federal pollution prevention policies, history matters. Congressional efforts in the late 1980s to promote an aggressive EPA pollution prevention program have gone against the grain of the EPA's institutional experience and evolved organizational culture. Further, semantic confusion of historic proportion—for which the EPA and Congress share much responsibility—confounds the implementation process (U.S. GAO 1987).

Irresolution and Resistance

During the years between 1970 and 1990, both Congress and the EPA repeatedly declared through statute and legislation that it was federal policy to promote pollution prevention, embodied in the term *waste minimization,* as the preferred method of hazardous waste treatment. As early as 1976, the EPA's "Statement of Preferred Options" spelled out the differences in waste treatment methods and the agency's commitment to waste minimization as the most desirable option (U.S. EPA 1976:61):

> The Agency believes that reuse, energy recovery and material recovery as well as treatment are desirable prior to ultimate disposal. Thus, the desired waste management options are (in order of priority):
> Waste Reduction
> Waste Separation and Concentration
> Waste Exchange
> Energy/Material Recovery
> Waste Incineration/Treatment
> Secure Ultimate Disposal

Moreover, almost all of the federal government's major statutory programs concerned with hazardous waste management have provided opportunities for waste minimization as a management strategy: the Clean Air Act, the Clean Water Act, the Resource Conservation and Recovery Act, the Toxic Substances Control Act, and Superfund among them.

Congress appeared to declare the primacy of waste minimization among all hazardous waste management strategies in the Hazardous and Solid Waste Amendments of 1984 (HSWA), which amended the Resource Conservation and Recovery Act (RCRA; U.S. Code Title 1982 6901 et seq.) by including a new preamble: "The Congress hereby declares it to be the national policy of the United States that, wherever feasible, the generation of hazardous waste is to be reduced or eliminated as expeditiously as possible. Waste nevertheless generated should be treated, stored, or disposed of so as to minimize the present and future threat to human health and the environment" (Resource Conservation and Recovery Act of 1976, section 1003 as amended). So much for intentions. By 1990, the EPA had done little to promote waste minimization as an important waste management technique and nothing to establish its primacy for that purpose. Nor was there evidence that a significant number of hazardous waste generators in the public and private sectors understood the meaning of "waste minimization" or practiced it. Consequently, a major institutional objective of later legislation was to assure that waste generators understood both the meaning of "waste minimization" and the congressional intention to make this the primary waste management technique.

Why had so little been accomplished? The U.S. Office of Technology Assessment provided one important explanation in the mid-1980s: "Federal law says that waste reduction is the preferred anti-pollution method; but government actions often send a different—or ambiguous—message to waste generators" (1986:6). The 1984 policy statement added to the RCRA does not clearly establish that waste minimization should be attempted before other methods, nor does it eliminate the confusion that exists among many waste producers about the meaning of "waste minimization." (Many producers assume that efficient management of hazardous waste after it is produced also constitutes "waste minimization" and, consequently, the 1984 legislation does not unambiguously establish the primacy of actions to reduce hazardous waste *before* it enters the waste stream.) Moreover, the 1984 legislation did not mandate waste minimization procedures, and waste generators were left to determine what activities constituted waste minimization.

The EPA's organizational culture appears to be another reason for pollution prevention's failure. The EPA's behavior assumes critical importance to the implementation of any federal pollution prevention strategies and the agency's support for pollution prevention was problematic—and still is. It is significant that none of the EPA's regulations implementing its 1976 policy statement or HSWA (1984) contain language repeating the primacy of waste minimization declared in both these enactments. Instead, regulatory language "appears to shift the emphasis of waste minimization from reducing the generation of hazardous waste to reducing land disposal as a hazardous waste management practice" (U.S. Office of Technology Assessment 1986:157). Also, the EPA assigned the responsibility for implementing the 1984 legislation to its Office of Solid Waste, where it assumed a priority so low as to be insignificant. Further evidence of the EPA's attitude appears in the congressional report the agency was required to provide according to the 1984 legislation. The report was supposed to advise Congress on possible methods for requiring the reduction of hazardous waste before it entered the waste stream and to identify the best management practices for those wastes that were generated. But the report largely defined "waste minimization" to mean the type of reduction practiced after waste is created by generators. Further, the report gave no emphasis to the primacy of waste reduction over other kinds of waste management, such as recycling.

The EPA's apparent reluctance to promote waste minimization seems to arise, in good measure, from the agency's "pollution control culture." From its inception in 1970, the EPA's regulatory programs have emphasized technological controls and end-of-the-pipe treatment of pollution together with reduction of residual wastes. The EPA's media offices—Air and Water, for example—live with statutory mandates, budgets, and program elements organized around pollution control, not pollution prevention. Most of the media offices' technical staff are professionally trained in pollution control as well. The agency's organizational experience over more than two decades is almost entirely derived from pollution control. Thus, extremely strong, durable organizational commitments to the philosophy and techniques of pollution control and historically strong disincentives to commit organizational resources to alternative management strategies are daily realities within the agency (Landy, Roberts, and Thomas 1990:chap. 1). Moreover, most state compliance programs are also organized by medium—air, water, or waste—as are many corporate environmental divisions as well. Thus, the effort to develop cross-media approaches to pollution management intrinsic to the pollution prevention strategy rubs against the grain of organizational and professional tradition that focuses environmental management upon media-specific activities in both private and public sectors (Geiser 1995:489).

It is against this background of congressional irresolution and congressional belief in the EPA's intractability that enactment of the Pollution Control Act of 1990 and later efforts to implement it must be evaluated. History, in effect, already provides a partial agenda of goals for evaluating the 1990 legislation's institutional impact:

1. The extent to which "pollution prevention" and "waste minimization" are clearly defined by statute and regulation and the clarity with which such definitions are communicated to waste-producing institutions, and used by them, are important institutional impact criteria.
2. The character of the EPA's administrative response to the new statutory mandates becomes a major institutional consideration in understanding and evaluating the law's impact on potential target institutions.

The Pollution Prevention Act of 1990

Congressional dissatisfaction with the EPA's pollution prevention programs was especially intense in the Senate Committee on the Environment and Public Works, which provided the initiative for the 1990 Pollution Prevention Act (PPA). The legislation is the most explicit and detailed congressional enactment concerning pollution prevention, the statutory framework that should set the design for any evaluation of the EPA's pollution prevention programs.

The legislation had several major objectives. It was supposed to send a message to the EPA that apathy about pollution prevention must end. The legislation required the EPA to establish the Office of Pollution Prevention (OPP) to carry out its statutory objectives and to review for the administrator the impact of proposed environmental regulations on source reduction. The Senate committee's report emphasized its high expectations for the new office: The EPA was to "establish an office independent of the Agency's single-medium program offices but shall have the authority to review and advise such offices on their activities to promote a multi-media approach to source reduction." Additionally committees intended "for this office [are] to be established within the Agency in a manner to reflect the importance, and multi-media significance of the functions of the new office." The committee also scolded the EPA: "The nation's pollution prevention efforts are scattered and uncoordinated, lack permanent institutional support . . . and for the most part pale in comparison to the established pollution control culture. . . . The EPA has not made much progress in using its current information collection systems to measure national waste reduction" (U.S. Senate Committee on the Environment and Public Works 1990:4; see also U.S. Senate Committee on the Environment and Public Works, Subcommittee on Environmental Protection 1990). To make its intention stupefyingly clear, the report added yet another admonition "that pollution prevention should be considered a high agency priority and should obtain the level of resources necessary to implement an effective source reduction policy and program" (U.S. Senate Committee on the Environment and Public Works 1990:6). Also: "The Committee is concerned that pollution prevention has not received the institutional support and priority necessary to effectively promote source reduction" (1990:7). This latter comment reflects a sentiment that resonated throughout congressional reviews of EPA pollution prevention programs during much of the 1990s (see, for example, U.S. House Committee on Science, Space, and Technology 1990:43–45).

The second statutory objective was to declare unambiguously the congressional intention that waste minimization should be the primary means of hazardous waste management. The act declared "the following pollution hierarchy as national policy; to reduce or eliminate pollution before it is created whenever feasible; to recycle pollutants which are generated in an environmentally safe manner, whenever feasible; to treat pollutants in an environmentally safe manner, whenever feasible; to treat pollutants which cannot be prevented or recycled in an environmentally safe manner" (section 2, PL 101-508). In a succeeding section *source reduction* was defined in a manner to distinguish it clearly from recycling and other waste management practices with which it is often confused.

The third major objective was to promote the voluntary reduction of hazardous wastes.

The fourth objective was to mandate specific pollution prevention activities at the EPA:

1. Promoting activities within the EPA that prevent the generation, emission, or discharge of hazardous wastes and other pollutants at their sources.
2. Making matching grants to states for technical assistance to businesses seeking to prevent the export of pollution.
3. Establishing a national clearinghouse on source reduction techniques.
4. Coordinating and improving the collection of information with respect to source reduction, recycling, and treatment practices under the Resource Conservation and Recovery Act and the Superfund Amendments and Reauthorization Act.
5. Establishing and maintaining a national database containing information on source reduction practices and opportunities for specific industries.

The act further defined in detail the nature of these activities, as if to leave the EPA as little room as possible for interpretation.

Altogether, the PPA, despite its brevity, constituted the most thorough, comprehensive statement of congressional expectations for the agency's pollution prevention activities and provided apparently clear guidelines for assessing the institutional impacts of the new legislation.

The EPA's Program Implementation and Self-Assessment

The Pollution Prevention Act became law in 1990 and by early 1991 the EPA was initiating its first efforts to comply with the new legislation. If one were to believe the agency's self-evaluations, the years since then are a tale of moderate but growing success in program implementation and impact—an opinion, however, not shared by most other program reviews (see, for example, U.S. GAO 1994; Geiser 1995). It is still too early to confidently describe, let alone evaluate, the full impact of all the act's mandated activities and other institutional objectives. However, it is possible to suggest in broad outline what the agency was attempting and what it appears to have accomplished so far. Table 3.2 identifies the initial six mandated activities upon which this analysis focuses because these activities were clearly considered essential for the program's integrity by the legislation writers.

Table 3.2. Evaluation Categories for Assessing the Institutional Impacts of the Pollution Prevention Act of 1990 (implementation FY 1992)

Mandated Activities	Outputs	Impacts
1. Establish pollution prevention office independent of EPA media offices with the authority to review and advise media offices on actions to promote source reduction	Office of Pollution Prevention internally reorganized; under authority of new EPA assistant administrator	
2. Review EPA activities and regulations to consider their effect on source reduction		
3. Establish standard methods for measuring source reduction, develop source reduction auditing procedures	Review of existing database; major change in methodology initiated; review of external data	
4. Facilitate adoption of source reduction techniques by business	Creation of 33/50 and Green Lights Programs.	
5. Identify measurable goals and establish milestones for the EPA to follow to implement the act	In preliminary stages	
6. Provide Congress with a report 18 months after law's enactment, and biannually thereafter, on various activities designated in the act	Report delayed until October 1992	

Program Implementation: Data Problems

Nothing better illustrates the salience of policy process in understanding policy outcomes than the strategic importance assumed by the EPA's data-gathering procedures in determining the program's impact on waste producers. Several of the PPA's major program objectives—the review of the EPA activities to assess their impact on source reduction and the promotion of source reduction by industry, for instance—depend upon the EPA's competent implementation of another PPA mandate to obtain accurate information about the quantity and type of wastes involved, their present modes of treatment, and related data. The initial data used by the EPA to implement the new law was obtained from its 1986 National Survey of Hazardous Waste generators required by RCRA. In 1986, long before the PPA's passage, it was already apparent to the EPA that the questionnaire was poorly designed and the acquired information wholly inadequate for the mandated waste survey (U.S. GAO 1991:1)—so deficient that the agency could not provide Congress in 1986 with legislatively required information on the need for mandatory waste minimization. The EPA redesigned its information system, using the revised procedure to gather further waste management data between 1987 and 1990. By 1991, however, the revised procedures had proven so flawed that the EPA had to abandon all the major analyses of waste minimization progress the agency had promised to provide to Congress.

Mistakes in the EPA's data system were egregious. For instance, 60 percent of the respondents did not report one or more of the data items needed to calculate changes in waste generation per unit of production between 1985 and 1986. Missing data was so pervasive in the EPA's surveys that one EPA analyst called it "the cancer of the generator survey" (U.S. GAO 1991:21). The GAO's evaluation of the data is damning: "The Agency will not be able to answer any of the central questions the Congress has asked about the extent of the progress, if any, that industry has made on minimizing waste generation. These questions require information on the amount of waste generated per unit of production and the industry characteristics (including voluntary efforts) that may affect the feasibility of waste minimization" (U.S. GAO 1991:7). The EPA officials, in fact, later indicated that their modified analyses would not even address these questions.

Enactment of the PPA has apparently done little to improve data quality. A later GAO study in mid-1992 found many of the same problems. For instance, the surveys used to obtain waste minimization information from respondents did not define the meaning of "waste minimization." This produced survey results suggesting, quite wrongly, that waste minimization was epidemic throughout the private sector. According to another GAO study, the questionnaire's ambiguity "caused officials of generating firms to attempt to define nearly every action on their part as a program to reduce the output of hazardous waste. Consequently, the reported number of existing waste minimization programs appears to be greatly inflated" (U.S. GAO 1992a:3). In brief, most waste producers did not understand the meaning, let alone appreciate the primacy, of waste minimization in federal law.

Another kind of data problem impeding the agency's efforts to demonstrate compliance with the PPA, for which it bears no responsibility, is the difficulty of measuring internal pollution prevention activities with an administrative or budget metric. These activities are inherently cross-media in administrative terms, imbedded within different media office activities and crossing bureaucratic office boundaries. The problems this poses for the EPA is evident in EPA administrator Carol Browner's response to a question posed by Congressman Estaban Torres (D-Calif.) during the agency's FY 1994 budget hearings in the House of Representatives:

> The budget request . . . doesn't reflect the total resources that are being devoted to integrating prevention into EPA's programs and policies. As the Agency's program offices develop their regulatory packages, they continue to devote more of their analyses and staff time to investigating the performance and feasibility of prevention approaches. The amount of resources that this time and energy reflects is difficult to break out because it is fully integrated into the regulatory development process. (U.S. House Committee on Appropriations 1993:581)

Implementation Problems: A Hostile Institutional Culture

The OPP has repeatedly evaluated and reevaluated its survey techniques and database for the pollution prevention program. Many of the EPA's persisting data problems and survey deficiencies revealed in these and other evaluations could have easily been anticipated, and perhaps could still be prevented, thereby leaving the agency vulnerable to criticism that it is incompetent, or worse, in its approach to pollution prevention. Some critics, especially those associated with relevant congressional committees and their staff, have suspected the EPA of deliberate program subversion (U.S. House Committee on Science, Space, and Technology 1990:26–31). It is more likely that these implementation difficulties arise from two conditions intrinsic to the EPA's regulatory role: the pervasive influence of the "pollution control culture" in shaping agency attitudes toward pollution prevention and the political tensions arising from regulatory relations with the states and waste-producing firms. Additionally, many agency offices and staff had little professional or bureaucratic investment in waste minimization prior to 1990 and many still do not despite increasing efforts by the EPA's leadership to promote a "pollution prevention mentality" among its regulatory personnel (Hirschhorn 1995).

Institutional Confusion

During the first two years of the PPA's implementation, the OPP, intended to coordinate and implement most of the act's provisions, endured continuing rumors of a major administrative relocation, congressional accusations of indifference to the PPA's mandate, and uncertainty about its administrative future. The OPP's precarious administrative status was another impediment to program implementation and was not eliminated until a major EPA reorganization in the fall of 1991.

The OPP was the responsibility of the assistant administrator in the Office of Policy Evaluation and Enforcement (OPEE) at the time of the PPA's passage. Congressional proponents of the act, however, long suspected that the OPEE lacked enthusiasm for pollution prevention—one reason why the act's mandates were so meticulously detailed. Proposals to place the OPP under a more sympathetic administrative venue within the EPA had been discussed even before the act was passed. The legislation had scarcely been enacted when congressional sources began to complain about the OPEE's management of the act and to suggest that the OPP should be moved. Twice in February 1991, for instance, congressional sources publicly criticized the OPEE's first major administrative actions under the new legislation. "Pollution prevention is only a priority in political rhetoric," asserted one critic (*Inside the EPA* 1991:1).

By March, rumors spread that EPA administrator William P. Reilly planned to move the OPP to the agency's Office of Pesticides and Toxics, where the leadership was presumably more congenial to pollution prevention. Uncertainty about the OPP's future status began to affect staff morale and raise large questions about its contemplated initiatives. The OPP's tenuous status further complicated an implementation problem already inherent in the nature of its mandate. Although the PPA required the OPP to create policy initiatives and to exert significant influence in the EPA's other major program offices, the OPP appeared to lack the administrative status required to exert strong influence on other EPA programs—it was, after all, only one office within the OPEE, not administratively superior to the program offices.

In early fall 1991, rumor became reality: the Office of Pollution Prevention was moved to the Office of Pesticides and Toxic Substances. To assure that pollution prevention would assume more visibility and status within the EPA, however, EPA administrator Reilly added to his own staff a special assistant for pollution prevention as well. While the OPP's administrative relocation ended uncertainties about its future, it did little to eliminate the OPP's status problems with other program offices. The OPP remained one program unit in one of the EPA's major media offices. The incongruity between its administrative status and its mandate to affect policy in the other program program offices remained. And it remained despite the addition of a position charged with pollution prevention in the administrator's own office.

Despite these reforms, congressional distrust of the EPA's commitment to pollution prevention remains. Moreover, many of the OPP's staff members share this mistrust. It is widely believed in the OPP that its 1992 appropriations contained language requiring the office to create "sector" strategies for all major EPA media offices because some influential staff members convinced the relevant congressional committees that the OPP would concentrate almost exclusively on industrial pollution prevention unless prevented by law.

Thus, the OPP's first eighteen months under the PPA were characterized by administrative and political problems that apparently impeded its program implementation and raised questions about the program's credibility. These problems were not necessarily all, or mostly, the fault of the OPP's leadership or staff. Responsibility for much of this difficulty can probably be attributed to the "pollution control" culture in the EPA and the difficulty of administratively implementing cross-media programs in powerful media program offices.

Regulatory politics is another nettlesome issue. Regulated firms were annoyed that the PPA required them to provide additional data about pollution prevention activities in their annual toxic release inventory report already required by Congress, a new task expected to cost 112,000 firms collectively an additional $50 million in 1992 and more than $36 million in each subsequent year (Grayson 1992:10392). Moreover, they suspected they were flirting with more grief. It was evident from the PPA's inception that the agency would use such data to determine whether a voluntary pollution prevention program would have to be replaced with enforced prevention. "Once EPA is in possession of sufficient data to demonstrate the viability of prevention-related technologies, it is possible that such options will be transformed into mandatory obligations imposed by regulation," warned one legal newsletter highly regarded among regulated firms in early 1992. "Since prevention-related technologies are often site specific," it noted ominously, "this scenario creates an enormous disincentive for industry compliance" (Grayson 1992:10395).

The states also contributed to the political complexities. The EPA depends on state reports for much of its program impact evaluation but the states have been persistently inconsistent in how they define and measure "pollution prevention," notwithstanding the clear wording of the PPA. The National Roundtable on State Pollution Programs, a voice for state pollution prevention officials, complained as early as 1993 that the state reports were too variable to establish meaningful measures or indexes of pollution prevention and that such unreliable data might be used to justify withdrawal of federal funding from states with apparently less effective programs (Geiser 1995). The EPA prefers not to enforce a uniform definition of "pollution prevention" in state reports and thus perpetuates its own data credibility problems. From the agency viewpoint this is seemingly the lesser evil. The alternative would likely provoke opposition from numerous states that suspect that uniform reporting requirements will invite invidious comparisons between states and penalties for those seemingly lackluster in their pollution prevention achievements (Blomquist 1995:393). Thus, the PPA has quickly become deeply embedded in the fabric of federal regulatory politics in ways that powerfully discourage the EPA from aggressively developing a coherent database for the PPA.

Not surprisingly, these administrative tribulations have done little to improve the credibility of the EPA's pollution prevention data with the public and private institutions who constitute its principal data consumers. And political tensions within the program, and between the agency and Congress, diminished confidence early among those state governments the EPA was attempting to influence. This is apparent if one briefly examines the EPA's so-called groundwater program of the early 1990s.

Implementation: An Evaluation from Target Institutions

In 1984, the EPA adopted a "groundwater strategy" to guide EPA and state officials in groundwater protection programs. The strategy was meant to promote the prevention of groundwater contamination as well as the cleanup of polluted waters. The strategy committed the EPA to providing financial and technical help to states to develop and implement groundwater protection programs. It also pledged the EPA to attain "agencywide consistency" in policies affecting groundwater—in effect, a coordinated administrative policy between major media offices. In 1988, well before the enactment of the PPA, the EPA's own evaluation of its groundwater strategy revealed that the states for which the strategy was intended did not understand or accept it. In 1990 the EPA adopted a "new" version of this groundwater strategy, intended to improve its credibility. The EPA's commitment to a more "aggressive" groundwater program was further emphasized in the strategy's final version in 1991.

In late 1991, however, a GAO study revealed that the EPA's groundwater program was ineffectual and unconvincing to most of the state agencies to which it was directed. State officials complained about inconsistent groundwater protection standards among different EPA media offices, lack of coordination among the offices, the pervasive preference for remediation rather than pollution prevention in EPA groundwater programs, and lack of resources committed to the new groundwater strategy. Most state officials doubted that the EPA was seriously committed to its pollution prevention program. One state official noted, as an example, that the EPA was currently providing his state with only $100,000 in federal assistance, barely enough to support an office with one staff member but not enough for a statewide program (U.S. GAO 1992b:22). The EPA's program budgets apparently sent a more persuasive message to the states than did its administrative pronouncements. State officials noted that between FY 1986 and FY 1991 the EPA had spent approximately $10 million on its Wellhead Protection Program, a groundwater pollution prevention activity, while at the same time spending an average of $25 million on cleanup of each hazardous waste site—an apparently convincing demonstration of the EPA bias for remediation rather than pollution prevention.

It seemed apparent that by mid-1992, about eighteen months after the passage of the PPA, the newly mandated programs were afflicted with a variety of implementation problems, many inherited from prior EPA pollution prevention activities. In light of these technical and administrative difficulties, it would be reasonable to conclude that the institutional impact of most programs, such as the source data collection system or state groundwater protection programs previously examined, had yet to be determined or, at best, was very modest in light of expectations. One would think quite differently, however, if the EPA's self-evaluations are to be trusted. Although none of the major programs had undergone even rigorous internal evaluation by mid-1992, the agency's public report, like the administrator's later congressional testimony, advertised many accomplishments.

Success by Acclamation: The EPA Evaluates Itself

In April 1992, the EPA issued *Securing Our Legacy,* a handsome, multicolor public report on its activities. The manner in which the agency evaluated its pollution prevention achievements is instructive. The report includes program activities that Congress excluded from the PPA, emphasizes several voluntary pollution prevention programs whose actual accomplishments cannot be measured with the existing EPA data resources, adds additional programs at very early implementation stages, and includes one clearly unsuccessful program.

"In the 1990s," begins the report, "pollution prevention has become a cornerstone of the EPA's work" (U.S. EPA 1992:20). The narrative then celebrates the accomplishment of two voluntary pollution prevention programs to which public and private institutions have pledged their cooperation. The Green Lights

Program claims a commitment from more than 400 corporations and many state and local governments to use energy-efficient lighting. The 33/50 Project involves a pledge from more than 700 companies to reduce the release or transfer of seventeen high-priority toxic chemicals by 33 percent at the end of 1992 and by 50 percent at the end of 1993. The program impacts seemed impressive. The EPA estimated that Green Lights would save $700 million yearly in electric bills and reduce air emissions of carbon dioxide, sulfur dioxide, and nitrogen oxide. The 33/50 Program, asserted the agency, would eliminate a "projected" release of 300 million pounds of air toxics by 1995. In 1995, the agency reported that the most recent data available from the toxic release inventory (covering the period through 1994) indicated the 1995 goal might be achieved two years early (U.S. EPA 1997). However, the accuracy of most EPA data and predictions used for pollution prevention reports and for Green Lights remains problematic. The EPA's reporting methods, based largely on state-provided information, do not reliably identify toxic release declines attributable only to pollution prevention, nor do they assure the reliability of estimates for "projected" toxic releases. And all Green Lights data are voluntary.

Thus, it is ultimately impossible to establish the validity of either program's alleged impacts with the EPA's current database. Nor is confirmation likely as long as both programs remain voluntary.

Several other reported program accomplishments seem inappropriate. A large portion of the pollution prevention report is concerned with recycling, an activity neither Congress nor most pollution prevention experts now include in the definition of *waste minimization*. Among "additional accomplishments" the EPA also lists the Wellhead Protection Program, whose lackluster history has been previously examined (U.S. GAO 1991). Most of the other reported activities, such as an ambitious multimedia initiative to control nonpoint pollution, consist largely of instruction projects, grants for research or demonstration, or new regulations and guidelines—that is, program outputs, not impacts.

Conclusions: Data Dilemmas, "Takeoffs," and Other Lessons to Be Learned

Viewed in the broad perspective of the EPA's total responsibilities, the OPP appears to be a small program with an attractive public image, an office with more potential than accomplishments, and a minor leaguer compared with the big-budget, high visibility, politically weighty media programs. Nonetheless, it has its lessons to teach of wider applicability than the constricted scope of the OPP's present influence.

Why the EPA's Evaluation Data Are Chronically Deficient

Any reliable evaluation of the OPP's implementation and impacts requires an enormous volume of high-quality empirical data on the character of materials involved in the production processes of all potential waste minimization practitioners. Here, impact indicators become critical to the agency's self-evaluation and to outside evaluators of agency performance. The needed information includes units of output and production, materials used in production, changes in waste streams over time, and much else that can be provided only by waste producers themselves. An evaluator has relatively little difficulty knowing what type of information to obtain. Getting it may be nearly impossible. The issue is unavoidable. Technical data that provide the basis for indicators to measure program impacts upon waste producers are so essential that they should be considered the sine qua non of any institutional evaluation. But the Office of Technology Assessment has called the data requirements "staggering" and warns, additionally, that waste producers possessing such information will resist its disclosure because they consider it a proprietary matter. Moreover, there is a well-founded fear that confidentiality cannot be guaranteed by the EPA (U.S. Office of Technology Assessment 1986:127).

This information problem partially explains the EPA's preference for voluntary waste minimization initiatives as a measure of program accomplishment. Many experts assert that a regulatory program for waste minimization—that is, a nonvoluntary approach—is predestined to founder in a morass of data deficiencies. For the policy analyst, the data problem poses a Hobson's choice between reliance upon voluntary reporting of program impacts, independent verification of reported impacts, or reliance upon currently flawed survey data. Significantly, the EPA has chosen to report voluntary information about program impacts in its public documents, particularly those addressed to the general public and public officials, without explaining, or even acknowledging, the serious problems involved in using this kind of data.

The EPA's difficulties with program indicators are not entirely its fault. The problem applies in varying degree to most environmental regulatory programs. Evaluating pollution prevention activities depends upon monitoring, indicator construction, and data management—among the least congressionally valued and supported administrative activities in any environmental regulatory agency. Because monitoring and related activities lack political "sex appeal," this situation is unlikely to change soon. But this state does not seem to trouble the OPP. And the OPP's apparently nonchalant approach to its waste minimization survey data is not peculiar to that office. The creation and use of high-quality evaluative data for program review is rare among the EPA program offices, in part because evaluation activities are permeated with political menace. The OPP cannot expect much organizational help even if it did choose to become more aggressive and articulate about its data management problems. The EPA's "pollution control culture" breeds a hostile organizational environment for the OPP in which institutional allies are hard to find. In short, there are not a lot of organizational incentives to improve data management within the OPP.

"Takeoffs" Are More Interesting than Landings

The EPA's leadership and the OPP are investing most of their bureaucratic resources (personnel, money, publicity, political capital) in what can be called the "takeoff" stage of implementation—new initiatives, output of regulations and guidelines, grants, and other activities that announce the program is underway. Relatively few resources are invested, or are likely to be invested, in careful impact assessment and review. This should not be surprising, if one is prepared to look at the politics of program development. From a political viewpoint, as Carol H. Weiss observes, agencies have far more to gain from advertising initiatives than from evaluating results: "It is often more important to a politically astute official to launch a program with great fanfare to show how much he is doing than to worry about how effectively the program serves the people's needs" (1987:55)

Organizational Culture as a Subversive Force

The impacts of the EPA's pollution prevention program cannot be properly evaluated without a sensitivity to the character of the implementation process and especially the nature of the EPA's organizational culture. One cannot, for instance, understand the EPA's plodding approach to pollution prevention, with the resulting failures of early program objectives, without recognition of the "pollution control culture," the political and administrative strength of the EPA's media offices in comparison to the OPP, and the problems of administratively implementing a cross-media policy mandate.

The OPP, in brief, is having, and will continue to have, considerable difficulty in program implementation because of its inhospitable organizational setting while at the same time it feels compelled to produce evidence to Congress and the public of "results." Powerful incentives are thereby created to follow the path of least resistance by producing evaluation documents that satisfy the OPP's need for some plausible proof of success in the face of organizational realities that continually frustrate such accomplishments. The path, in this instance, is papered with policy self-evaluations of dubious quality whose virtue, apart from their advertisement of program "success," is to avoid identifying the internal agency forces that frustrate better program implementation.

External Influence on Policy Evaluation

Finally, to understand the EPA's organizational behavior, pollution prevention achievements, or failures, one must also be sensitive to the administrative implications of congressional political agendas. Congressional distrust of the EPA's commitments to pollution prevention, and the resulting determination to micromanage the OPP's activities, resulted in organizational confusion and conflict that contributed substantially to delays in program implementation and to the OPP's difficulty in influencing the EPA media offices. Congressional desire to show "results" placed enormous pressure on the EPA to give operational priority and status to those mandates in the PPA that could most readily produce an appearance of accomplishment. Moreover, the EPA's leadership had powerful incentives to promote program elements whose "success" could be most easily demonstrated as a way of improving the agency's legislative stature.

If, as *Alice's Adventures in Wonderland* insists, stories must have a moral, this excursion into institutional self-evaluation leads to at least one challenge to the evaluator. In many ways, the most important variables in explaining how the EPA evaluated its pollution prevention program appear the least amenable to measurement or objectification. In fact, those agency officials most knowledgeable are often the most reluctant to discuss them. One must somehow get to these institutional forces in program implementation if one is to explain satisfactorily why agency evaluations turn out as they do. "Rational analysis carried on in an ignorance of political reality," observes Michael Carley, "may well end up so divorced from social reality as to be of little use to anyone" (1990:6–7).

REFERENCES

Alkin, M. C. 1990. *Debates on Evaluation*. Newbury Park, Calif.: Sage Publications.
Bartlett, R. V. 1990. "Comprehensive Environmental Decisionmaking: Can It Work?" In *Environmental Policy in the 1990s,* ed. N. J. Vig and M. E. Kraft. Washington, D.C.: Congressional Quarterly Press. 235–56.
Blomquist, R. F. 1995. "Government's Role Regarding Industrial Pollution Prevention in the United States." *Georgia Law Review* 29 (3): 349–448.
Brewer, G. D., and P. deLeon. 1983. *The Foundations of Policy Analysis*. Homewood, Ill.: Dorsey Press.
Carley, M. 1990. *Rational Techniques in Policy Analysis*. London: Heineman.
deLeon, P. 1988. "The Contextual Burden of Policy Design." *Policy Studies Journal* 17 (2): 297–309.

Dunn, W. N. 1981. *An Introduction to Public Policy Analysis.* Englewood Cliffs, N.J.: Prentice-Hall.
Geiser, K. 1995. "The Unfinished Business of Pollution Prevention." *Georgia Law Review* 29 (3): 473–91.
Gormley, W. T. 1992. *Political Methodology at the GAO.* Washington, D.C.: Georgetown University Public Policy Program.
Grayson, E. L. 1992. "The Pollution Prevention Act of 1990: Emergence of a New Environmental Policy." *Environmental Law Reporter* 22 (June): 10392–10396.
Gumm, J. G., and S. L. Wasby. 1981. *The Analysis of Policy Impact.* Lexington, Mass.: Lexington Books.
Hirschhorn, J. S. 1995. "Pollution Prevention Act of 1990: Emergence of a New Environmental Policy." *Georgia Law Review* 29 (3): 325–47.
Inside the EPA Weekly Report 12 (Feb. 8, 1991): 1.
Landy, M. K., M. J. Roberts, and S. R. Thomas. 1990. *The Environmental Protection Agency: Asking the Wrong Questions.* New York: Oxford University Press.
Mazmanian, D. A., and P. A. Sabatier. 1983. *Implementation and Public Policy.* Glenview, Ill.: Scott, Foresman.
Nagel, S. S. 1990. *Policy Theory and Policy Evaluation: Concepts, Knowledge, Causes, and Norms.* New York: Greenwood Press.
Nakamura, R. T., and F. Smallwood. 1980. *The Politics of Implementation.* New York: St. Martin's Press.
Palumbo, D. J., ed. 1987. *The Politics of Program Evaluation.* Newbury Park, Calif.: Sage Publications.
Pressman, J. L., and A. Wildavsky. 1973. *Implementation.* Berkeley: University of California Press.
Ripley, R. 1985. *Policy Analysis in Political Science.* Chicago, Ill.: Nelson-Hall Publishers.
Schneider, A. L., and H. Ingram. 1990. "Policy Design: Elements, Premises, and Strategies." In *Policy Theory and Policy Evaluation,* ed. S. S. Nagel. New York: Greenwood Press. 77–102.
U.S. Environmental Protection Agency. 1976. "Statement of Preferred Option." *Federal Register* 41 (CFR 35050) (Aug. 18): 61.
———. Office of Communications, Education, and Public Affairs. 1992. *Securing Our Legacy: The EPA Progress Report, 1989–1991.* Washington, D.C.: Environmental Protection Agency. EPA 175 R-92-001.
———. Office of Pollution Prevention and Toxics. 1997. "EPA's 30/50 Program." http://es.inel.gov/partners/3350/3350.html#progress (Feb. 3, 1997).
U.S. General Accounting Office. 1987. *Hazardous Waste: Uncertainties of Existing Data.* Washington, D.C.: GPO. GAO/PEMD 87-11BR.
———. 1991. *Waste Minimization: The EPA Data Are Severely Flawed.* Washington, D.C.: GPO. GAO/PEMD 91-21.
———. 1992a. *Waste Minimization: Major Problems of Data Reliability and Validity Identified.* GAO/PEMD 92-16.
———. 1992b. *Water Pollution: More Emphasis Needed on Prevention in the EPA's Efforts to Protect Groundwater.* GAO/RCED 92-47.

———. 1994. *Pollution Prevention: EPA Should Reexamine the Objectives and Sustainability of State Programs.* Washington, D.C.: GPO. GAO/PMED 94-8.

U.S. House Committee on Appropriations. Subcommittee on VA, HUD, and Independent Agencies. 1992. *Departments of Veterans Affairs and Housing and Urban Development and Independent Agencies Appropriations for 1993: Hearings.* 102 Cong., 2d sess. H. Rept. 102-710.

———. 1993. *Departments of Veterans Affairs and Housing and Urban Development and Independent Agencies Appropriations for 1994: Hearings.* 103 Cong., 2d sess. H. Rept. 103-555.

U.S. House Committee on Science, Space, and Technology. Subcommittee on Natural Resources, Agriculture Research, and the Environment. 1990. *Solid Waste Reduction, Recycling, Pollution Prevention: Hearings.* 101 Cong., 1st sess. H. Rept. 101-107.

U.S. Office of Technology Assessment. 1986. *Serious Reduction of Hazardous Waste.* Washington, D.C.: GPO. OYA-ITE 317.

U.S. Senate Committee on the Environment and Public Works. 1990. *Pollution Prevention Act of 1990: Report.* 101 Cong., 2d sess. S. Doc. 101-526.

———. Subcommittee on Environmental Protection. 1990. *Hearing: Pollution Prevention.* 101 Cong., 2d sess. S. Doc. 101-670.

———. 1995. *Environmental Protection Agency's Fiscal Year 1996 Budget: Hearing before the Committee on Environment and Public Works.* 104th Cong., 1st sess. S. Hearing 104-48.

Weiss, C. H. 1987. "Where Politics and Evaluation Research Meet." In *The Politics of Program Evaluation,* ed. D. J. Palumbo. Newbury Park, Calif.: Sage Publications. 47–71.

4

Environmental Program Evaluation in an Intergovernmental Context

WILLIAM R. MANGUN

Environmental management in the United States has been an intergovernmental affair at least since the Clean Air Act of 1970 and the Clean Water Act of 1972. Following these precedent-setting acts most federal legislation has required state and local governments to develop environmental legislation and accompanying programs to implement federal environmental policies. State and local governments, however, vary significantly in resources and willingness to comply with the intent of the federal legislation and the regulations established by the U.S. Environmental Protection Agency (EPA). Further, most federal environmental acts contain subsections that require states to manage different aspects of air and water quality in different ways. This jurisdictional and substantive fragmentation of federal environmental programs complicates the task of environmental program evaluation.

Due in part to the complexity of federal environmental programs, environmental program evaluation at both the state and federal levels tends to focus on process. Very few evaluations focus on actual changes in environmental quality. Perceptions differ considerably across levels of government, as well as from one state agency to the next, as to how environmental program evaluation should be conducted. Any assessment of environmental program evaluation efforts is shaped by the perspective of the analysis. For example, an analyst with a federal perspective would be interested in ascertaining the degree to which states have developed programs and initiated enforcement actions required by federal legislation.

State evaluation efforts also tend to focus on process, possibly because performance-based state budgets drive management activities. Annual budget justifications require information on program accomplishments measured in process terms: number of enforcement actions taken, permit applications processed, permits issued, reviews completed, contracts issued during the year under review, and figures compared to those of previous years.

However, as the quality of the environment continues to deteriorate, despite massive expenditures, some state agencies are being pressed to assess environmental quality outcomes. The basis of such concern presumably lies with expectations that program officials will eventually be required to associate program activities with changes in air and water quality. Arizona, California, Connecticut, Florida, Maryland, Oregon, and Washington have initiated substantial efforts toward this end, but it is still too early to assess how well they are succeeding in providing program officials with the information necessary to evaluate the outcomes of specific program activities. More states need to move in this direction if environmental quality is to improve significantly in the United States.

The purposes of this chapter are to describe how the EPA evaluates state responses to federal environmental initiatives; to describe how states evaluate their own environmental programs; and to offer suggestions for improving the evaluation of intergovernmental environmental programs. To address these issues, a brief questionnaire on evaluation efforts was sent to all fifty states. The questionnaires were followed up with interviews of officials in several of the more innovative states and in the EPA.

EPA Oversight of State Environmental Programs

States' responses to federal environmental initiatives vary considerably. As reported in table 4.1, for example, total per capita expenditures for environmental protection in 1988 ranged from $6.78 in Texas to $271.87 in Wyoming. State responses seem to vary according to economic resources, political culture, and severity of environmental problems, among other reasons. Analysts have attempted to identify variables that help to explain why different states use specific processes or develop particular institutional responses to federal environmental initiatives (Lester 1994; Lester and Lombard 1990; Blomquist 1991; Lombard 1994; Lowry 1992; Ringquist 1993). The influence of such variables on the development of environmental programs is difficult to determine with precision, but analysis of such variables provides some basis for explaining the differences in responses across states (Goggin, Bowman, Lester, and O'Toole 1990; Wood 1992). The greater a state's economic resource base or wealth, the more likely it will have the capacity to develop environmental programs (Game 1979; Lester and Keptner 1984; Lowry 1992; Ringquist 1993). A state's political culture has a substantial effect on the willingness of state legislators to develop new environmental programs (Lowry 1992). There is also a relationship between perceived severity of environmental problems and a state government's willingness to develop more or less stringent regulations (Wenner 1976; Lowry 1992; Lester and Lombard 1990).

Despite the range of state responsiveness, the EPA does not conduct specific program evaluations to determine whether individual state programs generate improvements in environmental quality. Instead, EPA scientists establish national standards deemed necessary to protect the health and welfare of humans, while state environmental agencies establish and implement control programs designed to ensure that federal standards are achieved. The EPA then conducts oversight through semiannual program reviews of state enforcement efforts to ensure compliance with federal environmental laws. Such assessments tend to be "bean counting." Instead of determining how well a state's actions have improved the quality of the air or water in that state, the EPA simply identifies the number of facilities or municipalities that are in significant noncompliance with federal environmental regulations in each program area. Statistics on compliance are typically presented in annual compliance reports that identify the number of significant noncompliers by program and by state.

Responsibility for the implementation of federal standards is blurred somewhat by the delegation of environmental management responsibilities to state governments. The federal government gives management responsibilities to states by granting primacy, delegation, or authorization. States typically want to "accept primacy" for program implementation to maintain greater control over regulatory affairs and, therefore, impacts within their respective jurisdictions.

Primacy means that the state agency has the primary responsibility for enforcing state and federal laws, as well as for managing and administering those programs for which primacy has been accepted. To accept primacy, a state government must agree to enact legislation as stringent or more stringent than federal regulations and establish specific programs capable of achieving implementation of federal standards (Crotty 1987). There is some assumption that the state governments will also fund and staff the programs sufficiently to guarantee enforcement of federal standards. The EPA and the respective state governments enter into formal program management agreements and the EPA assumes oversight responsibilities to determine whether state programs are in compliance with federal environmental laws.[1]

Federal delegation of authority means that a state environmental agency accepts responsibility for most of the major components of a federal program. The Alaska Department of Environmental Conservation (ADEC), for example, has accepted responsibility for the federal air quality control program (except for new source performance standards) within Alaska. In contrast, federal authorization of authority means that the EPA could authorize the ADEC to administer a federal program in Alaska that incorporates federal rules directly into state rules. Plans to seek authorization for the hazardous waste program have been postponed in Alaska because of budget constraints (Alaska Department of Environmental Conservation 1993).

Congress requires the EPA to issue periodic reports on the quality of air and water in all states. The Clean Air Act requires the EPA to issue an annual national air quality and emissions trends report and the Clean Water Act requires the EPA to prepare a national water quality inventory biennially. Like the compliance reports, neither of these reports actually provides an assessment of the capacity of state programs to improve air or water quality.

Because of the way the EPA collects data and the sometimes dubious quality of the data collected (as noted by Solomon in this volume),[2] it is easier to ascertain whether a particular state's environmental regulations are substantially in accordance with federal requirements than to determine whether the state's environmental management programs are actually improving environmental quality.

Results of interviews with EPA officials indicate varying levels of evaluation and review of state and local programs. Individual program offices within the EPA headquarters, in Washington, D.C., typically do not collect information on the effectiveness of state environmental programs. In fact, the U.S. General Accounting Office reports that EPA staff members tend to look for numbers first, not quality, and to focus on national priorities instead of environmental results (1988). Because of this, EPA officials in Washington are more likely to maintain records on whether an individual state has formally adopted a rule (or regulation) in the respective office's area of responsibility (e.g., safe drinking water standards or national pollution discharge elimination system pretreatment permitting process), rather than how well a rule has been implemented.

The ten regional offices of the EPA (Boston, New York, Philadelphia, Atlanta, Chicago, Dallas, Kansas City, Denver, San Francisco, and Seattle) are better sources of information about state program effectiveness than EPA headquarters because their responsibilities include direct oversight of state environmental programs within their respective jurisdictions. Actual concern about the effectiveness of state environmental management programs lies appropriately at the state level. However, since no uniform method or system of collecting data about program effectiveness exists from one state to another, comparative analysis is confounded considerably.

On the basis of recommendations contained in a National Academy of

Science report entitled *Rethinking the Ozone Problem in Urban and Regional Air Pollution* (National Research Council 1991), Congress amended the Clean Air Act to require states to prepare reasonable further progress reports to judge the effectiveness of their state implementation plans (SIPs). The basis of the evaluation is a periodic inventory conducted every three years to track emission reduction progress. Through this inventory collection process, the EPA is able to ascertain the rate of progress of each state toward achievement of the national ambient air quality standards. To help the states determine how well they are progressing, the Office of Air Quality Planning and Standards issued *Guidance on the Post-1996 Rate-of-Progress Plan and the Attainment Demonstration* (U.S. EPA 1994), which describes how state agencies are to calculate and present the base year inventory of volatile organic compounds and other special circumstances that should be considered during the development of the rate of progress plans. This guide explains that under the 1990 Clean Air Act, state agencies must produce a base year emissions inventory, periodic inventories, and annual aerometric information retrieval system facility subsystem inventories (U.S. EPA 1994:7). Through such guidance the EPA develops and utilizes emissions estimates and inventories in almost all of their air programs. In general, the SIP evaluation process consists of reviewing proposed or in-place programs and estimated emissions reductions expected from these programs (personal communication with Director Merrylin Zaw-Mon of the Maryland Air and Radiation Management Administration, Nov. 1993).

State Air Pollution Control Programs

Evaluation of state responses to federal mandates under provisions of the Clean Air Act is complicated because a state can accept primacy for implementation of an entire program requirement, a portion of that requirement, or none of a particular requirement. As noted earlier, through acceptance of primacy, states are given the right to create regulations for the control of air pollution within their jurisdictions. A state can accept primacy for the implementation of federal regulations for air pollution control programs dealing with, for example, new source review, prevention of significant deterioration, or national ambient air quality standards.

To determine whether a state has established appropriate regulatory programs, section 107 of the Clean Air Act requires states to develop state implementation plans that must be approved by the EPA. There are SIP requirements for many program elements. For example, there are SIPs for prevention of significant deterioration of each of the national ambient air quality standards (for carbon monoxide, lead, nitrogen oxides, ozone, particulate matter, and sulfur oxides). The SIP, then, serves as a primary means of monitoring state regulatory compliance in the implementation of federal regulations. The attainment status for each of the criteria pollutants is reported for each of the air quality maintenance areas in each state. The number of severe nonattainment areas in 1991 by state, for example, is reported in table 4.2.

Evaluation of state responses is complicated by the highly dynamic situation associated with SIPs; almost continual change occurs pursuant to changes required by the EPA after review of SIP submissions and state responses in the form of new regulatory initiatives. States that had initiated air pollution control programs for air toxics, emission fees, and ozone protection as of 1991 are reported in table 4.3.

Prevention of Significant Deterioration

To assure that the quality of the air does not deteriorate in those parts of the country where the air quality is good, section 163 of the Clean Air Act requires states to develop programs for the prevention of significant deterioration (PSD). As reported in table 4.4, only ten states (including California) had received full delegation of authority for prevention of significant deterioration by the fall of 1992. By this date, the EPA had approved PSD SIPs for thirty-six states, granted partial delegation authority to three states, delegated and approved SIP authority for selected districts in California, and approved no program transfer for South Dakota (U.S. EPA, personal contact, 1992).

Toxic Air Emissions

Title III of the Clean Air Act of 1990 requires states to establish programs to control hazardous air pollutants. Perhaps in anticipation of these provisions, as of 1991 thirty-five states had developed legislation to control toxic air emissions. As reported in table 4.5, the top ten states in the production of toxic air emissions (in thousands of tons) in 1991 were Texas (214.1), Ohio (135.3), Louisiana (135.2), Tennessee (134.2), Michigan (115.8), Indiana (110.7), Illinois (102.1), Pennsylvania (90.6), Utah (84.5), and Florida (50.5) (U.S. EPA 1990). Seven of the ten states generating the greatest amount of air toxics have established legislation to reduce the emission of air toxics. Only Tennessee, Indiana, and Utah do not have air toxics laws. However, out of the top ten toxic emission generators, only Illinois ranks in the top ten in expenditures for air pollution control, ranking fourth in the country.

State Operating Permits

Title V of the Clean Air Act of 1990 requires the EPA to establish a new source operating permit program. On July 21, 1992, the EPA promulgated the final rule for the program requiring states to establish operating permit programs. In anticipation of these requirements a number of states had passed legislation or had legislation pending. The preexisting authority for substantive permits was already adequate in seventeen states; of the thirteen states that enacted legislation, eleven had specifically created new legislation. However, the preexisting authority for program administration was adequate in only eight states; sixteen additional states had enacted new legislation for program administration. Twenty-four states either had adequate preexisting authority or had enacted legislation that imposed fees to offset the administrative costs of a permit program; nine states had legislation before the legislature. The overall status of state operating permit program submissions, as of July 1992, is reported in table 4.6.

State Water Pollution Control Programs

Water Quality Assessment

Under section 305 (b) of the Clean Water Act, each state is required to submit to the EPA an annual water quality assessment report. The assessments are compiled into a biennial report to Congress on the status of water quality in the United States. The 1990 national water quality inventory prepared by the EPA indicated that nontraditional sources of pollution such as runoff from agricultural fields and urban areas, primarily nutrients, are the leading reason for the impairment of surface waters regulated under the Clean Water Act. The Clean Water Act addresses surface waters and the Safe Drinking Water Act addresses both surface and underground drinking water supplies; other underground waters are addressed primarily by state and local authorities and only indirectly by the Clean Water Act. As a result, water pollution control initiatives vary extensively by state (see table 4.7).

The standard measure of water quality reported by the states is the degree to which waters support the uses for which they have been designated, such as high-quality cold water fishing, recreation, or drinking water supply (U.S. EPA 1992b:3). According to the 1990 national water quality inventory 95.4 percent of the river miles assessed by the states fully or partially supported a fishable goal and 92.4 percent a swimmable goal (see table 4.8); while 81.0 percent of acres in lakes and reservoirs fully or partially met a fishable goal and 93.1 percent met a swimmable goal (see table 4.9). Although siltation, nutrients, organic enrichment, and pathogens were reported as the greatest overall pollution causes of impairment of rivers, heavy metals presented major problems in certain areas.

National Pollutant Discharge Elimination System

Section 402 of the Clean Water Act of 1972 established the National Pollutant Discharge Elimination System (NPDES). The NPDES program was designed to eliminate or significantly reduce the discharge of wastes into water bodies from all point sources, especially from industrial and publicly owned wastewater treatment plants throughout the United States. NPDES is a preconstruction review permitting process to limit the amount and type of discharges to rivers, lakes, reservoirs, and estuaries. As reported in table 4.10, only twenty-four states had fully authorized NPDES programs by the fall of 1992 (i.e., programs for federal facilities, pretreatment standards, and general permits). However, the EPA had approved the NPDES permit programs of thirty-nine states for municipal and nonmunicipal facilities; thirty-four states had received EPA approval to regulate federal facilities; and twenty-seven states had approved pretreatment programs (U.S. EPA 1992b). Significant noncompliance with NPDES permits, as of fall 1992, is reported for major NPDES nonmunicipal sources in table 4.11 and for major municipal sources in table 4.12.

Nonpoint Source Pollution

Section 319 of the 1987 Clean Water Act Amendments requires state and local governments to address nonpoint source problems by establishing a comprehensive nonpoint source pollution control program. These programs are to identify waters with nontraditional source impacts, implement management plans to mitigate those impacts, and develop monitoring programs to study the effects of mitigation measures. By the end of 1990, all states had completed EPA-approved nonpoint assessments. The EPA also fully approved thirty-nine state nonpoint source management programs and portions of the remaining state management programs (U.S. EPA 1992b:xx) (see table 4.13 for more information on nonpoint source program status as of January 1990).

Groundwater Protection

The states have been very active in developing groundwater protection programs. As of 1991, forty-four states were in the process of adopting and implementing groundwater protection strategies. Seventeen states had EPA-approved wellhead protection programs and twenty-nine other states had programs under development. Thirty-seven states were in the process of enacting legislation requiring the development of comprehensive groundwater protection programs. Thirty-six states were promulgating groundwater protection regulations. Forty-one states were developing groundwater quality standards. Thirty-five states were developing groundwater monitoring programs (U.S. EPA 1992b:xix).

Public Water Supply Standards

The Safe Drinking Water Act of 1974 was passed in response to concerns about the quality of drinking water obtained from underground and surface water sources. This act led to the establishment of programs for public water supply standards, underground injection control, and sole source aquifer protection. The public water supply standards program is the keystone of the Safe Drinking Water Act. Public water plants are required to monitor drinking water supplies for the presence of selected toxic substances. As is clear from table 4.14, only Wyoming had not accepted primacy for the implementation of the original public water supply standards by the fall of 1991. New rule requirements were established under the Safe Drinking Water Act Amendments of 1986. As reported in table 4.15, forty-nine states had adopted rules on volatile organic compounds and thirty-eight states had adopted public notification in standard-setting as of September 1992, thirty-four states had promulgated rules on surface water treatment, thirty-seven states had adopted rules on total coliform, and thirteen states had adopted rules for phase two pollutants.

State Environmental Program Evaluation Initiatives

Environmental program evaluation at the state level varies widely from one state to another. Formal environmental program evaluations that attempt to relate program actions to changes in environmental quality typically are not conducted in most states. Because state governments are concerned about the manner in which funds are expended, however, environmental agencies typically have to prepare annual, semiannual, or quarterly reports. In these reports, program officials indicate how well their offices performed against objectives listed in annual work plans, such as the previous year's compliance record. A major problem with such "evaluations" is that they tend to address bureaucratic concerns of a "bean counting" nature, i.e., the number of enforcement actions taken, the number of endangered species listed, and the number of plans completed rather than information about the status of air or water quality or the rate of deterioration of environmental quality. Such process-based evaluations are used to reassign staff, adjust priorities, and lower unrealistic goals. The remainder of this chapter provides descriptions of various state environmental program evaluation activities.

Typical State Environmental Program Evaluation Activities

Arizona The following listing of the typical process-based program evaluation efforts conducted by the Arizona Department of Environmental Quality (ADEQ) is common among state environmental agencies. The ADEQ program evaluation efforts fall into five categories:

1. Those required for annual budget submissions (service measurements);
2. Those required by the EPA for program grants—(submitted quarterly and annually);
3. Program and media reporting requirements, e.g., annual reports and groundwater quality reports;
4. Audit evaluations—both internal and external, on an irregular basis;
5. Special analyses for specific purposes, e.g., incentive plans, customer surveys (personal communication with ADEQ Deputy Director William Wiley, Nov. 1993).

The ADEQ's incentive program concept is specifically designed to promote program effectiveness. The incentive program has three phases. In phase 1, ADEQ programs establish performance baselines in key areas. This phase has three parts: key outputs are determined for each program, such as number of permits issued or number of inspections performed; the current performance level relative to the key outputs is established; and the current external customer satisfaction level is determined. In phase 2 the programs develop and implement performance-based incentive (monetary reward) pilot programs. And in phase 3 the programs evaluate the effectiveness of the pilot programs at the end of the performance periods by looking at program productivity, key outputs, work quality, and cost savings. Among the Air Quality Program's goals under the incentive program, for example, are efforts to increase the quantity and quality of air quality permits issued; increase the timeliness of field inspections performed; increase the quality of responses to citizen inquiries and technical inquiries; increase the number and timeliness of SIP revisions processed; and increase the number and quality of amended rules adopted.

Most state program evaluation efforts probably would fall under the process-based categories that ADEQ conducts, but as of 1992 the ADEQ was in the process of developing and adopting an evaluation system based on environmental indicators. Until this system is fully developed, the agency will continue to conduct process-based program evaluations.

Maryland The director of the Maryland Air and Radiation Management Administration reports that many of the individual air programs operated at the state level in Maryland are reviewed by the state in conjunction with annual or biennial reviews by the EPA (personal correspondence, Nov. 1993). The reviews are held in concert with established quality assurance programs. The director points out that, generally speaking, any program for which the EPA has established a formal quality assurance program is reviewed regularly and improves substantially within a few years.

Examples of individual programs reviewed on a regular basis by the EPA include the Ambient Air Monitoring Program and vehicle inspection and maintenance programs. The state agency usually prepares for EPA reviews by conducting program-level evaluations. The director indicates that the evaluation of the Ambient Air Monitoring Program has several parts:

1. The collected data are reviewed for usefulness and the network siting is reviewed and confirmed. The monitor siting is evaluated by its contribution to the data-collection goals of the program. For some pollutants, appropriateness of the monitoring site can be determined through a saturation study.
2. A formal review of randomly selected data and quality assurance checks is performed by the EPA.
3. Two types of independent testing are performed during the year: cylinders of pollutant gases are shipped to the agency to determine the concentrations using network analyzers and an auditor does on-site accuracy audits for selected analyzers.
4. The data themselves are reviewed for completeness, accuracy, and reasonableness.

Maryland air program officials have found that monitoring programs and vehicle inspection and maintenance programs are easier to evaluate because they lend themselves to definitive accuracy testing. Other programs with multiple variables that affect their performance are more difficult to evaluate.

Innovative State Environmental Program Evaluation Efforts

Although bean counting is far easier than developing complex monitoring systems, conducting costly measurements, and maintaining essential databases, a few states have begun to develop comprehensive environmental quality evaluation initiatives with such requirements. The purpose of these evaluation efforts is to keep state officials apprised on how well environmental program efforts are succeeding in improving or maintaining environmental quality within their jurisdictional limits. Through these initiatives the states are attempting to get a handle on the complex problem of how to obtain meaningful measurements of environmental quality for program management purposes. Through such measures they hope to reach the elusive goal of attaining measurable improvements in environmental quality. Attainment of this goal is nearing reality in several innovative states; responses to letters sent to all state environmental agencies indicates that as of 1992 environmental quality evaluation systems were in place or under current development, at least, in Arizona, California, Connecticut, Florida, Oregon, and Washington.

Oregon In 1991, the Oregon Department of Environmental Quality (ODEQ) initiated an environmental management system based on benchmarks associated with measurable environmental goals. ODEQ establishes performance goals then measures how effective and efficient the programs are in their efforts to achieve the agency's mission. The ODEQ has measured and reported its performance since 1992. In fact, the air pollution control benchmark has become the lead quality of life benchmark in Oregon. In 1980 only 30 percent of Oregonians lived in clean air areas; by 1991 this figure had reached 50 percent. With the added emphasis on air pollution control growing out of the 1990 Clean Air Act Amendments and ODEQ efforts, 82.8 percent of Oregon citizens lived in clean air areas by 1993. The benchmark goal is 100 percent by the end of 1995 (ODEQ 1993).

The first year of the performance measurement program focused on gathering data to establish baselines from which to measure future progress. The system is supposed to emphasize results more than workload, with separate measures for effectiveness (how well they are doing) and efficiency (how well they are doing per full-time equivalent employee [FTE]). The ODEQ points out that

> performance measures rely on matrices that show the baseline, or average, performance and the potential, which is the best level that could possibly (but realistically) be achieved. Each quarter the actual achievement corresponds to a level on the scale between the baseline and potential. The actual result gets a score. Each measure is weighted, so that the results that are most important have a bigger impact on the overall score. Though the matrix is filled with numbers and may be hard to read at first, it provides a good snapshot of where we are, where we could be, and the relative value of the various results. (ODEQ 1993:1)

98   *An Intergovernmental Context*

To prevent one type of environmental protection from being weighed against another, ODEQ measures are broken down into forty-one separate program measures. For example, the air quality programs are measured against themselves on seven measures: an air pollution exposure index; an air pollution index; adherence to federal submittal deadlines (percentage successfully met); average waiting time above acceptable limits (frequency and magnitude); timeliness of permit processing (percent processed); number of permitted sources per FTE; and number of vehicles inspected per FTE per month (ODEQ 1993). The ODEQ performance system integrates environmental quality effectiveness concerns with bean-counting administrative efficiency concerns.

The following example of the performance measure for air pollution control provides insight into the nature of the ODEQ evaluation process:

AIR QUALITY
1. AIR POLLUTION EXPOSURE INDEX (APEI)
Definition: The index depicts how frequently federal air quality standards are violated, how severe the violations are, and the magnitude of people exposed. This product is figured for each exceedance day. The index is the summation of each product. This will be based on a 12 month moving average and reported quarterly. An index of 0.00 means there were no violations during the 12 month period prior to the reporting quarter.
Demonstrates: Effectiveness of control strategies and compliance assurance mechanisms such as the provision of technical assistance, rules, permits, inspections and enforcement actions.
Sections: Program Operations, Small Business Assistance, Planning, Technical Services, Vehicle Inspections and Field Burning.
Example: If during a previous 12 month period there were two exceedance days that were each 20% over the standard and each exposed 30% of Oregon's population, the index would be 0.72. This is calculated as follows:

$$\text{Exceedance 1} = (1.20 \times .30) + \text{Exceedance 2} = (1.20 \times .30) = 0.72$$

Note: This measure can also be used as a gauge of the environment, not just human health. Where able, data will be collected for attainment areas as well as non-attainment areas. We will also list what the APEI was in 1970 and 1980 to demonstrate where we have been and how far we have come.

2. AIR POLLUTION INDEX (API)
Definition: The API value is used to objectively assess ambient air quality. A lower value of 0 to 50 means the air quality is GOOD for that day. The higher the value, the worse the air is for the environment and public.
    The API values are already calculated daily. A 12 month moving average will be calculated and reported quarterly for the previous 12

month reporting period. An average index between 0 to 50 is Good, 51 to 100 is Moderate, 101 to 200 is Unhealthful, 201 to 300 is Very Unhealthful, and any value above 301 is Hazardous.

Demonstrates: Effectiveness of control strategies of compliance mechanisms such as the provision of technical assistance, rules, permits, inspections and enforcement actions. Unlike the APEI in #1, this index is able to provide a quantifiable indication of air quality regardless of whether or not that quality falls above or below the federal air quality standard.

Sections: Program Operations, Small Business Assistance, Planning, Technical Services, Vehicle Inspections and Field Burning.

Note: This measure can also be used as a gauge of the environment, not just human health. Where able, data will be collected for attainment areas as well as non-attainment areas. We will also list what the API was in 1970 and 1980 to demonstrate where we have been and how far we have come.

3. ADHERENCE TO FEDERAL SUBMITTAL DEADLINES

Definition: The percentage of submittal deadlines under the Clean Air Act Amendments that are successfully met.

Demonstrates: How well Oregon is meeting its federal responsibilities. This could further be compared to the national average. Indirectly, and most importantly, the deadlines established to ensure the quality of air is maintained or enhanced as quickly as possible. Meeting the established deadlines illustrates that Oregon is on track.

Sections: All Sections.

4. AVERAGE WAITING TIME ABOVE ACCEPTABLE FOR VEHICLE INSPECTIONS

Definition: The average waiting time above an "acceptable" waiting time. Frequency and magnitude.

Demonstrates: That the customer is served in a timely manner, builds credibility and a more positive image of state government.

Sections: Vehicle Inspection Program.

5. TIMELINESS OF PERMIT PROCESSING

Definition: Percent of permits processed in a timely manner. The ideal processing time will be determined for the type of source by the program. For example, timely processing will be determined for small, large, new, renewal and controversial sources. A 12 month moving average will be calculated based on the percentage of all of the types of permits that are processed in a timely manner. This percent will be reported quarterly for the previous 12 month period.

Demonstrates: That the customer is served in a timely manner, builds credibility and a more positive image of state government.

6. NUMBER OF PERMITTED SOURCES PER FTE
Definition: Number of industrial sources on Air Contaminant Discharge Permits relative to the FTE needed to manage the program. Activities include industrial source inspection, enforcement, compliance assurance, permitting, notice of construction approval, tax credit processing, complaint response, technical assistance to sources, public involvement and information, program administration. Includes all regional office air quality activities, headquarters air quality positions primarily involved in industrial source control and an additional 3 FTE to account for other air quality staff engaged part time in support of source control programs.
Demonstrates: Efficiency of industrial source control program.

7. NUMBER OF VEHICLES INSPECTED PER FTE PER MONTH
Definition: The number of vehicles inspected per the number of FTE in the Vehicle Inspection Program.
Demonstrates: The efficiency of the Vehicle Inspection Program. (ODEQ 1993:10–12)

The benchmark system that the ODEQ officials developed is dynamic. For example, they found the water quality benchmarks to be somewhat unwieldy and decided to revise them into a water quality index. Because of the difficulty of monitoring water quality in all of the navigable bodies of water throughout the state, the 305 (b) water quality assessment reports reflect the status of only 4 percent of Oregon's water bodies. In the revision to the water quality benchmark the ODEQ is attempting to develop an index that can facilitate the data collection effort and be calculated automatically so that substantial labor resources are not devoted to maintaining the index.

Washington The State of Washington has embarked upon a bold joint project with the EPA called Washington Environment 2010. Throughout 1990, the Governor's Office solicited input about environmental concerns from citizens across the state. Although Washington has some of the most beautiful natural resources in the country, citizens of that state are painfully aware of increasing environmental threats.

Based on these initial concerns a technical team of more than twenty-five experts from a variety of state agencies involved in environmental protection and natural resource management was formed to conduct a comprehensive evaluation of the condition of the state's environmental resources and important trends affecting those resources. The team gathered data on the current status of six key environmental resources—air, water, land, wetlands, fish and shellfish, and wildlife. It also reviewed historical data and information on trends in population growth, economic development, energy demand, and transportation needs. Then an assessment of the major existing and emerging threats to the environmental resources was conducted. Next it assessed the state's potential for managing those threats. Finally, the team established a preliminary list of priorities for environmental action.

The resource evaluation process was a major event for several reasons. It highlighted the need for accurate environmental and ecological data to assess the rate of deterioration or improvement in environmental resources. In the first state of the environment report, analysts indicate that environmental data in Washington State are limited. They found that in some instances, good information simply does not exist. For example, little information exists on levels of toxic air pollutants in ambient air. Information on other threats was not in usable form, often due to inconsistent collection practices. Perhaps the most important realization that came out of the analysis was what the team did not know about the environment (Washington Department of Ecology 1989:6). By collecting and assessing existing data, the State of Washington is now in a position to establish uniform data collection procedures that will minimize data deficiencies. Such procedures will help state officials better ascertain changes in air, water, and other environmental resources.

The environmental threats were ranked into five priority levels for action based on an analysis of their human health, ecological, and economic risks. The priority levels of action, based on the analysis of a technical advisory committee, were Priority Level 1: ambient air pollution, point source discharges to water, and nonpoint source discharges to water; Priority Level 2: drinking water contamination, uncontrolled hazardous waste sites, wetlands loss/degradation, nonchemical impacts on forest lands, and nonchemical impacts on agricultural lands; Priority Level 3: indoor air pollution, hydrologic disruptions, global warming and ozone depletion, regulated hazardous waste sites, nonhazardous waste sites, nonchemical impacts on recreational lands, and pesticides not covered elsewhere; Priority Level 4: indoor radon, radioactive releases, acid deposition, sudden and accidental releases, and nonchemical impacts on range lands; and Priority Level 5: nonionizing radiation, materials storage, and litter. All threats at each priority level are considered distinctly higher priorities than the threats included on the next level, but are not ranked within each priority level (Washington Department of Ecology 1990:2–3).

The extensive analysis and gathering of data on environmental problems that the State of Washington has conducted placed the Washington Department of Ecology in a good position to evaluate program effectiveness and to develop appropriate strategies to improve problem resolution. State resource managers also are in a much better position to ascertain rates of progress toward environmental quality goals. By communicating environmental conditions to the public through the comparative risk assessment and two state of the environment reports, the department has generated a great deal of support for its programs. Through all of these data collection efforts the department was able to issue its first report on environmental indicators, *Washington's Environmental Health 1995*.

Conclusion

The intergovernmental structure of environmental programs may facilitate a propensity of state and federal agencies to focus on administrative details rather than outcomes when evaluating environmental programs. Government agencies are constantly challenged to justify their existence in terms of how many units they produce.

When state governments accept primacy to implement federal environmental legislation, the EPA evaluates whether states have actually created legislation and programs that will allow the state agencies to achieve federal standards. A great deal of federal program oversight involves collecting data on program development.

The EPA also needs to know whether state environmental agencies are taking actions to ensure that the state environmental laws created in accordance with federal legislation are being implemented. The EPA, therefore, requires state agencies to report the number of abatement actions taken pursuant to the Clean Air Act, Clean Water Act, Hazardous and Solid Waste Act, and so forth. But problems occur because of vagueness in what constitutes an abatement action. For example, in 1993 South Carolina reported abatement actions 300 times greater than the number taken in California, which has substantially more polluting sources, agency personnel, and financial resources (Ringquist 1993: 133). Tighter definitions are needed for better comparability and greater accountability.

The requirements for reports to Congress on the quality of the nation's air and water provide some of the best stimuli for federal and state agencies to collect information on environmental quality that can be used for policy-making purposes. Each iteration of these reports provides more information about air and water quality and greater insight into the need for better data collection techniques. For example, the 1990 national water quality inventory identified several problems with state reporting on water quality. Most of the states provided most or all of the data requested in EPA guidelines. However, the EPA reported that the absence of data from some states limits its ability to analyze the data over time and creates gaps in understanding of water quality conditions nationwide. Inconsistencies between states in the collection of data due to different monitoring approaches, different pollution problems, and varying water quality standards is complicated by an overall lack of generally accepted assessment methodologies. Each state considers its needs and makes a judgment on how it can best use available monitoring resources. The monitoring problems may be reduced once state and EPA initiatives to improve water monitoring and reporting are completed and a computerized water quality data system is in place. This system is designed to manage state information on the causes, sources, and magnitude of degradation in individual bodies of water. The EPA and the states are also attempting to develop more cost-effective monitoring techniques. The EPA has already developed the section 305 (b) waterbody system (WBS) to assist the states in organization and analysis of water quality data. The WBS was used by the states in the development of data for the 1990 national water quality inventory.

The Washington Department of Ecology approach identifies key elements that need to be undertaken for state environmental program evaluation. Working groups or teams need to be identified early in the process to define the project scope, objectives, and expectations. These teams also can educate and train others on environmental indicator definitions and use. Once key indicators are identified the important tasks of data generation, data collection, and data quality assurance should be addressed. As the Washington experience shows, however, it is essential that the data collected be translated into policy relevant information for both citizens and program managers. An informed public is more likely to support difficult choices once it is educated about the necessity to take action.

As the Oregon and Washington examples illustrate, states are taking new and innovative approaches to both environmental management and environmental program evaluation. More states are likely to do so as responsibilities for environmental management continue to devolve to the states. Such devolution creates the potential for learning a great deal about the efficacy of alternative management approaches and alternative approaches for evaluating management efficiency. As a result, the prospects for environmental program evaluation to evolve from bean counting into a vigorous and richly diverse field are extremely bright.

APPENDIX

Table 4.1. State Environmental Protection Expenditures, 1988

	Air		Water		Spending on All Environmental Programs					
State	$ per Capita	Rank	$ per Capita	Rank	Total in $1000s	$ per Capita	Rank	State Budget	Rank	Score	Rank
Alabama	0.48	45	3.24	42	64,907	15.73	42	1.02	41	241	46
Alaska	3.03	2	14.33	18	131,684	256.69	2	4.00	4	35	2
Arizona	0.81	31	4.97	35	46,613	13.45	45	0.96	43	225	42
Arkansas	0.39	47	3.78	38	44,189	18.24	39	1.15	39	221	40
California	2.26	5	25.49	5	1,486,124	52.76	8	2.60	9	42	3
Colorado	1.96	7	5.01	34	76,150	23.15	34	1.65	24	149	24
Connecticut	2.64	3	5.02	33	61,996	19.13	38	0.77	46	169	32
Delaware	2.10	6	14.45	17	33,170	50.26	9	1.80	19	63	7
Florida	0.58	37	20.28	9	465,591	37.62	13	2.51	11	107	13
Georgia	0.55	40	1.61	48	93,344	14.58	44	1.07	40	236	45
Hawaii	1.58	12	10.61	20	27,832	61.50	29	0.85	44	163	27
Idaho	0.78	34	22.10	6	61,442	25.46	6	4.22	3	78	8
Illinois	2.38	4	21.08	8	392,844	34.03	18	2.26	14	105	12
Indiana	0.85	26	3.32	41	52,766	9.46	49	0.68	47	243	47
Iowa	0.38	48	16.38	14	88,065	31.07	22	1.44	32	201	35
Kansas	0.84	29	9.44	21	47,817	19.23	37	1.23	37	202	36
Kentucky	1.16	16	6.73	26	120,289	32.33	20	1.64	25	162	26
Louisiana	0.99	19	21.89	7	193,836	43.85	11	2.64	7	102	11
Maine	0.55	39	3.78	39	39,332	32.61	19	1.88	16	126	18
Maryland	0.99	20	8.57	23	150,091	32.32	21	1.60	26	151	25
Massachusetts	0.83	30	28.70	3	237,936	40.53	12	1.56	27	126	18
Michigan	0.84	28	11.01	19	221,425	23.81	32	1.42	33	174	33
Minnesota	0.78	33	4.56	36	126,236	29.32	28	1.46	31	146	23
Mississippi	0.50	43	1.96	46	54,154	20.61	35	1.40	34	223	41
Missouri	0.47	46	8.34	24	119,907	23.33	33	1.73	21	190	34
Montana	1.48	13	26.72	4	69,560	86.52	3	4.29	2	29	1
Nebraska	0.34	49	4.29	37	27,988	17.48	40	1.29	36	234	44
Nevada	0.52	42	18.68	11	36,487	34.42	17	2.57	10	139	22
New Hampshire	0.88	25	15.86	15	33,588	30.62	23	2.41	12	110	14
New Jersey	1.68	9	58.29	2	523,874	67.86	5	3.61	5	84	9
New Mexico	0.99	18	6.92	25	44,782	29.66	27	1.48	29	137	21
New York	0.90	24	5.11	32	236,484	13.21	46	0.59	50	213	38
North Carolina	0.50	44	1.91	47	96,943	14.85	43	1.00	42	246	48
North Dakota	1.66	10	17.90	13	125,669	49.06	10	2.32	13	99	10
Ohio	1.02	17	2.49	45	40,869	11.56	48	0.65	48	227	43
Oklahoma	0.60	36	1.58	49	35,524	15.52	47	0.79	45	262	49
Oregon	3.87	1	6.30	28	186,438	68.02	4	3.03	6	53	5
Pennsylvania	0.95	22	5.54	30	288,766	24.01	31	1.49	28	163	27
Rhode Island	0.78	32	18.04	12	35,879	36.06	15	1.86	17	120	17
South Carolina	0.26	50	2.52	44	71,124	20.36	36	1.21	38	219	39
South Dakota	0.53	41	6.35	27	21,264	29.74	26	1.85	18	168	31
Tennessee	0.58	38	5.61	29	81,180	16.50	41	1.34	35	203	37
Texas	0.68	35	1.05	50	113,797	6.78	50	0.60	49	278	50

Table 4.1, continued

	Air		Water		Spending on All Environmental Programs						
State	$ per Capita	Rank	$ per Capita	Rank	Total in $1000s	$ per Capita	Rank	State Budget	Rank	Score	Rank
Utah	1.24	14	9.01	22	51,419	30.41	24	1.80	20	126	18
Vermont	1.60	11	3.72	40	20,222	36.37	14	1.94	15	110	15
Virginia	0.93	23	5.30	31	152,149	25.38	30	1.47	30	164	29
Washington	1.22	15	14.73	16	246,873	53.45	7	2.63	8	57	6
West Virginia	0.85	27	3.11	43	56,189	29.82	25	1.68	23	165	30
Wisconsin	0.96	21	19.78	10	167,779	34.54	16	1.70	22	115	16
Wyoming	1.89	8	145.11	1	128,051	271.87	1	7.73	1	49	4
Average	1.15		12.47		146,613	29.89					

Source: Hall and Kerr 1992.

Table 4.2. Number of Severe Nonattainment Areas by State, 1991

State	CO	Lead	Ozone	PM-10	SO$_2$
Alabama	0	1	2	0	0
Alaska	2	0	0	2	0
Arizona	2	0	1	7	6
Arkansas	0	0	0	0	0
California	24	0	40	7	0
Colorado	11	0	6	6	0
Connecticut	9	0	10	1	0
Delaware	0	0	3	0	0
Florida	0	0	6	0	0
Georgia	0	1	13	0	0
Hawaii	0	0	0	0	0
Idaho	1	0	0	4	0
Illinois	0	0	13	4	3
Indiana	2	1	8	2	5
Iowa	0	0	0	0	0
Kansas	0	0	2	0	0
Kentucky	2	0	15	0	2
Louisiana	0	0	17	0	0
Maine	0	0	12	1	1
Maryland	3	0	14	0	0
Massachusetts	7	0	14	0	0
Michigan	3	0	37	1	0
Minnesota	9	1	0	2	2
Mississippi	0	0	0	0	0
Missouri	2	3	8	0	0
Montana	3	1	0	7	3
Nebraska	0	1	0	0	0
Nevada	5	0	1	2	1

Table 4.2, continued

State	CO	Lead	Ozone	PM-10	SO$_2$
New Hampshire	2	0	10	0	0
New Jersey	17	0	21	0	2
New Mexico	1	0	0	1	2
New York	8	0	22	0	0
North Carolina	4	0	9	0	0
North Dakota	0	0	0	0	0
Ohio	1	0	28	2	9
Oklahoma	0	0	0	0	0
Oregon	8	0	5	5	0
Pennsylvania	2	0	45	1	3
Rhode Island	0	0	5	0	0
South Carolina	0	0	1	0	0
South Dakota	0	0	0	0	0
Tennessee	1	2	7	0	6
Texas	1	1	17	1	0
Utah	3	0	2	2	4
Vermont	0	0	0	0	0
Virginia	2	0	29	0	0
Washington	6	0	4	7	0
West Virginia	0	0	6	0	1
Wisconsin	0	0	11	0	6
Wyoming	0	0	0	1	0

Source: U.S. EPA 1991.

Table 4.3. Air Pollution Control Initiatives by State, 1991

State	Control Air Toxics	Air Pollution Emission Fees	Ozone Protection
Alabama	X	X	
Alaska		X	
Arizona	X		
Arkansas	X	X	
California	X	X	X
Colorado		X	
Connecticut	X	X	X
Delaware		X	
Florida	X		
Georgia		X	
Hawaii			X
Idaho	X	X	
Illinois	X	X	X
Indiana		X	
Iowa	X	X	X
Kansas	X	X	
Kentucky	X	X	
Louisiana	X	X	
Maine			X

Table 4.3, continued

State	Control Air Toxics	Air Pollution Emission Fees	Ozone Protection
Maryland	X	X	X
Massachusetts		X	X
Michigan	X		
Minnesota		X	X
Mississippi	X	X	
Missouri	X		X
Montana	X	X	
Nebraska	X		
Nevada	X	X	
New Hampshire	X	X	
New Jersey	X	X	
New Mexico		X	
New York	X	X	X
North Carolina	X		
North Dakota	X	X	
Ohio	X	X	
Oklahoma	X	X	
Oregon	X	X	X
Pennsylvania	X		
Rhode Island	X	X	X
South Carolina	X	X	
South Dakota			
Tennessee		X	
Texas	X	X	
Utah			
Vermont	X		X
Virginia	X	X	
Washington	X		
West Virginia		X	
Wisconsin		X	X
Wyoming	X	X	
No. of states	35	37	15

Source: Hall and Kerr 1992.

Table 4.4. Prevention of Significant Deterioration Program Status by State, 1992

State	PSD SIP Status
Alabama	Approved PSD SIP
Alaska	Approved PSD SIP
Arizona	Approved PSD SIP
Arkansas	Approved PSD SIP
California	Delegated and SIP approved by district

Table 4.4, continued

State	PSD SIP Status
Colorado	Approved PSD SIP
Connecticut	Approved PSD SIP
Delaware	Approved PSD SIP
Florida	Approved PSD SIP
Georgia	Approved PSD SIP
Hawaii	Full delegation
Idaho	Approved PSD SIP
Illinois	Full delegation
Indiana	Full delegation
Iowa	Approved PSD SIP
Kansas	Approved PSD SIP
Kentucky	Approved PSD SIP
Louisiana	Approved PSD SIP
Maine	Approved PSD SIP
Maryland	Approved PSD SIP
Massachusetts	Partial delegation
Michigan	Approved PSD SIP
Minnesota	Full delegation
Mississippi	Full delegation
Missouri	Approved PSD SIP
Montana	Approved PSD SIP
Nebraska	Approved PSD SIP
Nevada	Full delegation
New Hampshire	Partial delegation
New Jersey	Full delegation
New Mexico	Approved PSD SIP
New York	Approved PSD SIP
North Carolina	Approved PSD SIP
North Dakota	Approved PSD SIP
Ohio	Full delegation
Oklahoma	Approved PSD SIP
Oregon	Approved PSD SIP
Pennsylvania	Approved PSD SIP
Rhode Island	Approved PSD SIP
South Carolina	Approved PSD SIP
South Dakota	No program transfer
Tennessee	Approved PSD SIP
Texas	Approved PSD SIP
Utah	Approved PSD SIP
Vermont	Approved PSD SIP
Virginia	Full delegation
Washington	Partial delegation
West Virginia	Approved PSD SIP
Wisconsin	Approved PSD SIP
Wyoming	Approved PSD SIP

Source: U.S. EPA, personal contact, 1992.

Table 4.5. Toxic Air Releases by State, 1991

State	Air Pounds	State	Air Pounds
Alabama	98,217,855	Nebraska	14,214,021
Alaska	31,324,172	Nevada	776,614
Arizona	17,557,458	New Hampshire	12,664,114
Arkansas	49,083,095	New Jersey	43,289,398
California	81,559,910	New Mexico	3,895,229
Colorado	10,168,740	New York	94,651,400
Connecticut	26,801,799	North Carolina	93,851,144
Delaware	6,463,334	North Dakota	828,862
Florida	50,505,521	Ohio	135,263,552
Georgia	91,681,105	Oklahoma	35,965,607
Hawaii	975,080	Oregon	19,576,823
Idaho	6,870,922	Pennsylvania	90,574,986
Illinois	102,145,620	Rhode Island	6,614,269
Indiana	110,653,475	South Carolina	64,068,010
Iowa	39,254,244	South Dakota	2,496,933
Kansas	24,896,642	Tennessee	134,213,411
Kentucky	44,840,930	Texas	214,110,207
Louisiana	135,202,095	Utah	84,533,765
Maine	15,874,624	Vermont	1,236,903
Maryland	18,432,845	Virginia	127,977,298
Massachusetts	31,256,802	Washington	29,183,116
Michigan	115,810,988	West Virginia	38,141,581
Minnesota	46,363,160	Wisconsin	49,040,922
Mississippi	55,900,737	Wyoming	2,505,827
Missouri	49,761,871	Total	2,575,968,562
Montana	2,750,456		

Source: U.S. EPA 1990.

Table 4.6. State Operating Permit Program Submissions, 1992

State	Permits Substantive	Permits Administration	Fees	Enforcement	Applicability
Alabama	E	E	E	E,G	E
Alaska	F	F	F	F	F
Arizona	B	B	B	B	B
Arkansas	E	A	A	A	A
California	D	D	E	A	D
Colorado	B	B	B	B,G	B,G
Connecticut	A	A,G	A	A,G	A
Delaware	E	C	C	D	E
Florida	E	A	A	E	E
Georgia	E	A	A	A	E
Hawaii	A	A	A	A	A
Idaho	E	E	E	F	E
Illinois	C	C	C	C	C
Indiana	D,E	D,E	D,G	D,G	D,G

Table 4.6, continued

	Permits				
State	Substantive	Administration	Fees	Enforcement	Applicability
Iowa	F	F	F	F	F
Kansas	F	F	F	F	F
Kentucky	E	E	A	E	E
Louisiana	B	B	A	B	B
Maine	D	D	D	D	D
Maryland	E	C	C	C	E
Massachusetts	D	D	D	D	D
Michigan	B,G	B,G	B,G	B,G	B,G
Minnesota	A	A	A	A	A
Mississippi	E	F	F	E	E
Missouri	A	A	A	A	A
Montana	D	D	D	D	D
Nebraska	A	A	A	A	A
Nevada	E	D	D	D	D
New Hampshire	E,G	E,G	E	F	E,G
New Jersey	B	B	B	B	B
New Mexico	A	A,G	A	A	A
New York	B	B	B	B	B
North Carolina	E	F	A	E	E
North Dakota	E	E	E	E	E
Ohio	D	D	D	D	D
Oklahoma	B	B	B	B	B
Oregon	A,G	A	A	D	A
Pennsylvania	B	B	B	D	B
Rhode Island	B	B	B	B	B
South Carolina	E	B	B	E	E
South Dakota	A	A	A	A	A
Tennessee	E	E	A,G	E	E
Texas	A	A	A	A	A
Utah	A	A	A,G	A	A
Vermont	F	F	F	F	F
Virginia	E	E	C	C	C
Washington	A	A	D	A	A
West Virginia	D	D	D	D	D
Wisconsin	A	A	A	A	A
Wyoming	A	A	A,G	A,E	A,G

Key: A = legislation enacted
B = legislation before the legislature
C = legislation prepared but not before the legislature
D = no legislation prepared
E = preexisting authority adequate
F = legislation prepared but not enacted by the legislature
G = region questions adequacy of legislation

Source: U.S. EPA, personal contact, 1992.

Note: The data reported above reflect reports from states and/or initial regional office checks. They do not reflect in-depth EPA review or conclusions as to their ultimate adequacy in supporting approval of state operating permit program submissions.

Table 4.7. Water Pollution Control Initiatives by State, 1991

State	Issue NPDES Permits	Pretreatment Program	Groundwater Protection	Underground Tanks	Groundwater Toxics	Wetland Protection	Phosphate Ban
Alabama	X	X					
Alaska					X		
Arizona			X	X	X		
Arkansas	X	X					
California	X		X	X	X		
Colorado	X		X	X	X		
Connecticut	X	X	X	X		X	
Delaware	X		X	X		X	
Florida			X	X	X	X	
Georgia	X	X	X		X		X
Hawaii	X	X	X	X			
Idaho			X		X		
Illinois	X		X	X	X		
Indiana	X		X	X			X
Iowa	X	X	X	X			
Kansas	X		X	X			
Kentucky	X	X	X	X			
Louisiana			X				
Maine			X	X	X	X	
Maryland	X	X	X	X	X		X
Massachusetts			X		X	X	
Michigan	X	X	X	X	X	X	X
Minnesota	X	X	X	X	X	X	X
Mississippi	X	X	X	X			
Missouri	X	X	X		X		
Montana	X			X	X		
Nebraska	X	X	X	X			
Nevada	X	X	X				
New Hampshire				X	X	X	
New Jersey	X	X	X	X	X	X	
New Mexico			X	X	X		
New York	X		X	X	X	X	X
North Carolina	X	X	X	X	X		X
North Dakota	X		X	X		X	
Ohio	X	X	X				
Oklahoma	X	X	X				
Oregon	X	X	X	X	X	X	
Pennsylvania	X		X			X	X
Rhode Island	X	X	X			X	
South Carolina	X	X		X	X		
South Dakota				X	X		
Tennessee	X	X			X		
Texas				X			
Utah	X	X		X			
Vermont	X	X	X	X		X	X
Virginia	X		X	X	X		X
Washington	X	X	X	X			

Table 4.7, continued

State	Issue NPDES Permits	Pretreatment Program	Groundwater Protection	Underground Tanks	Groundwater Toxics	Wetland Protection	Phosphate Ban
West Virginia	X	X					
Wisconsin	X	X	X	X	X	X	X
Wyoming	X		X	X	X		
No. of states	39	27	39	35	28	16	11

Source: Hall and Kerr 1992.

Table 4.8. Attainment of Clean Water Act Goals in Rivers and Streams by State, 1990

	Fishable Goal (miles)			Swimmable Goal (miles)		
State	Meeting	Partially Meeting	Not Meeting	Meeting	Partially Meeting	Not Meeting
Alabama	8,703	1,669	1,049	8,703	1,669	1,049
Arizona	662	2,312	2,061	1,810	566	645
Arkansas	9,426	1,535	349	9,236	0	2,074
California	9,069	1,492	887	7,947	1,620	896
Colorado	28,105	—	572	9,062	—	254
Connecticut	664	170	57	679	94	118
Delaware	498	71	74	98	56	365
Florida	6,750	268	897	6,750	268	897
Georgia	19,395	556	49	—	—	—
Hawaii	349	—	0	349	—	0
Illinois	11,476	1,101	546	1,144	2,003	424
Indiana	2,986	1,143	788	138	153	2,013
Iowa	90	6,620	428	158	1,456	223
Kansas	8,838	2,265	976	627	1,546	9,769
Kentucky	6,914	1,702	723	1,481	576	1,538
Louisiana	6,510	1,553	472	5,670	2,055	810
Maine	31,282	0	390	31,506	0	166
Maryland	15,618	1,281	101	16,998	0	2
Massachusetts	1,061	208	354	682	620	319
Michigan	35,632	—	718	36,086	—	264
Minnesota	2,292	681	2,306	2,045	548	1,628
Mississippi	8,023	7,350	350	13,086	2,126	511
Missouri	11,216	9,830	17	5,370	—	0
Montana	38,474	11,565	1,173	46,096	4,536	579
Nebraska	5,131	1,773	426	1,035	606	1,034
Nevada	793	583	83	735	560	57
New Hampshire	1,294	3	50	859	339	150
New Jersey	1,315	307	97	91	48	453
New Mexico	2,851	—	266	3,117	—	0
New York	69,300	—	0	69,200	—	800
North Carolina	23,820	12,551	922	23,820	12,551	922

Table 4.8, continued

State	Fishable Goal (miles) Meeting	Partially Meeting	Not Meeting	Swimmable Goal (miles) Meeting	Partially Meeting	Not Meeting
North Dakota	8,548	656	0	8,489	715	0
Ohio	2,524	1,482	2,868	—	—	—
Oklahoma	2,161	1,258	359	2,148	412	515
Oregon	26,197	—	1,542	26,773	—	966
Pennsylvania	19,137	1,982	2,713	19,137	1,982	2,713
Rhode Island	512	0	24	512	0	24
South Carolina	3,230	54	310	2,010	388	1,040
South Dakota	3,085	—	879	396	—	406
Tennessee	10,857	—	224	10,420	—	661
Texas	16,044	—	159	14,435	—	1,768
Utah	1,303	10,476	0	120	4,164	36
Vermont	4,468	608	190	4,854	13	140
Virginia	8,862	1,468	479	8,862	1,468	479
Washington	2,873	1,084	1,184	3,441	311	1,176
West Virginia	17,969	506	1,343	18,415	0	1,397
Wisconsin	8,272	4,334	220	8,235	4,269	212
Wyoming	19,430	—	7	947	0	0
Totals	524,009	92,498	29,682	433,772	47,718	39,493

Source: U.S. EPA 1992b.
Note: Data for Alaska and Idaho were not reported.

Table 4.9. Attainment of Clean Water Act Goals in Lakes and Reservoirs by State, 1990

State	Fishable Goal (acres) Meeting	Partially Meeting	Not Meeting	Swimmable Goal (acres) Meeting	Partially Meeting	Not Meeting
Alabama	286,039	133,635	63,548	286,039	133,635	63,548
Arizona	17,747	96,224	7,398	113,179	6,742	120
Arkansas	355,063	0	0	355,063	0	0
California	538,832	54,931	445,082	619,089	70,519	347,617
Colorado	146,494	—	1,862	112,539	—	0
Connecticut	20,304	2,875	0	23,179	0	0
Delaware	2,384	0	158	1,294	969	280
Florida	845,696	384	111,040	845,696	384	111,040
Georgia	404,704	13,000	26	—	—	—
Illinois	160,402	20,920	27,714	136,436	59,765	12,834
Indiana	97,372	63	101	97,372	63	89
Iowa	42,856	26,332	11,115	42,787	29,706	5,586
Kansas	173,236	0	366	173,199	0	403
Kentucky	214,642	219	0	214,642	219	0
Louisiana	455,869	137,000	0	434,237	137,960	20,672
Maine	796,394	164,189	0	911,493	49,090	0

Table 4.9, continued

	Fishable Goal (acres)			Swimmable Goal (acres)		
State	Meeting	Partially Meeting	Not Meeting	Meeting	Partially Meeting	Not Meeting
Maryland	17,255	3,641	106	20,999	0	2
Massachusetts	37,706	569	25,663	44,893	17,952	1,080
Michigan	472,587	—	16,846	489,433	—	0
Minnesota	225,838	0	1,964,482	1,290,496	68,088	177,601
Mississippi	339,358	160,596	46	492,453	7,501	46
Missouri	285,074	—	2,469	260,376	—	79
Montana	226,475	392,977	13,253	288,934	332,619	11,151
Nevada	149,042	159,083	0	43,075	264,100	950
New Hampshire	149,776	3,270	470	146,442	5,293	3,728
New Mexico	115,598	—	0	115,598	—	0
New York	560,000	—	190,000	678,000	—	72,000
North Carolina	295,221	—	10,124	303,134	—	2,221
North Dakota	367,484	21,855	221,734	376,767	12,248	221,734
Ohio	24,879	865	271	20,953	75,722	6,056
Oklahoma	219,855	200,493	28,428	285,197	123,310	25,924
Oregon	504,928	0	0	504,928	0	0
Rhode Island	15,553	100	487	15,553	100	487
South Carolina	175,069	72,713	13,120	197,142	20,525	0
South Dakota	682,034	—	0	682,034	—	0
Tennessee	483,675	—	56,355	521,473	—	16,849
Texas	1,543,884	—	14	1,536,346	—	7,552
Utah	0	133,754	0	2,329	131,425	0
Vermont	221,525	1,877	1,957	40,636	331	838
Virginia	150,158	10,482	0	160,640	—	0
Washington	62,551	5,277	115,020	180,512	660	1,635
West Virginia	12,988	4,152	1,730	18,870	0	0
Wisconsin	42,284	89,918	7,128	42,284	89,918	7,128
Wyoming	427,219	0	0	245,311	0	0
Totals	12,366,050	1,911,394	3,338,113	13,371,052	1,638,844	1,119,250

Source: U.S. EPA 1992b.
Note: Data for Alaska, Hawaii, Idaho, Nebraska, New Jersey, and Pennsylvania were not reported.

Table 4.10. State NPDES Program Status, 1992

State	Approved State NPDES Permit Program	Approved to Regulate Federal Facilities	Approved State Pretreatment Program	Approved General Permits Program
Alabama	10/19/79	10/19/79	10/19/79	06/26/91
Arkansas	11/01/86	11/01/86	11/01/86	11/01/86
California	05/14/73	05/05/78	09/22/89	09/22/89
Colorado	03/27/75	—	—	03/04/83
Connecticut	09/26/73	01/09/89	06/03/81	03/10/92
Delaware	04/01/74	—	—	—

Table 4.10, continued

State	Approved State NPDES Permit Program	Approved to Regulate Federal Facilities	Approved State Pretreatment Program	Approved General Permits Program
Georgia	06/28/74	12/08/80	03/12/81	01/28/91
Hawaii	11/28/74	06/01/79	08/12/83	09/30/91
Illinois	10/23/77	09/20/79	—	01/04/84
Indiana	01/01/75	12/09/78	—	04/02/91
Iowa	08/10/78	08/10/78	06/03/81	08/12/92
Kansas	06/28/74	08/28/85	—	—
Kentucky	09/30/83	09/30/83	09/30/83	09/30/83
Maryland	09/05/74	11/10/87	09/30/85	09/30/91
Michigan	10/17/73	12/09/78	06/07/83	—
Minnesota	06/30/74	12/07/78	07/16/79	12/15/87
Mississippi	05/01/74	01/28/83	05/13/82	09/27/91
Missouri	10/30/74	06/26/79	06/03/81	12/12/85
Montana	06/10/74	06/23/81	—	04/29/83
Nebraska	06/12/74	11/02/79	09/07/84	07/20/89
Nevada	09/19/75	08/31/78	—	07/27/92
New Jersey	04/13/82	04/13/82	04/13/82	04/13/82
New York	10/28/75	06/13/80	—	—
North Carolina	10/19/75	09/28/84	06/14/82	09/06/91
North Dakota	06/13/75	01/22/90	—	01/22/90
Ohio	03/11/74	01/28/83	07/27/83	08/17/92
Oregon	09/26/73	03/02/79	03/12/81	02/23/82
Pennsylvania	06/30/78	06/30/78	—	08/02/91
Rhode Island	09/17/84	09/17/84	09/17/84	09/17/84
South Carolina	06/10/75	09/26/80	04/09/82	09/03/92
Tennessee	12/28/77	09/30/86	08/10/83	04/18/91
Utah	07/07/87	07/07/87	07/07/87	07/07/87
Vermont	03/11/74	—	03/16/82	—
Virginia	03/31/75	02/09/82	04/14/89	05/20/91
Washington	11/14/73	—	09/30/86	09/26/89
West Virginia	05/10/82	05/10/82	05/10/82	05/10/82
Wisconsin	02/04/74	11/26/79	12/24/80	12/19/86
Wyoming	01/30/75	05/18/81	—	09/24/91
No. of states	39	34	27	33

Source: U.S. EPA, personal contact, 1993.
Note: Number of fully authorized programs (federal facilities, pretreatment, general permits) = 24.

Table 4.11. Major NPDES Nonmunicipal Significant Noncompliance by State, 1992

State	Major Nonmunicipals	Major Nonmunicipals in Significant Noncompliance	Percent Significant Noncompliance
Alabama	99	6	6
Alaska	2	0	0
Arizona	15	2	13
Arkansas	38	1	3
California	83	4	5
Colorado	36	3	8
Connecticut	65	9	14
Delaware	18	0	0
Florida	147	5	3
Georgia	53	1	2
Hawaii	13	0	0
Idaho	41	2	5
Illinois	92	5	5
Indiana	72	4	6
Iowa	31	2	6
Kansas	13	3	23
Kentucky	57	7	12
Louisiana	153	13	8
Maine	32	4	13
Maryland	45	4	9
Massachusetts	71	34	4
Michigan	95	9	9
Minnesota	26	1	4
Mississippi	34	0	0
Missouri	49	4	8
Montana	17	0	0
Nebraska	25	0	0
Nevada	3	0	0
New Hampshire	24	0	0
New Jersey	143	12	8
New Mexico	12	0	0
New York	144	11	8
North Carolina	93	2	2
North Dakota	7	0	0
Ohio	123	33	27
Oklahoma	34	2	6
Oregon	25	0	0
Pennsylvania	149	16	11
Rhode Island	14	0	0
South Carolina	89	5	6
South Dakota	4	0	0
Tennessee	43	1	2
Texas	217	31	14
Utah	12	2	17
Vermont	7	0	0
Virginia	58	2	3

Table 4.11, continued

State	Major Nonmunicipals	Major Nonmunicipals in Significant Noncompliance	Percent Significant Noncompliance
Washington	44	9	20
West Virginia	66	10	15
Wisconsin	52	0	0
Wyoming	14	3	21

Source: U.S. EPA, personal contact, 1992.

Table 4.12. Major NPDES Municipal Noncompliance by State, 1992

State	Major Municipals	Major Municipals in Significant Noncompliance	Percent Significant Noncompliance
Alabama	107	11	10
Alaska	2	0	0
Arizona	16	1	6
Arkansas	62	4	6
California	154	4	3
Colorado	69	3	4
Connecticut	64	9	14
Delaware	14	0	0
Florida	145	6	4
Georgia	119	7	6
Hawaii	10	0	0
Idaho	27	2	7
Illinois	180	8	4
Indiana	107	25	23
Iowa	79	4	5
Kansas	41	2	5
Kentucky	63	5	8
Louisiana	92	14	15
Maine	65	9	14
Maryland	44	0	0
Massachusetts	97	6	6
Michigan	92	15	16
Minnesota	53	5	9
Mississippi	51	3	6
Missouri	80	10	13
Montana	26	0	0
Nebraska	41	1	2
Nevada	6	1	17
New Hampshire	41	1	2
New Jersey	135	25	19
New Mexico	24	1	4
New York	237	30	13
North Carolina	134	7	5

Table 4.12, continued

State	Major Municipals	Major Municipals in Significant Noncompliance	Percent Significant Noncompliance
North Dakota	16	0	0
Ohio	169	42	25
Oklahoma	55	5	9
Oregon	46	1	2
Pennsylvania	248	23	9
Rhode Island	19	1	5
South Carolina	86	9	10
South Dakota	26	0	0
Tennessee	85	7	8
Texas	678	55	15
Utah	27	3	11
Vermont	28	0	0
Virginia	56	1	2
Washington	45	9	20
West Virginia	39	3	8
Wisconsin	85	4	5
Wyoming	16	1	6

Source: U.S. EPA, personal contact, 1992.

Table 4.13. Nonpoint Source Program Status, 1990

State	Approval Status	Approval Date	Agriculture	Silviculture	Construction	Urban Runoff	Mining	Land Disposal	Hydrological Modification
Alabama	Full	08/89	X	X	X	X	X		X
Alaska	Full	09/90							
Arizona	Full	01/90	X	X	X	X	X	X	
Arkansas	Partial[a]		X						
California	Full	01/90	X	X		X	X	X	
Colorado	Full	12/89	X	X	X	X	X		X
Connecticut	Full	06/89	X	X	X				
Delaware	Full	10/89	X	X	X	X		X	
Florida	Full	07/89	X			X		X	X
Georgia	Full	01/90	X	X	X	X	X		
Hawaii	Full	01/90	X		X	X			
Idaho	Full	12/89	X	X		X	X	X	X
Illinois	Full	01/90	X		X	X			X
Indiana	Full	01/90	X		X	X	X	X	
Iowa	Partial	12/89	X						
Kansas	Full	12/89	X			X	X	X	X
Kentucky	Full	11/89	X	X	X	X	X	X	X
Louisiana	Partial[a]	09/89	X			X		X	X
Maine	Full	01/90	X	X	X	X	X	X	
Maryland	Partial	10/89	X	X	X	X	X	X	X
Massachusetts	Full	01/90	X	X	X	X	X	X	X
Michigan	Full	01/90	X	X	X	X	X		
Minnesota	Full	01/90	X	X	X		X	X	X
Mississippi	Full	08/89	X	X	X	X	X	X	X
Missouri	Full	12/89	X				X		
Montana	Full	06/89	X	X	X			X	X

Table 4.13, continued

State	Approval Status	Approval Date	Agriculture	Silviculture	Construction	Urban Runoff	Mining	Land Disposal	Hydrological Modification
Nebraska	Full	01/89	No specifics						
Nevada	Full	01/90	X	X		X		X	X
New Hampshire	Full	01/90			X	X		X	
New Jersey	Full	01/90	X		X	X	X	X	
New Mexico	Partial[a]	09/89	X	X	X		X	X	
New York	Full	01/90	X		X	X	X	X	X
North Carolina	Full	08/89	X					X	
North Dakota	Full	06/89	X				X	X	X
Ohio	Full	01/90	X	X	X	X	X	X	X
Oklahoma	Partial[a]	09/89	X						
Oregon	Partial[b]	11/89	X						
Pennsylvania	Partial[b]		X						
Rhode Island	Full	04/89	X		X	X		X	
South Carolina	Full	08/89	X	X	X	X	X	X	X
South Dakota	Full	04/89	X	X	X	X	X	X	
Tennessee	Full	09/89	X	X	X				X
Texas	Partial[a]	08/89	X	X		X		X	
Utah	Full	08/89	X	X	X	X			X
Vermont	Full	03/89	X	X	X	X		X	X
Virginia	Full	08/89	X	X	X	X			
Washington	Full	10/89	X	X					
West Virginia	Partial		X	X	X	X	X	X	
Wisconsin	Full	01/90	X				X		
Wyoming	Partial	09/89	X	X	X	X	X		X
Totals			47	29	31	33	26	29	22

Source: U.S. EPA 1992b.

a. Region 6 approved only the portions of states' management programs that relate to program implementation.
b. Agriculture only. Only the grazing section of Wyoming's management program remains to be approved.

Table 4.14. States with Primacy under Original PWSS Program, 1992

State	Date	State	Date
Alabama	07/10/77	Montana	03/29/78
Alaska	09/22/78	Nebraska	06/22/77
Arizona	09/24/78	Nevada	03/29/78
Arkansas	07/10/77	New Hampshire	08/18/78
California	06/02/78	New Jersey	12/30/79
Colorado	05/06/78	New Mexico	04/01/78
Connecticut	05/06/77	New York	09/09/77
Delaware	04/01/78	North Carolina	03/14/80
Florida	02/17/78	North Dakota	02/17/78
Georgia	08/06/77	Ohio	03/15/79
Hawaii	10/22/77	Oklahoma	04/29/77
Idaho	03/29/78	Oregon	02/24/86
Illinois	09/28/79	Pennsylvania	02/04/85
Indiana	06/28/91	Rhode Island	11/22/78
Iowa[a]	09/11/77	South Carolina	09/30/77
Kansas	03/29/78	South Dakota	01/09/84
Kentucky	10/27/77	Tennessee	09/30/77
Louisiana	05/07/77	Texas	01/29/78
Maine	10/07/77	Utah	02/28/80
Maryland	02/12/78	Vermont	04/25/80
Massachusetts	12/01/77	Virginia	09/09/77
Michigan	02/12/78	Washington	03/29/78
Minnesota	09/26/77	West Virginia	04/01/78
Mississippi	06/19/77	Wisconsin	03/14/78
Missouri	09/22/79		

Source: U.S. EPA, personal contact, 1992.
a. Iowa assumed primacy on 9/11/77 and returned it voluntarily on 7/01/81. It reassumed primacy on 8/01/82.

Table 4.15. State Public Water Supply Standards, 1992

State	Phase 1 (VOCs) Rule Adopted	Phase 1 (VOCs) Primacy Approved	Public Notification Rule Adopted	Public Notification Primacy Approved	Surface Water Treatment Rule Adopted	Surface Water Treatment Primacy Approved	Total Coliforms Rule Adopted	Total Coliforms Primacy Approved	Phase 2 Rule Adopted
Alabama	01/89	11/89	01/89	11/89	10/90	06/91	10/90	06/91	
Alaska	06/91	06/92	06/91	06/92					
Arizona	06/89	01/91	06/89	01/91	08/91		08/91		
Arkansas	07/87	02/90	10/87	02/90	12/90	01/92	12/90	01/92	07/91
California	12/88	04/92		05/91		07/92			
Colorado	04/91	07/91	04/91	07/91	04/91		04/91		
Connecticut	01/92	12/90		01/92		01/91			
Delaware	06/89	05/90	05/90	05/90	03/91		03/91		
Florida	01/89	10/91	01/89	10/91	11/90	05/91	11/90	05/91	07/92
Georgia	06/89	10/91	06/89	10/91	10/90	05/91	10/90	05/91	07/92
Hawaii	02/92	02/92							
Idaho	11/89	11/89							07/92
Illinois	08/90	08/90		08/90		08/90			
Indiana	02/91	05/91	02/91	05/91	02/91		02/91		
Iowa	03/90	08/91	03/90	08/91	11/90	08/92	11/90	08/92	
Kansas	08/92	08/92				08/92			
Kentucky	07/90	10/91	11/90	10/91	04/91		04/91		
Louisiana	04/88	11/89	04/88	11/89	04/91		07/91		
Maine	08/91	08/91		11/91		11/91			
Maryland	03/91	10/90	10/90	10/90					
Massachusetts	06/89	12/89	06/89	12/89	10/90		10/90		
Michigan	09/89	10/90	09/89	11/90	11/91		11/91		
Minnesota	05/89	05/90	05/89	05/90	02/91		02/91		
Mississippi	05/89	10/91	05/89	10/91	10/90	12/91	10/90	12/91	07/92

Missouri	08/88	03/92	08/88	07/92	12/91	08/92	07/91	08/92	07/92
Montana	06/91	04/91	06/91	04/91	06/91		06/91		
Nebraska	09/89	09/89		09/91	09/91				
Nevada	11/88	06/91	11/88	06/91	10/91		06/92		06/92
New Hampshire	06/90	06/90		03/91		06/90	06/90	06/92	
New Jersey	01/89	08/91	04/89	08/91	06/90		06/90		06/92
New Mexico	12/88	11/89	12/88	11/89	04/91		04/91		
New York	01/90	10/91	01/90	10/91	12/91		06/91		
North Carolina	11/89	10/91	11/89	10/91	08/90	05/91	08/90	05/91	05/92
North Dakota	11/91	11/91				11/90	03/92		
Ohio	05/89	01/90	05/89	01/90	12/90		12/90		
Oklahoma	04/88	11/89	04/88	11/89	02/90	01/92	02/90	01/92	03/92
Oregon	11/89	06/90	11/89	06/90	12/90	12/91	12/90	12/91	07/92
Pennsylvania	06/89	05/90	12/91	03/89			05/92		
Rhode Island	10/89	12/89	10/89	12/89	12/90		12/90		06/92
South Carolina	11/88	05/89	11/88	05/89	10/90	04/91	10/90	04/91	
South Dakota	02/89	06/90	03/90	06/90	05/91	02/92	05/91		
Tennessee	11/88	10/91	05/89	10/91	11/90	05/91	11/90	05/91	06/92
Texas	04/88	11/89	08/89	11/89	10/90	01/92	10/90	01/92	
Utah	03/89	09/90	12/90	04/91			04/91	01/92	
Vermont	09/92	09/92		09/92		09/92		09/92	
Virginia	04/91	06/92	04/91	06/92	06/92		06/92		
Washington	04/91	06/92	11/89	06/92			01/92		
West Virginia	10/90	04/91	10/90	04/91	03/91		03/91		
Wisconsin	08/89	02/90	08/89	02/90	03/91		03/91		
No. of states	49	49	38	44	34	23	37	16	13

Source: U.S. EPA, personal contact, 1992.

NOTES

1. For a good description of the oversight responsibilities of the U.S. Environmental Protection Agency, see Stever and Dolin 1992:22–145 and Ruckelshaus 1984.

2. Even in this one area, in which "better" statistics are available, problems of data reliability exist, as is evident in the following caution to users of the data in the U.S. Environmental Protection Agency's "FY 1991 State-by-State Enforcement Data Summaries": "Users of these data should recognize that it is a subset of information on program performance, and without the use of more detailed information, or without dialogue with appropriate EPA program enforcement personnel and State officials, accurate conclusions about program performance cannot be developed. In addition, due to the dynamic nature of the automated and manual databases that form the basis of this summary, and also due to the timing of data retrievals for this report, some data may not precisely match data reported in other EPA reports. The data reported in STARS [Strategically Targeted Activities Reporting System] (and taken from the national enforcement data bases) does not include all enforcement activities, especially activities undertaken by State agencies.... Administrative action totals for EPA and the States for all Programs may include actions taken to address violations other than violations defined as SNC [Significant Noncompliers] and, therefore, it is not possible to draw a direct relationship between enforcement activity levels and SNC.... Not all State administrative actions are included in the state columns since not all state actions are entered into the national database.... Data on State criminal enforcement activities for most programs are not available and are not included" (1992a:3).

REFERENCES

Alaska Department of Environmental Conservation. 1993. *Alaska State/Environmental Protection Agency Agreement (SEA): Fiscal Years 1994 and 1995, Vol. II.* Juneau: Alaska Department of Environmental Conservation.

Blomquist, W. 1991. "Exploring State Differences in Groundwater Policy Adoptions, 1980–1989." *Publius* 21 (Spring): 101–15.

Crotty, P. M. 1987. "The New Federalism Game: Primacy Implementation of Environmental Policy." *Publius* 16 (Spring): 53–67.

Game, K. W. 1979. "Controlling Air Pollution: Why Some States Try Harder." *Policy Studies Journal* 7 (1979): 728–38.

Goggin, M. L., A. O'M. Bowman, J. P. Lester, and L. J. O'Toole Jr. 1990. *Implementation Theory and Practice: Toward a Third Generation.* Glenview, Ill.: Scott, Foresman/Little, Brown.

Hall, B., and M. L. Kerr. 1992. *1991–1992 Environmental Index: A State-by-State Guide to the Nation's Environmental Health.* Washington, D.C.: Island Press.

Lester, J. P. 1994. "A New Federalism?: Environmental Policy in the States." In *Environmental Policy in the 1990s,* ed. N. J. Vig and M. E. Kraft. 2d ed. Washington, D.C.: Congressional Quarterly Press. 51–68.

Lester, J. P., and P. M. Keptner. 1984. "State Budgetary Commitments to Environmental

Quality under Austerity." In *Western Public Lands,* ed. J. G. Francis and R. Ganzel. Totowa, N.J.: Rowman and Allanheld. 193–214.

Lester, J. P., and E. N. Lombard. 1990. "The Comparative Analysis of State Environmental Policy." *Natural Resources Journal* 30 (Spring): 301–19.

Lombard, E. N. 1994. "Determinants of State Air-Quality Management: A Comparative Analysis." *American Review of Public Administration* 23 (1): 57–73.

Lowry, W. R. 1992. *Dimensions of Federalism: State Governments and Pollution Control Policies.* Durham, N.C.: Duke University Press.

National Research Council. 1991. *Rethinking the Ozone Problem in Urban and Regional Air Pollution.* Washington, D.C.: National Academy Press.

Oregon Department of Environmental Quality. 1993. "Memorandum to Environmental Quality Commission on Agenda Item I, September 10, 1993 EQC Meeting Work Session Discussion: Environmental Performance Measures." Aug. 16.

Ringquist, E. J. 1993. *Environmental Protection at the State Level: Politics and Progress in Controlling Pollution.* Armonk, N.Y.: M. E. Sharpe.

Ruckelshaus, W. D. 1984. "Agency Policies on Delegation and Oversight: Making the State/EPA Partnership Work." Memorandum to assistant administrators. Apr. 4.

Stever, D. W., and E. A. Dolin, eds. 1992. *Environmental Law and Practice: Compliance/Litigation/Forms.* Vol. 3. Deerfield, Ill.: Clark, Boardman, and Callaghan.

U.S. Environmental Protection Agency. 1990. *Toxics in the Community: National and Local Perspectives.* Washington, D.C.: GPO. EPA 560/4-90-017.

———. 1991. "Designation of Areas for Air Quality Planning Purposes." *Federal Register* 56 (174) (Nov. 6): 56694–56858.

———. 1992a. *FY 1991 State-by-State Enforcement Data Summaries.* Washington, D.C.: Office of Enforcement.

———. 1992b. *National Water Quality Inventory: 1990 Report to Congress.* Washington, D.C.: U.S. Environmental Protection Agency, Office of Water. EPA 503/9-92-006

———. 1994. *Guidance on the Post-1996 Rate-of-Progress Plan and the Attainment Demonstration.* Research Triangle Park, N.C.: Office of Air Quality Planning and Standards.

U.S. General Accounting Office. 1988. *Environmental Protection Agency: Protecting Human Health and the Environment through Improved Management.* Washington, D.C.: GPO. GAO/RCED-88-101.

Washington Department of Ecology. 1989. *The State of the Environment Report.* Olympia: Washington Department of Ecology.

———. 1990. *Toward 2010: An Environmental Action Agenda.* Olympia: Washington Department of Ecology.

———. 1995. *Washington's Environmental Health 1995.* Olympia: Washington Department of Ecology.

Wenner, L. 1976. *One Environment under Law: A Public Policy.* Pacific Palisades, Calif.: Goodyear Publishing.

Wood, B. D. 1992. "Modeling Federal Implementation as a System: The Clean Air Case." *American Journal of Political Science* 36 (1): 40–67.

PART 3

ENVIRONMENTAL SCIENCE AND ENVIRONMENTAL CHANGE

5

The Role of Chemical Indexes in Environmental Program Evaluation

ROGER A. MINEAR AND MARK A. NANNY

All environmental programs, regardless of their specific goals, are concerned with three general aspects of environmental quality: the assessment of environmental quality, the means for affecting or preventing changes in the monitored environmental system, and monitoring to ensure that programs to cause or prevent change are effective. Successful evaluation of an environmental program, in light of these aspects, depends upon appropriate measurements of the monitored environmental system. It is clear that the proper selection of indexes is a crucial part of an environmental program. Thus, selection of parameters or indexes for measurement needs to be a function of the environmental program's goals and also related to its definition of environmental quality. It is also crucial that the parameters or indexes selected must accurately respond to changes in the environmental system being monitored. Selection of appropriate indexes may be a simple task, such as measuring the concentration of a specific pollutant to determine its absence or presence, or it may be difficult and complex, such as quantitatively determining the quality of an environmental system.

Not only is the selection of indexes to be measured essential for successful monitoring but so is the methodology employed. Methodology pertains to how the parameters or indexes are measured. This includes the sampling method, the storage and handling of samples, quality assurance, quality control, and the instrumentation and assays used for analysis. The methodology employed determines the detection limits, the percent error in a measurement, reproducibility, precision, and accuracy of the measured parameters. Since important decisions, often involving political, economic, and social ramifications, are dependent upon these measurements, it is crucial that the methodology be well defined and reliable.

Because of the great dependence of an environmental program assessment upon measurements, which in most cases are measurements of chemical indexes, many aspects of environmental programs have been the result of or driven by changes in analytical chemistry. Problems that were not previously imagined or understood have come to the attention of the environmental scientific community as a result of new developments in analytical chemistry and their application to "environmental" chemistry. These developments have been in both qualitation and quantitation. The latter have resulted in dramatic decreases in lower detectable quantities in complex matrixes while the former have resulted in ever-increasing capabilities in reliable multiple speciation of individual components in complex mixtures.

In this chapter we present key historical developments in analytical chemistry and the impact these developments have had on environmental program evolution, function, and evaluation. We will also examine the role of chemical indicators in environmental program evaluations by addressing the following questions: How does one choose environmental indicators? What is the proper response to the specific environmental variable being assayed? What constitutes environmental improvement? How can environmental chemistry contribute to a more effective environmental program evaluation? Finally, we will examine the likely critical aspects of environmental programs in the future.

Evolution of Analytical Methods

Early Techniques and Gas Chromatography

Before the early 1960s, analysis of organic compounds in the environment was severely limited due to the lack of adequate techniques. Organic compounds as a composite could be evaluated only by measuring indirect or surrogate parameters, e.g., the biochemical oxygen demand (BOD) and the chemical oxygen demand (COD), both of which measured the extent the receiving water's dissolved oxygen level was depleted by the organic content. Other indirect or surrogate parameters available for measurement were total organic carbon (TOC), organic nitrogen, and oil and grease. Only a few methods existed for the detection of specific compounds such as cyanides, tannin and lignin, phenol, and furfural present in water and wastewater. In general, individual organic compounds present at low concentrations in water could be analyzed only by passing extremely large volumes of water (five thousand gallons or more) through activated carbon (Rosen 1976). The sorbed organic material was then eluted with organic solvents such as ether, chloroform, and ethanol. The solvent was evaporated and compounds in the solid residue, after extensive isolation procedures, could be analyzed with functional group characterization techniques (Cooper and Young 1984), ultraviolet-visible spectroscopy (Cooper and Young 1984), and infrared spectroscopy (Kawahara 1984). This methodology was obviously very difficult and labor intensive, raising concerns about the integrity of the sample during the extensive processing. This was a severe hindrance to effective and rapid analysis of organic compounds present at low concentrations in the environment. The development of gas chromatography (GC) changed this situation dramatically and quickly. GC provided the necessary factors for the analysis of trace amounts of organic compounds: the ability to separate and detect fifty to one hundred individual compounds present in a 1 to 2 mg sample (Heller, McGuire, and Budde 1975). A major limitation of GC methodology was that it was limited to volatile compounds or those for which volatile derivatives could be generated.

Pesticides were one of the first groups of organic compounds to be of environmental concern (Rosen and Kraybill 1966). Their deleterious effect upon the environment was brought to the public's attention in 1962 by Rachel Carson's book *Silent Spring*. The ability to detect and quantify pesticide residues in soils, sediment, foods, and plant and animal tissues was facilitated by the gas chromatograph. The use of GC provided immense knowledge regarding the fate and transport of pesticides that eventually led to the realization that chlorinated pesticides were resistant to chemical and biological degradation and therefore bioaccumulated in the food chain (Metcalf 1971).

GC had been commercially available since 1955 for the analysis of gases and organic liquids at percent levels of constituents. The extensive research into pesticides was made possible by the development of highly and in some cases element or chemical compound class sensitive GC detectors, notably the flame ionization detector and the electron capture detector. GC coupled with sensitive detectors provided detection of "trace" concentrations of pesticides, typically at the parts per million level. Besides advances in detector sensitivity, the separation efficiency of the GC columns increased, primarily resulting from the development of capillary columns (Mullin and Filkins 1981; Trussell, Lieu, and Moncur 1981). This allowed separation and detection of even more individual compounds present in a single sample.

Although GC dramatically enhanced the abilities to detect and analyze trace organic compounds, individual species identification was dependent upon the retention time of the organic compound, which had to be calibrated against standards, i.e., pure known compounds. The influence of the sample matrix upon the retention time was an unknown factor. Another problem with using standards for environmental samples was that often a suitable standard was unavailable. This is illustrated with the analysis of Araclors, which are mixtures of the 209 different polychlorinated biphenyl (PCB) congeners (i.e., distinct compounds) (Alford-Stevens 1986). Different Araclors are used for different industrial and manufacturing purposes. The concentration of individual PCB congeners in an Araclor can vary from batch to batch, thus the analyst is reduced to matching the unknown Araclor GC trace with that of the best matching standard GC trace. Because of the possible difference in concentration of the individual PCBs in the sample Araclor and the standard Araclor, the GC analysis results could appear different. This difference could lead to misidentification of the specific Araclor present in the environmental sample.

During the early to mid-1970s an explosive growth of analytical capabilities occurred, permitting the analysis of an even wider range of trace organic compounds present in natural water, air, and soil samples; air and water waste discharges; and treated drinking water. This was due not only to the advances in GC methodology but also to the linkage or interfacing of mass spectroscopy (MS) with GC (Hrutfiord, Christman, and Horton 1973); the use of high-speed computers to handle, analyze, and store the huge volume of data collected; and the development of new isolation and concentration techniques. Instrument automation was also a factor in this explosive growth because it permitted the routine analysis of large numbers of samples while enhancing and ensuring reproducibility among the sample analyses (Allen 1971; Ciaccio, Cardenas, and Jeris 1973). The combination of these systems made it possible to identify hundreds of compounds in complex environmental samples at ever-decreasing concentrations. Detection limits went from parts per million to parts per billion to, in several cases, parts per trillion.

Gas Chromatography–Mass Spectroscopy

One major problem with GC was specificity of detector response. For example, electron capture detectors, although very sensitive, respond only to highly electronegative compounds, principally halogenated compounds (i.e., organic compounds containing fluorine, chlorine, bromine, or iodine), therefore, polyhalogenated species such as chlorinated pesticides or PCBs are suitable for analysis, but hydrocarbons (i.e., such materials as oils, tars, kerosene, benzene, and toluene) and other nonhalogenated organic species are not. Because MS responds to all compounds, its use as a GC detector greatly enhances analytical capabilities. Not only are the organic compounds separated by GC but the mass spectrometer also provides information regarding molecular weight and ion fragmentation, which results in structural information and compound identification (Heller, McGuire, and Budde 1975). Furthermore, advances in MS instrumentation, such as the development of different types of ionization sources (Willard, Merritt, Dean, and Settle 1981) permitted a large diversity of organic compounds to be analyzed by MS.

During a GC-MS analysis of an environmental sample, several thousand mass spectra are generated that need to be effectively handled, analyzed, and stored. Fortunately, during the development of GC-MS, advances in computers during the 1970s made it possible to meet these demands. Computers also aided in the identification of mass spectra by comparing the results with electronically stored reference mass spectra (Heller, McGuire, and Budde 1975). The coupling of GC, MS, and a high-speed computer allowed the rapid analysis, by one person, of a complex environmental sample that previously may have required several months and many people. This created an explosive growth in the research of PCBs (Alford-Stevens 1986), polychlorinated dioxins and polychlorinated dibenzofurans (PCDDs and PCDFs, respectively) (Choudhary, Keith, and Rappe 1983; Rappe 1984; Kimbrough 1990), and polyaromatic hydrocarbons (PAH) (Hase and Hites 1976; Menzie, Potocki, and Santodonato 1992).

Purge and Trap Methodology

In the mid-1970s, analytical capacities were enhanced further through the development and use of isolation and concentration methods such as purge and trap techniques. This led to the assay of natural systems and demonstrated the virtually ubiquitous presence of volatile organic compounds (VOCs) (Bellar and Lichtenberg 1974). VOCs are organic compounds that have a high volatility and a low water solubility; trihalomethanes (THMs)—which include chloroform, trichloroethylene (TCE), trichloroethane (TCA), perchloroethylene (PCE), and carbon tetrachloride—are examples of VOCs. Because of VOC detection with purge and trap technology, the chlorination of drinking water containing dissolved organic carbon was recognized as a possible health hazard (U.S. EPA 1985; Westrick 1990).

The definition of VOCs is operational in that VOCs are compounds that are collected and concentrated on activated carbon or other hydrophobic material (Brass 1990; Hertz and Suffet 1990). After collection, GC or GC-MS is used to analyze the activated carbon.

Since the use of purge and trap methods numerous other techniques for VOC collection and concentration have evolved: solid phase extraction, micro- and large volume liquid-liquid extraction, continuous liquid-liquid extraction, and steam distillation.

Inorganic Analyses

Beside advances in organic compound analysis, similar developments in inorganic analyses occurred, even though metal analyses were far more routine in the 1950s and 1960s for many of the toxic metals of environmental concern (Minear and Murray 1973). The introduction of atomic adsorption spectroscopy (AA) in the 1960s allowed highly sensitive, element-specific natural analyses to be performed (Hume 1967; Platte 1968; Ellis and Demers 1968). At that time AA methodology was thought to be essentially independent of interferences from the sample and thus adaptable for trace metal analysis of any environmental sample. This was later found to be untrue and AA methodology was put into proper perspective. Advances in AA, such as graphite tube furnace methods and the introduction of inductively coupled plasma instrumentation (ICP), overcame some of the sample interference difficulties encountered with AA (Fassel 1984; Rains 1984). ICP also permitted the simultaneous analysis of multiple elements, making the analysis of metals in a wide array of environmental samples even more feasible. Even lower concentrations ($< 1 \times 10^{-8}$ M) of dissolved toxic metals in aquatic samples could be quantified and speciated using electrochemical methods such as anodic stripping voltammetry (Burrell and Lee 1975; Johnson 1984). Even so, these methods were often found to be of much more limited use than the AA methods because of additional sample interferences.

Future Analytical Trends

There is little doubt that the future direction of analytical capabilities is going to continue to focus on higher sensitivity and increasing ability to analyze complex samples, especially samples that are not amiable to GC-MS methods. Compounds that are aqueous, polar, and nonvolatile and also high-molecular-weight humic-derived complexes are examples of compounds that are not readily amenable to GC-MS analysis. Their identification and characterization need further development, especially since the chemistry of dissolved humic material in water produces by-products that may be a health concern. Liquid chromatography and high performance liquid chromatography coupled with MS are beginning to be used for such compounds (Brown 1990).

Another important area is the development of highly sensitive, but rugged and portable instrumentation that can be used directly in the field, allowing rapid identification of samples on site. This would eliminate having to collect, store, and transport samples, all of which could jeopardize sample integrity. This would also greatly reduce analysis time, thus reducing response time. This is important for situations that demand immediate attention or are quickly changing, such as an oil or hazardous waste spill.

Remote sensing devices represent another important area in environmental analytical instrumentation. Remote sensing devices can be used to examine changes over a large area. These devices are especially important in global and atmospheric studies, which provide a wide-scale evaluation of sources, transport, and the fate of chemicals or environmental processes, such as acid rain, ozone depletion, and the movement of air-borne pollution. These devices are also used for monitoring difficult to reach places, such as smokestack plumes. In 1994 they were used to monitor exhaust emissions of moving vehicles in traffic (Cadle and Stephens 1994). This provides a direct method to characterize and minimize exhaust emissions.

Impact of Changing Analytical Capabilities on Environmental Programs

The result of improving instrument analytical capabilities was that more analyses could be conducted at lower detection limits. This allowed previously undetected organic compounds to be found throughout the environment. Because of this, there was a great interest in determining what organic chemicals were present in our drinking water and the possible health risks they might pose (U.S. EPA 1975; National Research Council 1977). It also meant that as more information became available through better analytical resolution, previous theories and understanding were significantly revised. This often led to a "pollutant of the month" syndrome, especially in the case of toxic or heavy metals such as lead and mercury at trace levels. For many metals, the problems at high concentration levels were known but the dangers of trace concentrations, previously undetectable, were unknown. The environmental research community has tended to overpublicize, overemphasize, or overstate problems to justify increased research support, although this was not necessarily conscious or premeditated. After heavy metals, specific subsets of organic chemicals, such as PAHs, PCBs, TCE, and TMHs, were focused on. Each had its rise and fall as a favorite research focus.

Since environmental programs are driven by or generated because of new discoveries of environmental contaminants, the driving force for these programs will likely increase as analytical methods of measurement, data handling, and data processing continue to improve. The primary focus, however, will be on making necessary decisions about the meaning of the chemical data relative to evaluating the success of programs. Clearly, the chemical measurements offer objective criteria for evaluation, but the context for setting these criteria seems to be dynamic and driven by other forces.

Improvements in analytical capabilities also strongly influenced legislation concerning numerous organic compounds that had been in extensive use for many years. PCBs and TCE are prime examples. PCBs have been in use since 1929, finding wide applications as coolants, dielectric fluids in transformers and capacitors, heat transfer fluids, inks, and paints (Alford-Stevens 1986). PCBs were first reported in environmental samples in 1966 and by 1974 had become one of the most discussed pollutants (Hutzinger, Safe, and Zitko 1980). Because of this, in 1976 Congress banned the manufacture, processing, distribution, and use of PCBs except in enclosed containers (Alford-Stevens 1986). TCE has been widely used since the 1940s as a degreasing agent (Leroy 1952), dry cleaning solvent, extraction agent for decaffeinated coffee, and even as a general anesthetic (U.S. EPA 1979). Because it is quite volatile and was considered to be nontoxic, general military practice was to dispose of TCE by pouring it directly on the ground (Schaumburg 1990). It was not until the late 1970s, when analytical methods were able to routinely detect trace amounts of TCE in aqueous samples, that it was discovered that many groundwater sources had been contaminated by TCE (Schaumburg 1990). This surprising information in combination with new information regarding the possible health effects of TCE and its degradation by-products inspired specific laws and regulations in the late 1970s and early 1980s (Schaumburg 1990).

The influence of GC-MS and purge and trap methods on the direction and development of regulations is demonstrated in EPA laboratory and analytical procedures for the analysis of organic contaminants in wastewater and drinking water (Hites and Budde 1991). These EPA methods were developed to meet the growing need for standardization of analytical procedures. The majority of the EPA analytical methods in the series use GC, MS, GC-MS, and purge and trap methods in one form or another, thus heavily emphasizing the analysis of volatile, nonpolar organic compounds. Another chromatographic methodology, high performance liquid chromatography, came into use later and offered advantages in that nonvolatile substances were amenable to this analysis. However, its detection methods are much less sensitive. As a result, only a few of the methods employ high performance liquid chromatography for more polar organic species.

As a result of this enhanced analytical sensitivity and increased awareness of contaminants in the environment, billions of dollars are now being spent on cleanup of environmental contamination at Superfund sites and other groundwater contamination sites for chemicals previously unmeasurable at most levels, let alone at the parts-per-billion levels that are now being measured. Large amounts of research are funded to examine the distribution, transport, and fate of these chemicals in addition to the capital required to change or modify water and wastewater treatment facilities, landfill and waste disposal sites, and industrial processes to maintain compliance with new regulations.

Because advances in analytical methodology have aided in creating much more specific criteria for regulation and enforcement, criteria that encompass ever-expanding numbers of specific compounds at ever-decreasing concentrations, a very perplexing question arises: How clean is clean? Stated in a more rigorous manner, What is the highest concentration of a chemical species in the environment that is permissible in light of human health, ecological integrity, and reasonable economics? Clearly, this question will have to be answered in relation to program evaluation using tools such as risk assessment and risk management. Attempts to achieve zero contaminant levels will become completely meaningless as analytical techniques continue to improve because, eventually, a little bit of everything will be detected everywhere. Obviously, this problem calls for a better interface between the chemical sciences and the social, economic, and policy aspects of environmental programs. While chemical measurements provide a quantitative means for assessing the environmental state and criteria for monitoring change resulting from particular programmatic actions, there must be a value system superimposed upon the numbers obtained (Kimbrough 1990).

Role of Chemical Measurements

Chemical measurements can serve as indicators of process function, whether it be a natural ecosystem or a water, wastewater, air, or soil treatment system. Since appropriate measurements are essential to a well-devised environmental program, the question of which environmental indicators are chosen has to be addressed. Obviously, basic requirements of any indicator include that it accurately responds to specific and well-understood changes in the environmental system under assessment. The objectives and goals of the environmental program determine which changes in the environment system are to be monitored. Depending upon these changes or parameters, a proper method is selected. As mentioned previously, if the objectives are to monitor a pollutant or set of pollutants, the indicator of change is the concentration of the pollutants. Often however, the objectives of an environmental program require measurement of the quality or status of an environmental system. For example, the primary goal of the Clean Water Act of 1972 was to restore and maintain the chemical, physical, and biological integrity of the waters of the United States. Its interim goal was the restoration of the nation's waters to "fishable and swimmable" conditions. These are obviously difficult parameters to define, much less quantitatively measure with specific indexes. More often than not, it is the intended use of the water system that guides which parameters are monitored. A system used as a drinking water source will have many more specific parameters monitored than one used only for agricultural irrigation or for industrial purposes, such as a shipping channel.

Typically, key but gross system characteristics or surrogate measurements are used. Surrogates can also provide a practical alternative to the detection and identification of specific pollutants that may require expensive and sophisticated equipment (American Water Works Association Research Foundation and Keuringsinstituut Voor Waterleidingartikelen 1988). They can also be used effectively as a screening tool or for rapid determination of changes in the quality of an environmental system (American Water Works Association Research Foundation and Keuringsinstituut Voor Waterleidingartikelen 1988). A simple example of the use of surrogates is the monitoring of a water system that receives a mixture of organic, nonhazardous compounds. Instead of identifying each individual compound present, which would be a difficult and labor-intensive task, it is easier to monitor the loss of dissolved oxygen resulting from decomposition of the organic compounds. This way, the total quantity of compounds degrading the system is monitored rather than the presence and concentration of each individual compound. BOD and COD are surrogates that are used for such measurements.

The use of surrogates has also been influenced by changes in analytical capabilities. For example, indexes of water contamination that ultimately reduce or destroy dissolved oxygen in receiving waters have revolved around the five-day biochemical oxygen demand (BOD_5), which measures the potential impact on oxygen reserves by giving rate data. The major disadvantage is that the analysis requires five days and therefore it is not a real time method. Replacement or surrogate analysis was developed and continued to evolve, resulting in the introduction of COD and TOC measurement procedures. These took three hours and three minutes, respectively, for results but sacrificed specifics regarding real biologically based processes. Correlations that were system specific served to bridge the gap between real time data and real system behavior. More recent developments in analytical capabilities include the reported use of an optical fiber microbial sensor for the detection of BOD_5 (Preininger, Kilmant, and Wolfbels 1994). Typical response time is reported to be five to ten minutes.

The next question that needs to be addressed is, What is a proper response to the measurements obtained? When the program is indexed on chemical criteria, the linkage is obvious. The numbers obtained for appropriate criteria serve as a direct scale for evaluation, and if contamination decreases, the program success is directly indicated. However, when environmental standards are indexed on the lower detection limits of a method, there are a number of problems. The first is the reasonableness of constantly revising standards downward because lower values can be detected. As detection limits have decreased and as more compounds have been found in the environment, regulation seems to have followed in kind. There seems to be the sense that if we can measure it, it is potentially bad. Risk assessments based on extrapolation of animal test data at high-exposure concentrations to humans at low-exposure concentrations are invoked whenever there is a potential for human exposure, such as when chemicals threaten groundwater sources. Even when human exposure is marginal, very costly cleanup or remediation procedures are initiated on such assessments. This has been referred to as the "tyranny of analytical chemistry." Quantitative aspects of using an analytical method's lower detection limits as the legal limits for contamination is when enforcement (or, worse yet, litigation) is based on numbers just barely exceeding the detection limit value. This is especially true when only single analyses are used. The uncertainty of such an analysis is as great as 40 percent, a fact that the legal and lay public usually do not understand.

Another problem is what constitutes environmental improvement. Chemical measurements as isolated data on one or a few parameters do not necessarily suffice as an indicator of improvement. Defensible and meaningful criteria must be established for which chemical measurements can be definitive indexes of improvement. Of particular importance is full knowledge of the fate of specific chemicals and the possible transformation pathways (Kimbrough 1990). In the past, the treatability of a specific compound was assessed based solely on its presence and decline in concentration within the aqueous phase of a treatment system. Uptake of the compound, either unchanged or as a more toxic degradation product, into the biomass or removal into the air were ignored. For example, TCE degradation in groundwaters was not initially understood until recently. In some situations, the by-products were found to be more toxic than the original compound, notably the formation of vinyl chloride from TCE degradation.

The issue of what constitutes environmental improvement is brought up by the call from several environmental groups for a total ban on the use of chlorine and chlorinated compounds. Chlorinated compounds are of concern because of their persistence in the environment and accumulation in the fatty tissues of living organisms. For these reasons, the use of chlorinated pesticides such as DDT were banned in the United States several decades ago. Concern for the same reasons has arisen over the many other chlorinated compounds that are used or formed in industrial processes, such as pulp bleaching, drinking water disinfection, and the manufacture of plastics, pharmaceuticals, and a wide variety of other important chemicals. Would a total ban on chlorinated compounds be effective in decreasing the concentration of these compounds in the environment? Current research has identified about two thousand chlorinated and other halogenated compounds that are naturally discharged into the environment by plants, marine organisms, insects, bacteria, fungi, and mammals. These compounds also result from natural processes such as forest fires and volcanic eruptions (Gribble 1994). Many of the chlorinated compounds regulated by the EPA are produced in the natural environment: chlorophenols, PCDDs, PCDFs, and chloromethane. Emission of chloromethane from marine and terrestrial biomass is estimated to be 5 million tons per year while anthropogenic emissions are only 26,000 tons per year (Gribble 1994). Two research groups have concluded that forest and brush fires are the major sources of PCDDs and PCDFs in the environment (Nestrick and Lamparski 1982; Sheffield 1985). It is estimated that Canadian forest fires annually release ten times the amount of PCDDs than that released in the 1976 Seveso plant accident.

Obviously an outright ban on compounds will not cleanse the environment of their presence, therefore trying to reach a "zero contaminant level" is not feasible or realistic. Even so, their ubiquitous nature does not give license for their irresponsible and indiscriminate use. It is becoming clear that more than the mere detection of chlorinated organic compounds in the environment is required to fully understand their impact. Their origin, transport, fate, and effect upon the environment and living organisms need to be addressed more adequately before effective environmental programs regarding these compounds can be created.

New Dimensions

These traditional analytical aspects, however, represent only a small subset of the importance of the impact of environmental chemistry on environmental programs. New dimensions include the integration of biological and chemical methodologies and a shift from the command and control approach to prevention.

Bioassays have been used to set limits for specific chemicals in ambient air or water and in discharge from point sources into the environment. These techniques are neither new nor chemical assays per se because known and predetermined levels of a chemical are placed in solution or maintained in the air for exposure to a test organism. By varying the concentrations of exposure, the test organisms (animals, plants, or other life forms) are exposed to a range of concentrations and the physiological response is evaluated, including time of death of a fraction of the population. Typically, the target death fraction is 50 percent and the time of exposure is commonly twenty-four to ninety-six hours. Parallel to this approach has been a focus on bioaccumulation of specific chemicals in organismal tissue. Initially focused on pesticides, this has included heavy metals as well as specific organic chemicals.

However, in the past ten years work has focused on more subtle indicators of potential harm. These methods combine chemical methods with biochemical processes and methodologies. Specific immunological responses to target chemicals have opened new avenues for assaying toxic chemicals at sublethal and chronic levels and, in some cases, subaccute physiological response levels. Related to this approach is the use of biomarkers in target organisms. The organisms may be the actual recipient of environmental exposure or may be a test species intentionally introduced for exposure evaluation.

The National Academy of Sciences (National Research Council 1989) conducted a workshop on the biomarkers concept as related to air pollution stress on forests in which many of these principles were illustrated. In addition to physiological changes, biochemical markers included measurements of changes in specific enzyme levels, changes in secondary metabolites in internal biochemical processes, accumulation of metabolic metals in specific locations within organisms, indicators of free radical stress on organisms via analysis of accumulation of antioxidants in tissue, changes in photosynthesis rates and related metabolites, and nutrient uptake and use patterns.

With the recent advent of DNA and RNA magnification procedures (referred to as PCR, or polyerase chain reaction) subtle damage to DNA and RNA can be evaluated and assigned to specific chemical exposures. Specific chemical changes in the form of addition products or adducts to the nucleic acid structure can be evaluated and related to individual chemicals or groups of chemicals. In some cases, these changes can be related to specific organism damage.

Integral to the shift from command-and-control regulatory approaches to prevention has been the evolution of risk assessment evaluation (National Research Council 1975). Figure 5.1 depicts aspects of both human health and ecological risk assessment that relate to chemical indexes and chemical understanding that go beyond analytical chemistry but rely heavily upon it. Environmental chemistry is used in risk assessment for more than just measuring the presence and concentration of individual chemical species. Risk assessment includes determining exposure and dose. The risk of exposure requires the evolution of the probability that the target species, human or otherwise, can come into contact with a chemical judged to be detrimental by some defined endpoint. The dose that causes a defined endpoint may not be a simple function of concentration because it may also be affected by time or duration of exposure (Rodericks 1992; Calabrese 1994).

While human health issues are in themselves quite complex, all the elements of human risk assessment are embodied in ecological risk assessment but in a far more complex way because of the great interdependency of the ecological compartments (National Research Council 1981; McCarty and MacKay 1993). Figure 5.1 is not intended to minimize these complexities by omission of distinct elements. Complex chemistry can be involved for many chemical species in that more than one chemical form can exist for which a wide variation in toxicity may be exhibited. Also, the specific chemical form may dramatically affect environmental mobility and thus the likelihood of exposure (Forstner 1987). If this were not complex enough, the relationship between exposure and manifestation of an effect, i.e., the actual transferred dose to the receptor, may also be related to the specific chemical form, its physical state (i.e., liquid, gas, or solid) and the mode of exposure (i.e., inhalation, ingestion, dermal contact, and so forth).

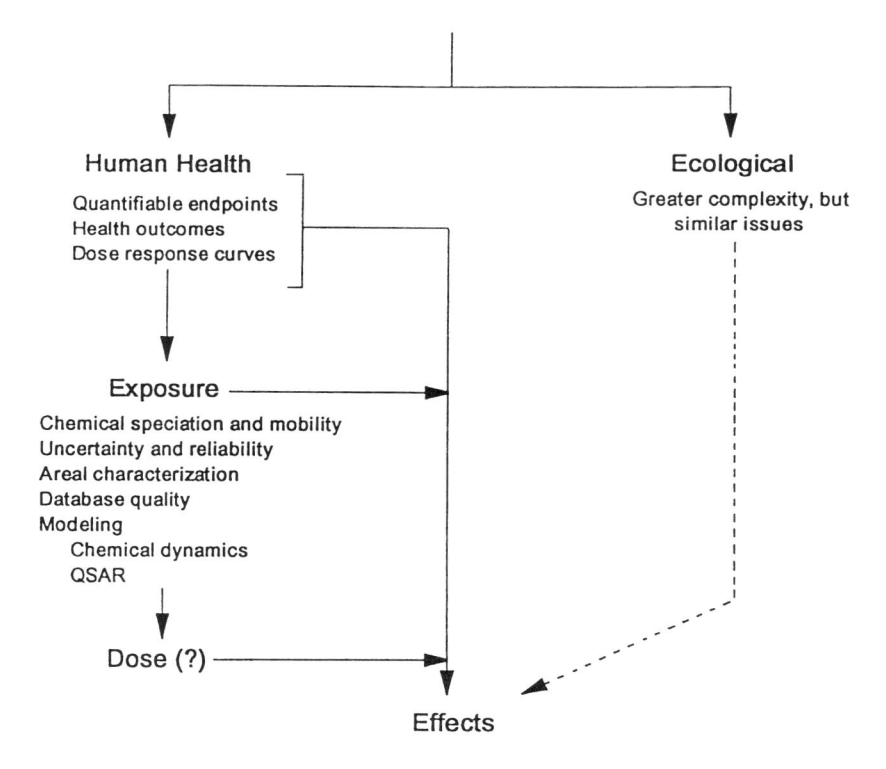

Figure 5.1. Chemistry of Risk Assessment

Environmental programs engaged in risk assessment evaluation or application must rely upon an understanding of chemical fate, including the intricacies of transport and transformation. Because the situations and controlling variables are complex and represent an almost unmanageable magnitude of possible combinations, a complete array of empirical evaluations is not feasible. Modeling of chemical dynamics in fate and transport in conjunction with risk assessment has become mandatory. Linkage of chemical transformations and biologically mediated changes in chemical species with physical models of fluid movement (atmosphere, surface waters, groundwater, or large geographical areas when dealing with atmospheric transport) has been realized for some time but has advanced only with the rapid growth in computer capabilities since 1980. However, these advances in computational power have outstripped the rate at which empirical data can be acquired for the enormous number of chemicals of potential environmental concern. The use of quantitative structure activity relationships has evolved as a means of predicting both fate and transport behavior as well as toxicity of chemicals for which elaborate and expensive test data are not available at times when decision makers cannot or will not wait for such data to be obtained experimentally (Nirmalakhandan and Speece 1988; Blum and Speece 1990).

Related to the need for reliable predictions of environmental behavior of chemicals is the issue of the reliability of existing databases (U.S. EPA 1988; Jensen 1989). Attempts have been made to develop uniform analytical methodologies and associated well-defined quality assurance and quality control protocols. But even with high confidence in specific analyses sets, there is the added difficulty in extrapolation of discrete analyses of discrete samples taken from large areas. Risk assessments require inputs of representative concentration values when determining whether there is acceptable risk at an alleged contamination site. Areal characterization of large sites from limited data is frequently dictated by cost considerations. Surface exposure estimates at times end up being based on composite samples taken over depths of 1 to 2 feet below the land surface. Much remains to be done in this area, yet major cleanup decisions of high cost and toxic tort litigation proceedings are underway using these limited data.

In many cases, court decisions are based on imperfect scientific data but when rendered, such decisions set precedent for future actions. Reinforcing the role of science in the decision process can be difficult. One possible outcome from the enormous and continually increasing costs involved in risk-based decisions is a need to deal with the level of risk society can afford and how clean is necessary at the expense of other societal benefits. Emotional arguments frequently are invoked that truly challenge the use of science in what many perceive as an increasingly antiscience society. This is an expanding challenge in developing and evaluating new environmental programs.

Environmental chemistry and related disciplines will continue to play a role in the evaluation of the true benefits of changes in the environment, and chemical indexes will be used to assess the success of new environmental programs. Negative biological responses to new chemicals and products will still be related to causative agents sooner or later, and there will be a chemical assessment component. One hopes the process is rational and reasoned, rather than responsive to new frontiers of analytical detection capabilities.

REFERENCES

Alford-Stevens, A. L. 1986. "Analyzing PCBs." *Environmental Science Technology* 20 (12): 1194–99.

Allen, H. E. 1971. "Advanced Chemical Analyses." In *Instrumental Analysis for Water Pollution Control,* ed. K. J. Mancy. Ann Arbor: Ann Arbor Science Publishers. 135–64.

American Water Works Association Research Foundation and Keuringsinstituut Voor Waterleidingartikelen. 1988. *The Search for a Surrogate.* Denver: American Water Works Association.

Bellar, T. A., and J. J. Lichtenberg. 1974. "Determining Volatile Organics at the µg/L Level in Water by Gas Chromatography." *Journal of the American Water Works Association* 66 (12): 739–44.

Blum, D. J. W., and R. E. Speece. 1990. "Determining Chemical Toxicity to Aquatic Species." *Environmental Science Technology* 24 (3): 284–93.

Brass, H. J. 1990. "Analytical Methods for Volatile Organic Compound Determination." In *Significance and Treatment of Volatile Organic Compounds in Water Supplies,* ed. N. M. Ram, R. F. Christman, and K. P. Cantor. Chelsea, Mich.: Lewis Publishers. 57–72.

Brown, M. A., ed. 1990. *Liquid Chromatography/Mass Spectroscopy: Applications in Agricultural, Pharmaceutical, and Environmental Chemistry.* ACS Symposium Series 420. Washington, D.C.: American Chemical Society.

Burrell, D. C., and M.-L. Lee. 1975. "Critical Review of Analytical Techniques for the Determination of Soluble Pollutant Heavy Metals in Seawater." In *Water Quality Parameters.* ASTM Special Technical Publication 573. American Society for Testing and Materials. Philadelphia. 58–70.

Cadle, S. H., and R. D. Stephens. 1994. "Remote Sensing of Vehicle Exhaust Emissions." *Environmental Science Technology* 28 (6): 258A–265A.

Calabrese, E., ed. 1994. *Biological Effects of Low Level Exposures: Dose-Response Relationships.* Chelsea, Mich.: Lewis Publishers.

Carson, R. 1962. *Silent Spring.* Cambridge, Mass.: Riverside Press.

Choudhary, G., L. H. Keith, and C. Rappe. 1983. *Chlorinated Dioxins and Dibenzofurans in the Total Environment.* Woburn, Mass.: Butterworth Publishers.

Ciaccio, L. L., R. R. Cardenas Jr., and J. S. Jeris. 1973. "Automated and Instrumental Methods in Water Analyses." In *Water and Water Pollution Handbook*, vol. 4, ed. L. L. Ciaccio. New York: Marcel Dekker. 1431–1556.

Cooper, W. J., and J. C. Young. 1984. "Chemical Nonspecific Organic Analysis." In *Water Analysis*, vol. 3, *Organic Species*, ed. R. A. Minear and L. H. Keith. New York: Academic Press. 41–82.

Ellis, D. W., and D. R. Demers. 1968. "Atomic Fluorescence Spectroscopy." In *Trace Inorganics in Water*, ed. R. A. Baker. ACS Advances in Chemistry 73. Washington, D.C.: American Chemical Society. 326–36.

Fassel, V. A. 1984. "Atomic Emission Methods." In *Water Analysis*, vol. 2, *Inorganic Species*, ed. R. A. Minear and L. H. Keith. New York: Academic Press. 111–54.

Forstner, U. 1987. "Changes in Metal Mobilities in Aquatic and Terrestrial Cycles." In *Metals Speciation, Separation, and Recovery*, ed. J. W. Patterson and R. Passino. Chelsea, Mich.: Lewis Publishers. 3–26.

Gribble, G. W. 1994. "The Natural Production of Chlorinated Compounds." *Environmental Science Technology* 28 (7): 310A–319A.

Hase, A., and R. A. Hites. 1976. "On the Origin of Polycyclic Aromatic Hydrocarbons in the Aqueous Environment." In *Identification and Analysis of Organic Pollutants in Water*, ed. L. H. Keith. Ann Arbor: Ann Arbor Science Publishers. 205–14.

Heller, S. R., J. M. McGuire, and W. Budde. 1975. "Trace Organics by GC/MS." *Environmental Science Technology* 9 (3): 210–13.

Hertz, C. D., and I. H. Suffet. 1990. "Research Methods for Determination of Volatile Organic Compounds in Water." In *Significance and Treatment of Volatile Organic Compounds in Water Supplies*, ed. N. M. Ram, R. F. Christman, and K. P. Cantor. Chelsea, Mich.: Lewis Publishers. 39–56.

Hites, R. A., and W. L. Budde. 1991. "EPA's Analytical Methods for Water: The Next Generation." *Environmental Science Technology* 25 (6): 998–1006.

Hrutfiord, B., R. F. Christman, and M. A. Horton. 1973. "Advanced Instrumental Techniques for the Identification of Organic Materials in Aqueous Solutions." In *Trace Metals and Metal-Organic Interactions in Natural Waters*, ed. P. C. Singer. Ann Arbor: Ann Arbor Science Publishers. 43–56.

Hume, D. N. 1967. "Analysis of Water for Trace Metals: Present Capabilities and Limitations." In *Equilibrium Concepts in Natural Water Systems*, ed. Werner Stumm. ACS Advances in Chemistry 67. Washington, D.C.: American Chemical Society. 30–44.

Hutzinger, O., S. Safe, and V. Zitko. 1980. *The Chemistry of PCBs*. Boca Raton, Fla.: CRC Press.

Jensen, A. A. 1989. In *Halogenated Biphenyls, Terphenyls, Naphthalenes, Dibenzodioxins, and Related Products*, ed. R. Kimbrough and A. A. Jensen. Amsterdam: Elsevier/North-Holland and Biomedical Press. 345–80.

Johnson, J. Donald. 1984. "Electrochemical Methods: Amperometric Analysis." In *Water Analysis*, vol. 2, *Inorganic Species*, ed. R. A. Minear and L. H. Keith. New York: Academic Press. 250–322.

Kawahara, F. K. 1984. "Infrared Spectrophotometry of Pollutants in Water Systems."

In *Water Analysis*, vol. 3, *Organic Species*, ed. R. A. Minear and L. H. Keith. New York: Academic Press. 381–443.

Kimbrough, R. D. 1990. "Environmental Protection: Theory and Practice." *Environmental Science Technology* 24 (10): 1442–45.

Kimbrough, R., and A. A. Jensen, eds. 1989. *Halogenated Biphenyls, Terphenyls, Naphthalenes, Dibenzodioxins, and Related Products.* Amsterdam: Elsevier/North-Holland and Biomedical Press.

Leroy, J. M. 1952. *Modern Metal Degreasing.* Sarna, Canada: Dow Chemical of Canada.

McCarty, L. S., and D. MacKay. 1993. "Enhancing Ecotoxicological Modeling and Assessment." *Environmental Science Technology* 27 (9): 1719–28.

Menzie, C. A., B. B. Potocki, and J. Santodonato. 1992. "Exposure to Carcinogenic PAHs in the Environment." *Environmental Science Technology* 26 (7): 1278–84.

Metcalf, R. L. 1971. "The Chemistry and Biology of Pesticides." In *Pesticides in the Environment*, vol. 1, ed. R. White-Stevens. New York: Marcel Dekker. 1–144.

Minear, R. A., and B. B. Murray. 1973. "Methods of Trace Metals Analysis in Aquatic Systems." In *Trace Metals and Metal-Organic Interactions in Natural Waters*, ed. P. C. Singer. Ann Arbor: Ann Arbor Science Publishers. 1–42.

Mullin, M. D., and J. C. Filkins. 1981. "Analysis of Polychlorinated Biphenyls by Glass Capillary and Packed-Column Chromatography." In *Advances in the Identification and Analysis of Organic Pollutants in Water*, ed. L. H. Keith. Ann Arbor: Ann Arbor Science Publishers. 1:187–96.

National Academy of Sciences. 1975. *Principles for Evaluating Chemicals in the Environment.* Washington, D.C.: National Academy of the Sciences.

National Research Council. 1977. *Drinking Water and Health.* Vol. 1. Washington, D.C.: National Academy of Sciences.

———. Commission on Natural Resources, Committee to Review Methods for Ecotoxicology. 1981. *Testing for Effects of Chemicals on Ecosystems.* Washington, D.C.: National Academy Press.

———. 1989. *Biologic Markers of Air Pollution Stress and Damage in Forests.* Washington, D.C.: National Academy Press.

Nestrick, T. J., and L. L. Lamparski. 1982. "Isomer-Specific Determination of Chlorinated Dioxins for Assessment of Formation and Potential Environmental Emission from Wood Combustion." *Analytical Chemistry* 54 (13): 2292.

Nirmalakhandan, N., and R. E. Speece. 1988. "Structure-Activity Relationships." *Environmental Science Technology* 22 (6): 606–15.

Platte, J. A. 1968. "Analysis of Industrial Waters by Atomic Adsorption." In *Trace Inorganics in Water*, ed. R. A. Baker. ACS Advances in Chemistry 73. Washington, D.C.: American Chemical Society. 247–52.

Preininger, C., I. Kilmant, and O. S. Wolfbels. 1994. "Optical Fiber Sensor for Biological Oxygen Demand." *Analytical Chemistry* 66:1841–46.

Rains, T. C. 1984. "Atomic Adsorption Spectrophotometry." In *Water Analysis*, vol. 2, *Inorganic Species*, ed. R. A. Minear and L. H. Keith. New York: Academic Press. 62–110.

Rappe, C. 1984. "Analysis of Polychlorinated Dioxins and Furans." *Environmental Science Technology* 18 (3): 78A–90A.

Rodericks, J. 1992. *Calculated Risks.* New York: Cambridge University Press.

Rosen, A. A. 1976. "The Foundations of Organic Pollutant Analysis." In *Identification and Analysis of Organic Pollutants in Water,* ed. L. H. Keith. Ann Arbor: Ann Arbor Science Publishers. 3–14.

Rosen, A. A., and H. F. Kraybill, eds. 1966. *Organic Pesticides in the Environment.* ACS Advances in Chemistry 60. Washington, D.C.: American Chemical Society.

Schaumburg, F. D. 1990. "Banning Trichloroethylene: Responsible Reaction or Overkill." *Environmental Science Technology* 24 (1): 17–22.

Sheffield, A. 1985. "Sources and Releases of PCDD's and PCDF's to the Canadian Environment." *Chemosphere* 14:811.

Trussell, A. A., F.-Y. Lieu, and J. G. Moncur. 1981. "Part-per-Trillion Analysis of Volatile, Base/Neutral, and Acidic Water Contaminants on a Single Fused-Silica Capillary Column." In *Advances in the Identification and Analysis of Organic Pollutants in Water,* ed. L. H. Keith. Ann Arbor: Ann Arbor Science Publishers. 1:171–86.

U.S. Environmental Protection Agency. Office of Toxic Substances. 1975. *Preliminary Assessment of Suspected Carcinogens in Drinking Water: Report to Congress.* Washington, D.C.: Environmental Protection Agency.

———. 1979. *Status Assessment of Toxic Chemicals: Trichloroethylene.* Washington, D.C.: GPO. EPA-600/2-79-210M.

———. 1985. "National Primary Drinking Water Regulations; Synthetic Organic Chemicals, Inorganic Chemicals, and Microorganisms; Proposed Rule. *Federal Register* 50 (219) (Nov. 13): 46936–47025.

———. 1988. *Availability, Adequacy, and Comparability of Testing Procedures for the Analysis of Pollutants Established under Section 304 (h) of the Federal Water Pollution Control Act: Report to Congress.* Cincinnati: Office of Research and Development, Environmental Monitoring Systems Laboratory. EPA/600/9-87/030.

Westrick, J. J. 1990. "National Surveys of Volatile Organic Compounds in Ground and Surface Waters." In *Significance and Treatment of Volatile Organic Compounds in Water Supplies,* ed. N. M. Ram, R. F. Christman, and K. P. Cantor. Chelsea, Mich.: Lewis Publishers. 103–25.

Willard, H. H., L. L. Merritt, J. A. Dean, and F. A. Settle. 1981. *Instrumental Methods of Analysis.* New York: Van Nostrand.

6

Biological Integrity: A Long-Neglected Aspect of Environmental Program Evaluation

JAMES R. KARR

Unbridled population growth and technological expansion threaten the integrity of the biosphere, including human welfare (Karr 1993, 1995b). Although the scale in modern times is unique, the threat is not unprecedented. Human history documents numerous civilizations that developed, prospered, and eventually collapsed because they depleted and degraded their natural resource base (Ponting 1991). Water and resources associated with water illustrate for modern society the kinds of challenges faced many times in human history.

Water of sufficient quality and quantity is critical to human welfare, indeed to all life. Modern society is devoting more and more attention to protecting adequate supplies of water. Although the perception of biological degradation stimulated state and federal legislation on water resource quality in the United States, the focus on living systems was lost as environmental managers concentrated on easily measured physical and chemical surrogates (water pollution was simply defined as chemical contamination). As the downward trend in the condition of biological resources has become more obvious, interest in biological monitoring has grown. We can no longer behave as if concepts such as "biological integrity" and "unreasonable degradation" were not central to implementation of the Clean Water Act. Assessing the quality of water resources by sampling biological communities in the field (ambient biological monitoring) offers a promising monitoring approach.

Several federal agencies and many states are calling for implementation of direct biological monitoring programs. New philosophies guide these efforts and signal major shifts that will play key roles in restoring and maintaining the biological integrity of the nation's waters, the explicit mandate of the Water Pollution Control Act Amendments of 1972 (PL 92-500) and amendments. The

U.S. Environmental Protection Agency (EPA) has called for including biological criteria (indicators of biological condition) in its water quality standards program, restructuring existing monitoring programs to document the impact of regulatory programs, evaluating and controlling nonpoint pollution, coordinating chemical sampling with biological surveys, assessing ecological risk, incorporating "good science" at all levels of water resource policy, and adopting narrative and numerical biological criteria into state water quality standards. The EPA and others call for developing biological criteria to protect terrestrial wildlife from the negative impact of human activities on water resources. The U.S. Geological Survey, the U.S. Forest Service, and the Tennessee Valley Authority are also expanding their uses of biological monitoring.

Specific tactics involved in biological monitoring programs vary among states. Some states have adopted legal biological criteria (Florida, Ohio, Vermont), biologically based use designations (e.g., excellent warm-water habitat: Maine, Arkansas), and biological criteria in assessments and monitoring. After a detailed, statewide program to evaluate ambient (field) biological monitoring, Ohio incorporated biological monitoring into regulations aimed at attaining the goals of the Clean Water Act (Yoder and Rankin 1995). Some states (Colorado) retain a focus on toxics and effluent, while others (Ohio) incorporate a broader goal for biological integrity. These advances emerged from a recognition that water resource problems involve biological as well as physicochemical and socioeconomic issues.

Philosophical shifts within state and federal agencies suggest that it is possible to overcome shortsighted and incomplete approaches to water resource management (e.g., approaches assuming that water resource problems can be solved by cleaning up the water). Replacing such approaches with sophisticated, quantitative evaluations based on biological and ecological principles will provide better assessments of the effectiveness of water management programs. As of 1996 forty-two states used multimetric biological assessments of biological conditions and six states were developing biological assessment approaches; only three states used multimetric biological approaches in 1989 (U.S. EPA 1996).

Although the foundations for these advances have existed for perhaps three decades, three factors have contributed to rapid progress during the last fifteen years: the development of integrative biological indexes, the development of an ecoregion approach, and a recognition of the importance of assessing cumulative impacts at regional scales. The challenge for basic and applied ecologists in the beginning of the twenty-first century will be to ensure that ecological principles are used to evaluate water quality programs and to protect and manage the nation's aquatic systems (Karr 1995a, 1997).

Why Has It Taken So Long?

Although the declining ability of our waters to support biological activity was the first sign of a problem, society embraced approaches to improve water quality that, in the main, ignored biology. At least eight factors contributed to this trend.

Dominance of Reductionist Viewpoints

The dominant reductionist views of several disciplines involved in state and federal water management have been, and remain, a major impediment. First, few knowledgeable ecologists are willing to either help develop and implement biological criteria or challenge the conceptual underpinnings of chemical criteria. "Standards" are legally established rules consisting of two parts: designated uses and criteria. "Designated uses" are the purposes or benefits to be derived from a water body (e.g., aquatic life, irrigation water, or drinking water), and "criteria" are the conditions presumed to support or protect the designated uses. Dissolved oxygen may not fall below 5 mg/L, for example, if the "designated use" is a cold-water fishery. In addition, the "antidegradation" goal added to the Clean Water Act reauthorization in 1987 is often considered a component of water quality standards. Second, engineers often fail to consider biotic impairment. Third, politicians implement programs based on local interests and short time scales. Finally, planners naively attack problems as if ecosystem dysfunctions were unimportant or easily reversible.

Limited Legal and Regulatory Programs

Water law within the American legal system is a complex integration of federal and state constitutions (fundamental law), statutes and ordinances (acts at local, state, and federal levels), administrative regulations (formulated and implemented by agencies), executive orders (orders by state and federal chief executives), and common law court decisions (Goldfarb 1988). As a result, responsibility for regulating, protecting, and developing water resources is vested in a patchwork of local, state, national, and international agencies. Although protecting water resources is the primary goal of water law, the law is not adequate for real protection. For example, ground and surface waters are treated as if they were unconnected, despite scientific evidence to the contrary. This artificial distinction highlights the fundamentally different approaches used in legal and scientific circumstances. Historically, courts could impose effluent controls only with proof, based on a "preponderance of evidence," that an effluent was degrading a water body. Science is more concerned with risk management, which involves evaluating probabilities of damage, than with such "burden of proof."

Another common, but technically indefensible, dichotomy firmly ingrained in water law separates the legal framework for water quality from that of water quantity (McDonnell 1990). For example, water rights laws (the prior appropriation doctrine for allocating water) dominate in the West and Southwest (Dufford 1995), where water supplies are limited (6 percent of the U.S. supply but 31 percent of its use); and water quality laws dominate in the East, where water supplies generally exceed demand (37 percent of supply and 8 percent of use) ("A Comparison" 1990). Toxicological approaches (water quality) dominate efforts to protect water resources in the East, whereas the need to protect in-stream flows (water quantity) dominate in the West (Bain and Boltz 1989).

The evolution of federal water quality legislation (and of the regulations that support it) illustrates additional constraints on the use of biology in protecting water resources. The first water quality act was probably the Refuse Act of 1899, created to treat the growing problem of disease and oil pollution in navigable waters. Since the 1940s, a series of laws and amendments has been passed under the general rubric Water Pollution Control Act (WPCA; also Water Quality Act or Clean Water Act). Amendments passed in 1948, 1956, 1961, 1965, 1966, and 1970 established several trends: more money for construction and technology development, expanded lists of pollutants to treat, and increased enforcement to control point sources of pollution. These efforts successfully controlled disease and, secondarily, reduced discharge of suspended solids (especially particulate organic carbon) that produced high biochemical oxygen demand (BOD) near wastewater treatment outflows. On the other hand, the growing array of chemicals from industrial plants; urban runoff; and agricultural sediments, nutrients, and pesticides were inadequately treated.

The burden of proof in documenting ecosystem degradation and establishing the causes of specific discharges fell on the government (Ward and Loftis 1989). Establishing cause was difficult, however, and enforcement actions rarely succeeded. First, no rigorously defined water quality criteria were available. Second, few tools existed to accurately and effectively portray the results of regulatory programs.

By 1972, Congress recognized the need to revamp water resource programs. The WPCA Amendments of 1972 (PL 92-500), which came on the heels of the first Earth Day and heightened environmental awareness at the national level in the 1960s, contained far-reaching provisions, including stronger enforcement, increased federal involvement in water resource programs, and strict deadlines to end pollution by 1985. These provisions were to be implemented primarily by achieving technology-based limits for point source effluents (Ward and Loftis

1989). For the first time, water quality standards covered intra- and interstate waters. Two visionary phrases in the act dealt with keeping water bodies "fishable and swimmable" and the charge to "restore and maintain the physical, chemical, and biological integrity of the Nation's waters." These phrases explicitly called attention to the need to protect "all forms of natural aquatic life (the ultimate goal of water quality management)" (Meybeck and Helmer 1989:304).

Although the new emphasis on technology-based controls was heralded as revolutionary, implementation programs were usually limited to establishing effluent limitations, an improved discharge permit system (the National Pollutant Discharge Elimination System), performance standards for new plants and industries, and the call for sewer and waste treatment plants in all municipalities in the United States. Regulations continued to stress rules and standards for effluents instead of measuring the biological effects in the receiving water body because regulators feared a return to the battles over burdens of proof. Thus, the focus on chemical parameters continued, or, when a biological perspective was taken, the emphasis was placed on acute and, later, chronic effects of chemical pollutants on laboratory organisms. Many have expressed disappointment that such a visionary law was applied so inadequately ("WPCF Roundtable" 1981; "Congressional Staffers" 1981; "Changing" 1983).

Although the call for protecting biotic integrity was explicit in the 1972 Clean Water amendments, point source effluents remained the primary target of regulation for at least three reasons: biological integrity was only one of several factors explicitly protected; politically and logistically, point sources were easier to clean up; and numerical chemical contaminant standards were thought to be not only legally defensible but also sufficient to protect water resources.

Success controlling point sources of pollution made the effects of nonpoint source problems more obvious. But programs to control nonpoint source pollution were (and remain today) largely unsuccessful because it is difficult to apply point source approaches to diffuse nonpoint source problems (Karr 1990) and because society remains unwilling to limit private land rights for the public good.

The next major legislative action came in 1977 with passage of the Clean Water Act. As a result, emphasis shifted from conventional pollutants (e.g., fecal coliform and BOD) to the growing list of toxic chemicals released into the nation's waters. Although a wider perspective was integral to the 1972 and 1977 legislation, the primary regulatory approach of both the EPA and the states focused on technology-based controls to limit point source pollutants discharged into bodies of water. All too frequently, efforts to measure progress toward water quality goals counted permits issued or point sources regulated rather than assessing real environmental results. An inability to associate standards based on water quality with biological integrity also limited the success of efforts to protect water resources, especially given the combined (synergistic, antagonistic, and additive; Risser 1988) effects of numerous pollutants and other human impacts (cumulative impacts). Chemical and physical approaches may be legally defensible (Mount 1985), but they cannot measure complex attributes such as ecological health or "biotic integrity."

Because of widespread public support, Congress passed the Water Quality Act of 1987, overriding a presidential veto. When combined with the regulations developed in 1983 and 1985, this act changed the emphasis from technology-based controls with simple chemical water quality standards to protection of specific water bodies (Plafkin 1989).

A major shift in philosophy came in the 1980s, when earlier approaches were recognized as inadequate in a report entitled *Surface Water Monitoring: A Framework for Change* (U.S. EPA 1987). At last, ambient monitoring of biological integrity was regarded as a direct, comprehensive indicator of ecological conditions and, thus, the quality of a water body. Although some argue that "the water quality criteria approach has served the science and the needs of society well" (Kimerle 1986:114), continuing degradation stimulated evolution in three areas of EPA policy: efforts to document environmental variability across landscapes and thereby to develop appropriate regional adjustments in standards, efforts to develop and implement approaches for direct assessment of biotic integrity, and recognition of the need to assess and mitigate cumulative impacts of human society.

Imprecise Definitions of Biological Integrity

The EPA convened a symposium on the integrity of water soon after passage of the Clean Water Act, but no clear definition of *biotic integrity* emerged. Many authors advocated use of a holistic perspective. James R. Karr and Daniel R. Dudley argued that "integrity" encompasses all factors affecting aquatic ecosystems. A place has biological integrity if it is able to support and maintain "a balanced, integrated, adaptive community of organisms having a species composition, diversity, and functional organization comparable to that of natural habitat of the region" (1981:56).

A biological system can be considered healthy when its inherent potential is realized, its condition is stable, its capacity for self-repair when perturbed is preserved, and minimal external support for management is needed. This umbrella goal motivates virtually all environmental legislation (Karr et al. 1986; Angermeier and Karr 1994).

These concepts establish broad biological goals to replace narrow chemical criteria, and their use depends on developing biological criteria based on ecological principles. Unfortunately, most theoretical ecologists have been reluctant to apply their knowledge (Schindler 1987), and a cynical attitude about the utility of ecologists dominates in some quarters (Wilk 1985; Kareiva 1990). Single-species bioassays, complicated models, and impact statements have been singularly unsuccessful at predicting the effects of anthropogenic stress on biological systems (Schindler 1987). Studies of population dynamics, food web organization, and taxonomic structure in communities have been more successful (Schindler 1987). Besides defining *biological integrity,* success at incorporating biotic integrity into program evaluation depends on an appropriate, cost-effective procedure for evaluating biotic impairment.

Inadequate Indexes to Assess Biological Integrity

Early efforts to develop biological indexes concentrated on detecting a narrow range of variation in biological integrity, yielded indexes sensitive to only a few types of degradation (low dissolved oxygen, selected toxins, and so on), or provided only a binary evaluation (degraded versus not degraded). Some indexes evaluated fecal contamination (Geldreich 1970), and others focused on effects of chemical stress on organisms (Ford 1989). Although valuable for measuring selected anthropogenic effects, early indexes were less useful for screening all types of degradation, including complex cumulative impacts.

The system developed in Europe based on classification of taxa according to their tolerance of low oxygen levels (Slàdaček 1973) focuses on BOD. Dominance of these tolerant organisms indicates high levels of oxygen-demanding wastes or sedimentation resulting from soil erosion. Even more complex community-based indexes, like the Hilsenhoff index (Hilsenhoff 1987) for benthic insects and other invertebrates (macroinvertebrates), are primarily sensitive to domestic effluents. Coliform counts can identify inputs of untreated sewage, although contamination from wildlife and livestock may also affect bacterial counts (Dudley and Karr 1979). Finally, many existing biological indexes may apply to only a narrow geographical area (e.g., lake trout stocks from Lake Ontario to Superior). When specific narrow impacts are known to be the only human influence at a site, these approaches are valuable to judge the magnitude of their effect. But effective assessment requires a more comprehensive approach.

One successful tactic developed in the 1980s is to combine two or more biological metrics into a single multimetric index. Each biological metric provides information about the sampling site and also the region (Steedman 1988). This effort is akin to using metrics such as blood pressure, urine analysis, white blood count, and temperature to evaluate a person's health or housing starts, unemployment rates, and gross national product to track economic health. It is important to note, however, that good health—human, economic, or ecological—is not a simple function of those metrics (Karr et al. 1986).

The ideal index would be sensitive to all stresses placed on biological systems by human society but would have limited sensitivity to natural variation in physical and biological environments (see figure 6.1). An array of indicators

Ecological Impact of Human-Induced Alterations →

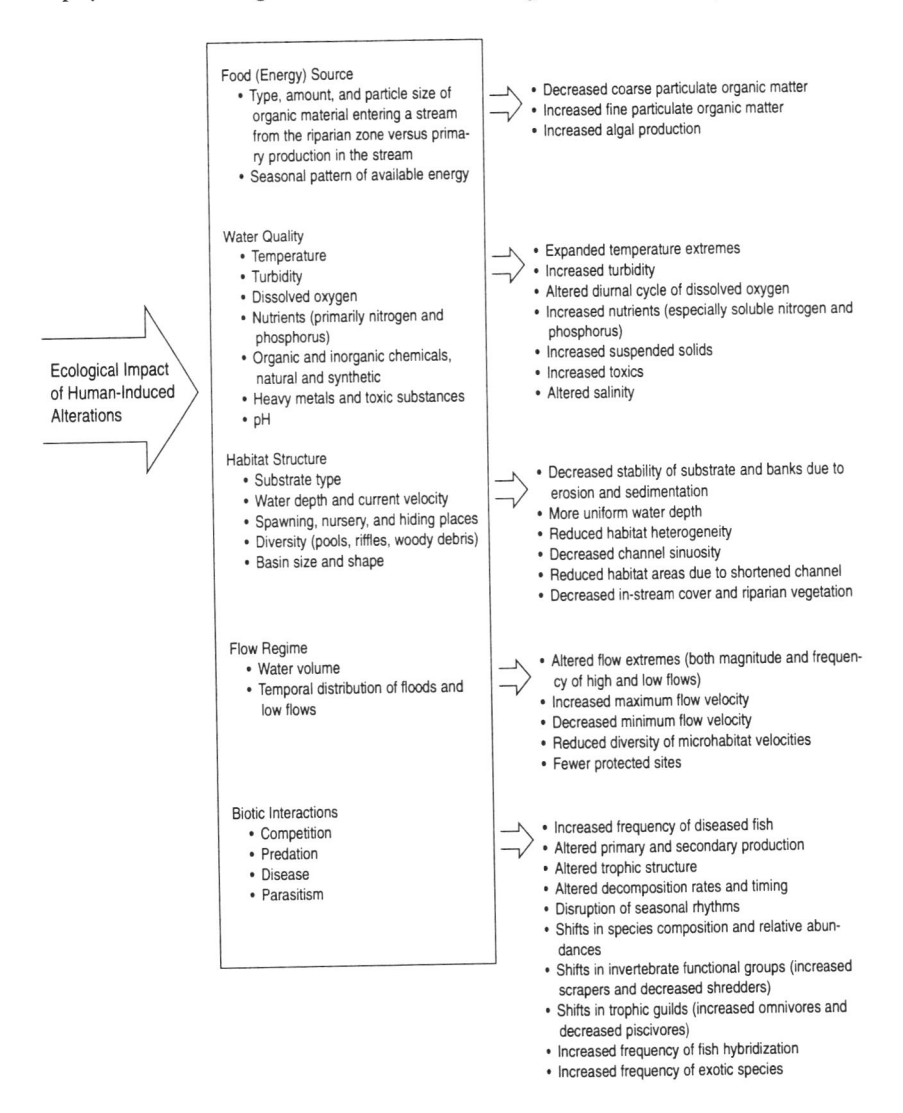

Food (Energy) Source
- Type, amount, and particle size of organic material entering a stream from the riparian zone versus primary production in the stream
- Seasonal pattern of available energy

→
- Decreased coarse particulate organic matter
- Increased fine particulate organic matter
- Increased algal production

Water Quality
- Temperature
- Turbidity
- Dissolved oxygen
- Nutrients (primarily nitrogen and phosphorus)
- Organic and inorganic chemicals, natural and synthetic
- Heavy metals and toxic substances
- pH

→
- Expanded temperature extremes
- Increased turbidity
- Altered diurnal cycle of dissolved oxygen
- Increased nutrients (especially soluble nitrogen and phosphorus)
- Increased suspended solids
- Increased toxics
- Altered salinity

Habitat Structure
- Substrate type
- Water depth and current velocity
- Spawning, nursery, and hiding places
- Diversity (pools, riffles, woody debris)
- Basin size and shape

→
- Decreased stability of substrate and banks due to erosion and sedimentation
- More uniform water depth
- Reduced habitat heterogeneity
- Decreased channel sinuosity
- Reduced habitat areas due to shortened channel
- Decreased in-stream cover and riparian vegetation

Flow Regime
- Water volume
- Temporal distribution of floods and low flows

→
- Altered flow extremes (both magnitude and frequency of high and low flows)
- Increased maximum flow velocity
- Decreased minimum flow velocity
- Reduced diversity of microhabitat velocities
- Fewer protected sites

Biotic Interactions
- Competition
- Predation
- Disease
- Parasitism

→
- Increased frequency of diseased fish
- Altered primary and secondary production
- Altered trophic structure
- Altered decomposition rates and timing
- Disruption of seasonal rhythms
- Shifts in species composition and relative abundances
- Shifts in invertebrate functional groups (increased scrapers and decreased shredders)
- Shifts in trophic guilds (increased omnivores and decreased piscivores)
- Increased frequency of fish hybridization
- Increased frequency of exotic species

Figure 6.1. Ecological Impact of Human-Induced Alterations

would be combined into one or more simple indexes and could be used to detect degradation, identify its cause, and determine if improvement resulted from management actions. In the best of all worlds, these could be used in a regulatory context to prevent degradation and thus preserve high-quality water resource systems. Indicators for general use must be applicable in a wide range of water resource systems and measure attainment of the biological integrity goals of the Clean Water Act (Ohio EPA 1988).

Regional Specificity of Quantitative Definitions of Ecological Health

The idea of chemical-specific toxicological criteria and water quality standards involves defining contaminant levels above which one can expect negative effects on water resources (Levin, Harwell, Kelly, and Kimball 1989). But standardized values for chemical criteria fail to recognize natural geographic variation in water chemistry. Natural heavy metal concentrations in western rivers often reach well above EPA standards, and dissolved oxygen often falls below standards established in water quality regulations. Many efforts in the 1980s to develop biological criteria call for defining specific expectations according to region (Fausch, Karr, and Yant 1984; Hughes, Larsen, and Omernik 1986) and stream size (Karr 1981; Ohio EPA 1988; Plafkin et al. 1989).

Many states have adopted Omernik's (1987) ecoregion concept as a framework for refining biological expectations. Ecoregions are geographic areas within which stream communities are relatively homogeneous. Nevertheless, their boundaries should not be flatly accepted because other boundaries, including those of river basins and physiographic provinces, may also play important roles. Accounting for natural geographic variation in the ecological features of undisturbed aquatic systems is essential, something that has been overlooked for several decades in efforts to set rigid nationwide chemical and physical standards.

Incomplete Standardization of Field Methods

Standardization of methods (quality assurance/quality control) is a fundamental prerequisite for any monitoring program. Without standardized methods, the utility of environmental monitoring data can and will be challenged (Plafkin et al. 1989). The first step is to define standards for biological sampling. James R. Karr and colleagues (1986), Ohio EPA (1988), James L. Plafkin and colleagues (1989), and Wayne S. Davis and Thomas P. Simon (1995) provide examples of efforts to define acceptable methods for sampling biological communities in the field with minimal effort. They also attempt to formalize analytic procedures.

Weak Links between Field Measurements and Enforceable Management Options

The national framework for water quality management provided by the 1965 Water Quality Act (PL 89-234) required policies and goals (standards) against which in-stream water quality would be evaluated. Problems developed when it became obvious that knowledge of the connections between in-stream conditions and the action of dischargers was inadequate. Under the assumption that point source discharges were causing problems, the 1972 act shifted from a focus on in-stream conditions to one on effluents. Enforcement targeted violators of discharge permits.

By the early 1980s, many people recognized that ignoring in-stream conditions, especially their effects on living organisms, did nothing to curtail continuing degradation of water resources. Although improved water quality did result in some areas from limiting discharge through permits, money was sometimes spent on wastewater treatment that did not improve in-stream conditions. Simply put, the shift to effluent monitoring was a high-cost program that failed to protect the overall quality of many key water resources from continuing impacts by unregulated sources (e.g., nonpoint sources).

One extreme view on this issue (Mount 1985) contends that the best reason to depend on testing based on toxicity is that such testing is decisive; toxicity-based criteria provide clear standards for estimating impacts on water quality. But decisiveness does not overcome the many deficiencies of toxicity testing in relation to ecological goals. Most chemical standards have no meaning outside the legal or regulatory context, and they protect only those environmental values explicitly included in the standard-setting process. In addition, toxicity-based criteria are not adequate early-warning detectors of degradation. Finally, poorly understood but important biological mechanisms and effects are not incorporated into the standard-setting process (Suter 1990). Single-species toxicity testing may, in specific situations, be well informed and decisive, but in many circumstances such decisiveness may be misleading and even detrimental for the resource as a whole.

In short, for nearly two decades, a narrow perspective on standards was presumed to be effective because it was decisive, legally defensible, and enforceable. Although this approach is viable to control point source discharges, sole reliance on it cripples society's ability to detect, much less reverse, degradation caused by nonpoint sources of pollution, habitat destruction, modifications in flow, spread of exotic organisms, and changes in the food supply (energy base) for aquatic organisms.

Need for Cost-Effective Approaches to Biological Monitoring

Water resource managers have long argued that the costs of ambient biological monitoring are too high (Loftis, Ward, and Smillie 1983). A compilation by the Ohio EPA (see table 6.1), however, shows that biological monitoring is not prohibitively expensive compared with other approaches, although costs may vary among agencies and circumstances. Accounting must go beyond the cost of data collection and analysis to include costs of building and operating expensive and potentially unnecessary or poorly designed treatment plants and costs of poor management decisions (e.g., see Karr, Heidinger, and Helmer 1985). The mandate in the Clean Water Act of 1987 to reduce emphasis on construction suggests more widespread recognition of this issue.

Although progress toward reducing the effects of these eight constraints varies, all have now been widely recognized, and considerable energy is being expended to overcome them.

Table 6.1. Comparative Cost Analysis for Sample Collection, Processing, and Analysis for Evaluating the Quality of a Water Resource

	Per sample[a]	Per evaluation[b]
Chemical/physical water quality		
4 samples/site	$359	$1,653
Bioassay		
Screening (acute—48-h exposure)	1,191	3,573
Definitive (LC50[c] and EC50[d]—48-h and 96-h exposure)	1,848	5,544
7-d (acute and chronic effects, single sample)	3,052	9,156
7-d (acute and chronic effects, composite sample collected daily)	6,106	18,318
Macroinvertebrate community[e]	824	824
Fish community[f]	340	740
Fish and macroinvertebrates (combined)	1,164	1,564

Source: Yoder and Rankin 1995

a. The cost to sample one location or one effluent; standard evaluation protocols specify multiple samples per location.

b. The cost to evaluate the impact of an entity; this example assumes sampling five stream sites and one effluent discharge site.

c. Dose of toxicant that is lethal to 50 percent of the organisms in the test conditions at a specified time.

d. Concentration at which a specified effect is observed in 50 percent of organisms tested; e.g., hemorrhaging, dilation of pupils, cessation of swimming.

e. Using invertebrate community index (ICI) (see text and table 6.4).

f. Using index of biotic integrity (IBI) (see text and table 6.2).

How to Measure Biotic Integrity

Human activities may alter the physical, chemical, or biological processes associated with water resources and thus modify the resident biological community. Biological criteria are valuable for assessing these alterations because they "directly measure the condition of the resource at risk, detect problems that other methods may miss or underestimate, and provide a systematic process for measuring progress resulting from the implementation of water quality programs" (U.S. EPA 1990:vii). They do not replace chemical and toxicological methods, but they do increase the probability that a program evaluation will detect degradation caused by anthropogenic influences or improvement resulting from remedial actions.

During the 1980s, five primary sets of variables were identified that, when affected by human activities, result in ecosystem degradation (see figure 6.2). Many individual studies demonstrate correlations (if not cause and effect) between degradation and some biological indicator (e.g., species richness, changing abundance of an indicator species, ratio of production to respiration; see Taub 1987 and Ford 1989 for reviews). Few attempts have been made, however, to integrate more than one of those indicators into a single index. The index of biotic integrity does so by using a set of attributes that measure the organization and structure of fish or insect assemblages.

The Index of Biotic Integrity

The index of biotic integrity (IBI) was conceived to provide a broadly based and ecologically sound tool to evaluate biological conditions in streams and human effects on a stream and its watershed (Karr 1981). The IBI incorporates many attributes of fish assemblages, covering the range of ecological levels from the individual through population, community, and ecosystem. Although initially developed for use with fishes, the ecological foundation of the IBI can be used to develop analogous indexes that apply to other taxa (Kerans and Karr 1994; Fore, Karr, and Wisseman 1996) or even to incorporate multiple taxa in a more comprehensive assessment of biotic integrity such as that used by the Ohio EPA (Yoder and Rankin 1995).

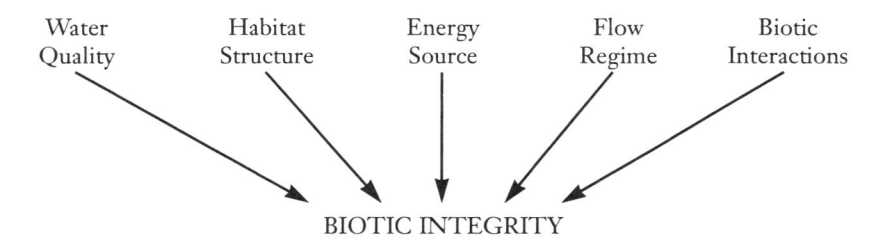

Water Quality → Habitat Structure → Energy Source → Flow Regime → Biotic Interactions → BIOTIC INTEGRITY

Figure 6.2. Influences on Biotic Integrity

Calculating a fish IBI for a stream reach requires a single sample that represents fish species composition and relative abundances (see Karr et al. 1986; Ohio EPA 1988; and Plafkin et al. 1989 for detailed sampling and data-handling protocols). Applying the IBI requires carefully standardizing procedures.

The value for each IBI metric is based on a comparison with a regional reference site that has been subjected to little or no influence from humans (Fausch, Karr, and Yant 1984). Expected values are based on that reference site, and observed values from the study stream are compared with expected values. For each metric, an index score of 5 is assigned if the study site deviates only slightly from the reference site, 3 if it deviates moderately, and 1 if it deviates strongly from the undisturbed condition. These assessments require experienced biologists to set standards based on knowledge of the regional biota (organisms living in a region or at a site) and stream size. For example, more fish species are expected in a river in Tennessee than in a stream of similar size in Nebraska and in a large stream than in a small stream. Thus, assessment of biotic integrity explicitly incorporates biogeographic variation into the evaluation.

To calculate the IBI as originally developed, twelve attributes (see table 6.2) of a fish community are rated. The sum of those ratings (5, 3, or 1) provides an IBI value, an integrative and quantitative assessment of local biological integrity (see table 6.3). The IBI uses three groups of metrics: species richness and composition, trophic composition, and fish abundance and condition.

Species Richness and Composition Metrics The first group of six metrics evaluates the extent to which the sample area is degraded as reflected in reduced species richness and altered species composition. Because richness varies as a function of region, stream size, elevation, and stream gradient, all sites must be evaluated against the expected richness from a similar undisturbed site (or, regionally, minimally disturbed site). For the Midwest, the IBI includes five metrics of species richness (table 6.2): three are taxon specific (Catostomidae [suckers]; Etheostomatinae, Percidae [darters]; and Centrarchidae [sunfish]); one assesses the presence of species intolerant of human activities; and one assesses total species richness (excluding nonnative species, or exotics). The three taxa represent groups that consume benthic (bottom-dwelling) invertebrates (suckers and darters) or drifting and terrestrial invertebrates (sunfish) as their primary food. The presence or absence of these taxa confirm whether their needed spawning habitat and food are available. Other taxa with similar ecological attributes should be substituted in regions where these families are not abundant (see "Adaptability of the IBI" below and table 6.2 for modification of IBI metrics in regions outside the Midwest).

The metric of intolerant species is based on the common observation that a

Table 6.2. Metrics Used to Assess Biological Integrity of Fish Communities Based on IBI

Metrics	Rating of Metric[a]		
	5	3	1
Species richness and composition			
1. Total number of fish species[a] (native fish species)[b]			
2. Number and identity of darter species (benthic species)	Expectations for metrics 1–5 vary with stream size and region.		
3. Number and identity of sunfish species (water column species)			
4. Number and identity of sucker species (long-lived species)			
5. Number and identity of intolerant species			
6. Percentage of individuals that are green sunfish (tolerant species)	< 5	5–20	> 20
Trophic composition			
7. Percentage of individuals that are omnivores	< 20	20–45	> 45
8. Percentage of individuals that are insectivorous cyprinids (insectivores)	> 45	45–20	< 20
9. Percentage of individuals that are piscivores (top carnivores)	> 5	5–1	< 1
Fish abundance and condition			
10. Number of individuals in sample	Expectations for metric 10 vary with stream size and other factors.		
11. Percentage of individuals that are hybrids (exotics or simple lithophils)	0	> 0–1	> 1
12. Percentage of individuals with disease, tumors, fin damage, or skeletal anomalies	0–2	> 2–5	> 5

a. Original IBI metrics for midwestern United States (Karr 1981).
b. Generalized IBI metrics (see Miller et al. 1988).

few species are especially sensitive to human disturbance (intolerants). Many species cannot tolerate siltation, but species intolerant of other factors may also be present. The "intolerant class" should be restricted to the 5–10 percent of species most susceptible to degradation; the designation should not be taken as equivalent to "rare" or "endangered" (Karr et al. 1986). The fifth metric of species richness–species composition is the total species richness of the community; the sixth metric relates solely to species composition. Green sunfish (*Lepomis cyanellus*) increase in relative abundance in degraded streams of the Midwest, reflecting the extent to which disturbance permits a single species to dominate the community. Other species used in this metric in other regions include common carp (*Cyprinus carpio*), white sucker (*Catastomus commersonii*), and gardon (*Rutilus rutilus*) (Miller et al. 1988; Oberdorff and Hughes 1992).

The number of species expected in an undisturbed stream increases with stream size. A "maximum species richness line" is determined by plotting the number of fish species collected in samples from streams against stream size (stream order, watershed area, or flow). A plot of data from a watershed (figure

Table 6.3. Total IBI Scores, Integrity Classes, and the Attributes of Those Classes

Total IBI Score[a]	Integrity Class of Site	Attributes
58–60	Excellent	Comparable to the best situations without human disturbance; all regionally expected species for the habitat and stream size, including the most intolerant forms, are present with a full array of age (size) classes; balanced trophic structure
48–52	Good	Species richness somewhat below expectation, especially because the most intolerant forms have been lost; some species are present at less than optimal abundances or size distributions; trophic structure shows some signs of stress
40–44	Fair	Loss of intolerant forms, fewer species, highly skewed trophic structures (e.g., increasing frequency of omnivores and green sunfish or other tolerant species); older age classes of top predators may be rare
28–34	Poor	Dominated by omnivores, tolerant forms, and habitat generalists; few top carnivores; growth rates and condition factors commonly depressed; hybrids and diseased fish often present
12–22	Very poor	Few fish present, mostly introduced or tolerant forms; hybrids common; disease, parasites, fin damage, and other anomalies regular
[b]	No fish	Repeated sampling finds no fish

a. The sum of the twelve metric ratings. Sites with values between classes are assigned to the appropriate integrity class following careful consideration of individual metrics by informed biologists.
b. No score can be calculated where no fish were found.

6.3) or ecoregion typically yields a right triangle whose hypotenuse approximates the upper limit of species richness (Fausch, Karr, and Yant 1984; Karr et al. 1986; Ohio EPA 1988).

Trophic Composition Metrics All organisms require reliable sources of energy, and major efforts have been made to measure the many dimensions of productivity. Measuring productivity directly is costly and time consuming, especially if attempted at several trophic levels (e.g., primary producers, primary consumers, predators), and interpreting the results may be ambiguous or even misleading (Schindler 1987). Thus, several metrics that measure divergence from expectation were developed to assess energy flow through the community (trophic structure). In degraded streams, the proportion of omnivores increases, and the proportion of insectivorous minnows (family Cyprinidae) and top car-

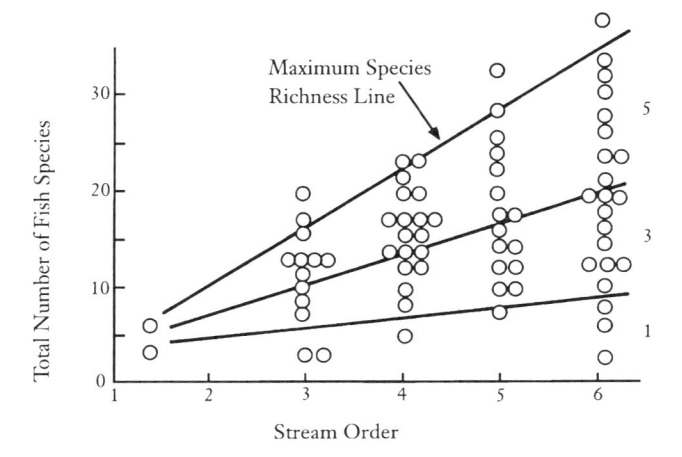

Figure 6.3. Number of Fish Species and Stream Order

nivores decreases. The proportion of individuals in the sample is used to rate stream reaches for each of these metrics. Scoring criteria for functional metrics have been remarkably consistent throughout North America, suggesting a general pattern for stream fishes.

Fish Abundance and Condition Metrics Three metrics evaluate population density and fish condition. The total number of individuals in the sample is an important parameter because severely disturbed areas often have fewer fish. The metric for number of individuals must be based on catch per unit of sampling effort, such as the number per sampled area or time period. The last two metrics evaluate the frequency of hybridization—apparently a function of habitat destruction and mixing of gametes in the best available spawning areas (Greenfield, Abdel-Hameed, Deckert, and Flinn 1973)—and the proportion of individuals with disease, tumors, fin damage, and major skeletal anomalies. Such anomalies increase in degraded areas, especially those with major toxic contamination (Brown et al. 1973). The metric for anomalies incorporates only those anomalies that can be found externally, without sacrificing the fish.

Each metric reflects the attributes of the fish assemblage that respond to variations in different aspects of the aquatic system. The relative sensitivity of the metrics varies from region to region (Angermeier and Karr 1986; Karr et al. 1986; Steedman 1988), in part because metrics are differentially sensitive to perturbations (siltation, flow alteration, toxins). In addition, natural variation among watersheds means that indexes based on one or a few metrics are unlikely to provide reliable assessments over a wide geographic area. In one study

(Karr, Heidinger, and Helmer 1985), for example, the total number of fishes decreased and trophic structure of the community shifted in areas exposed to municipal outfall. Sedimentation and other habitat alteration reduced the number of fishes feeding on benthic invertebrates (e.g., darters). In the most degraded sites, many IBI metrics reflect the serious degradation, reinforcing the strength of inferences about the status of biological systems (degree of biotic integrity) at the site. Finally, IBI metrics have differential sensitivity along the gradient from undisturbed to degraded (figure 6.4).

IBI scores can be used to evaluate current conditions at a site; determine trends at a site over time, with repeated sampling; compare sites from which data are collected more or less simultaneously; and, to some extent, identify the cause of local degradation (Karr et al. 1986). More than forty-five states and provinces and several federal agencies have used the IBI (or modifications of it; see "Adaptability of the IBI" below). At least four states and the Tennessee Valley Authority have incorporated the IBI into their standards and monitoring programs (Miller et al. 1988), and many others are expanding use of the IBI and conceptually similar approaches into their routine monitoring programs.

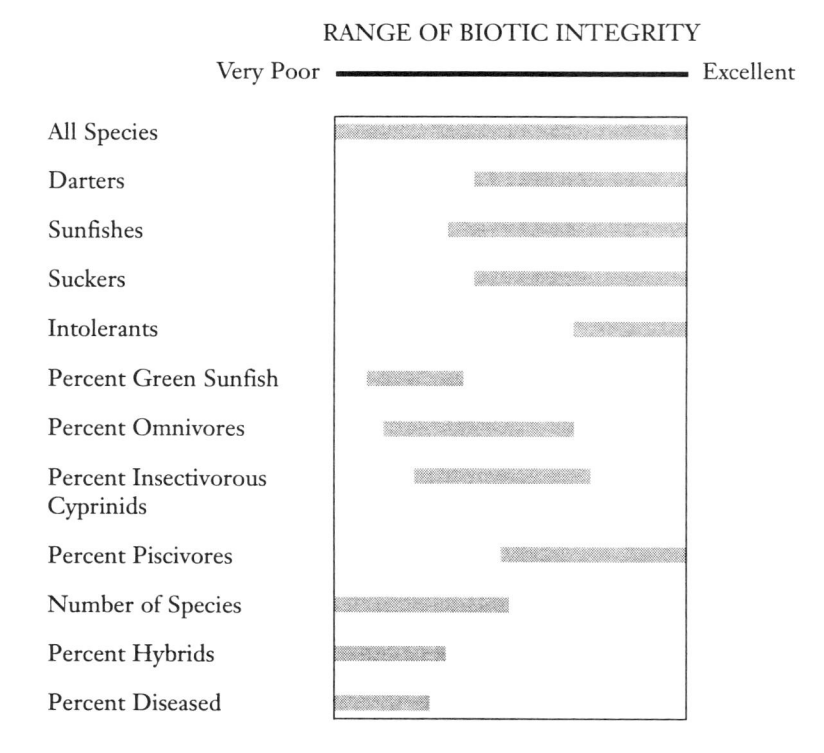

Figure 6.4. Range of Primary Sensitivity for Each Metric in the IBI

The IBI is now being used in Europe, South and Central America, Africa, Asia, and Australia.

Many advantages of the IBI have been cited (Miller et al. 1988; Plafkin et al. 1989; Fausch, Lyons, Karr, and Angermeier 1990), including: it is quantitative; it gauges a stream against an expectation based on minimal disturbance in the region; it reflects distinct attributes of biological systems, including temporal and spatial dynamics; there is no loss of information from constituent metrics when the total index is determined because each metric contributes to the total evaluation; and professional judgment is incorporated in a systematic and ecologically sound manner.

The IBI does not serve all needs of detailed biological monitoring (Karr et al. 1986; Fausch, Lyons, Karr, and Angermeier 1990) and certainly cannot be advocated as a replacement for physical and chemical monitoring or toxicity testing. Nevertheless, ecologically sophisticated biological monitoring provides direct information about conditions at a sample site compared with a site subject to little or no human influence or with criteria for a designated biological use (e.g., high-quality warm-water fishery).

Some Successful Applications of the IBI

Successful applications of the IBI in varied contexts (effects of mine drainage, sewage effluent, habitat alteration, and so forth) and in diverse geographic areas prove the usefulness of the IBI concept (Karr et al. 1986; Steedman 1988; Lyons 1992; Oberdorff and Hughes 1992; Deegan, Finn, Ayvazian, and Ryder 1993; Kerans and Karr 1994).

The influence of chemical contaminants on biotic integrity have been directly evaluated; the IBI declines as residual chlorine concentration increases in three watersheds in east-central Illinois (Karr, Heidinger, and Helmer 1985). In another study, general watershed conditions and wastewater treatment outflows were evaluated in the Scioto River in Ohio (Yoder and Rankin 1995). IBI values were low throughout the river in 1980, especially downstream of wastewater discharge from large sewage treatment plants. After ten years of effort to control the plants' effluents, downstream biotic integrity finally complied with water quality standards, although regional nonpoint pollution and habitat degradation kept the IBI below optimal levels.

In southern Ontario, the IBI was strongly associated with two independently derived measures of watershed condition (Steedman 1988): urbanized area and the condition of a riparian corridor. Biotic integrity declined sharply for Toronto-vicinity streams under conditions ranging from 75 percent removal of riparian vegetation at sites with no urbanization to 0 percent removal of riparian vegetation at 55 percent urbanization.

Finally, between-stream variation in IBI arose from the mobility of fish and the presence of nearby habitat refuges (sources of colonists) in Jordan Creek (Karr, Yant, Fausch, and Schlosser 1987), suggesting that cumulative regional impacts may also be important in determining local biotic integrity.

In short, the IBI satisfies three traits named by Schindler (1987) for useful monitoring programs: it is inexpensive, simple, and highly sensitive to changes in ecosystems.

Adaptability of the IBI

No single index or set of metrics can be expected to detect all water resource problems. Still, the IBI is very successful as a broadly based approach to assess the quality of a water resource. The IBI can be modified to incorporate other aspects of the fish community, such as species composition within major taxa, population structure (e.g., size frequency distribution), growth rates, and relative health of individuals within populations of selected species. James R. Karr (1981) mentioned all of these, but did not incorporate any into the index because the necessary information was not easily obtained, especially from historical databases, which furnished the primary information for initial IBI development and testing.

Adapting the IBI to geographic regions outside the midwestern United States requires modifying, deleting, or replacing selected IBI metrics (see table 6.2). David L. Miller and colleagues (1988) review changes needed to reflect regional differences in biological communities and fish distribution. The kind of flexibility offered by the IBI results from an integrative framework with a strong ecological foundation. Places as diverse as the streams of Colorado, New England, northern California, Oregon, southeastern Canada, France (Seine River), and Appalachia and coastal environments in Louisiana, the Chesapeake Bay, and New England have been evaluated with the IBI's conceptual approach.

In California, the principal attributes that must be accommodated are reduced species richness, high endemism among watersheds, absence of midwestern taxa such as darters and sunfish, and many salmonid species (salmon, trout, steelhead). Modifications in the IBI needed for use in estuarine regions of Louisiana include variation in salinity and the landscape context of estuary sites. New IBI metrics reflect aspects of fish residency, presence of nearshore marine fishes and large freshwater fishes, and seasonal variation in community structure. Other special considerations include the importance of stream gradient (steepness) in Appalachia and geographic variation in tolerances within some species. For example, the creek chub (*Semotilus atromaculatus*) varies appreciably in its tolerance of stream degradation and in food habits from Colorado to Illinois to the New River drainage of Virginia.

Modifications adopted by the Ohio EPA include replacement of several of the original IBI metrics with alternatives applicable to large rivers. The agency proposes substituting round-bodied suckers for darters in large rivers sampled with electrofishing gear—which stuns fish with an electric shock—mounted on boats, an excellent suggestion in a situation where darters are likely to be undersampled. Agency biologists have also field-tested and evaluated other IBI aspects.

The Tennessee Valley Authority has shown the value of IBI in assessing declining biotic integrity (Saylor and Scott 1987; Kerans and Karr 1994; Jennings, Fore, and Karr 1995). In one case, release of cold water limited fish communities in a stream segment immediately downstream of a reservoir, and in another case, low-flow periods left much of the channel dry. In both cases, the IBI detected this degradation while general reviews of habitat conditions and water quality did not.

Perhaps the most innovative use of the IBI is the work of Robert J. Steedman (1988) in southern Ontario. He sampled twenty-one fish species at 209 tributary sites on the northwestern shore of Lake Ontario near Toronto. His ten-metric IBI accommodated both cold- and warm-water reaches, for example, by combining taxonomic groups in selected metrics: sculpins plus darters, salmonids plus sunfish, and suckers plus catfishes.

Steedman's analysis of threshold effects in riparian degradation raises a persistent but still unanswered question about what threshold of destruction in riparian vegetation within a watershed causes major disruption of biotic integrity (Karr and Schlosser 1978). His use of the IBI may answer that question directly and deserves considerable study in many geographic areas.

David L. Miller and colleagues (1988) encouraged modifying the IBI to make it suitable for a wide range of geographical areas. I suggest three cautions. First, avoid modifications unless they yield significant improvement in the index's utility (Angermeier and Karr 1986). Paul M. Leonard and Donald J. Orth (1986), for example, modified many IBI metrics for studying streams in West Virginia. Reanalysis of their data, however, showed that the modified metrics did not improve the IBI's ability to detect degradation. Second, modifications should be done only by experienced fish biologists familiar with local fish faunas, watershed conditions, and the conceptual framework of the IBI.

Finally, efforts should be made to develop indexes similar to the IBI for other environments, such as wetlands, lakes, and terrestrial ecosystems. Successful applications of the IBI have been made in lakes in Minnesota (Heiskary, Wilson, and Larsen 1987), reservoirs (Jennings, Fore, and, Karr 1995), and wetlands (Brooks and Hughes 1988). Attempts have been made, with limited success, to apply IBI concepts to birds of forest islands; for example, the number of omnivorous birds increases as the size of a forest island declines (Karr 1987).

Apparently, the altered food base in remnant forest islands parallels changes in species richness and abundance of omnivores in disturbed headwater streams.

Assessment of Biotic Integrity with Other Taxa

The framework of the fish IBI has been adopted by invertebrate biologists to develop robust measurements of degradation based on benthic invertebrates (Ohio EPA 1988; Plafkin 1989; Kerans and Karr 1994; Fore, Karr, and Wisseman 1996). One extensively tested integrative effort is the invertebrate community index (ICI) developed by the Ohio EPA (1988), a ten-metric index that emphasizes structural attributes of invertebrate communities. The Ohio EPA took this approach because the findings are easy to interpret, the index requires only simple derivation, and the method is widely accepted. Metric 10 is scored from a qualitative field sample; metrics 1–9 are based on artificial-substrate sampling.

As part of its effort to establish biological metrics, the EPA has also supported development of a hierarchy of methods for biological monitoring. Its rapid bioassessment protocol III (RBP III) (Plafkin et al. 1989) is similar to the ICI but has only eight metrics, both structural and functional. The RBP III combines sampling invertebrates from a riffle or run habitat and from clumps of dead leaves (leaf packs) at each sampling site.

Finally, based on work with the Tennessee Valley Authority, the benthic index of biotic integrity (B-IBI) was developed for eastern rivers (Kerans and Karr 1994) and is being tested in the Pacific Northwest (Fore, Karr, and Wisseman 1996) and Japan (Rossano 1996). These efforts strengthen the role of biology in the assessment of water resource quality. Both the B-IBI and the ICI are robust additions to the arsenal of assessment tools. The RBP III has been less extensively tested, and many validation studies remain to be done.

Fish and invertebrate IBI approaches are a major improvement over past programs in river and stream environments. Ecologists are now in a position to develop suites of metrics that integrate taxa (fish, invertebrates, protozoa, and diatoms) and levels of ecological organization (population, community, ecosystem, and landscape).

The Future of Biological Monitoring

Growing dissatisfaction with present water resource programs and appreciation of the potential contribution of improved biological monitoring have stimulated nationwide interest in biological monitoring to help attain the goals of the Clean Water Act (Davis and Simon 1995). The solution of water resource problems will not come from better regulation of chemicals or better tools to detect degradation caused by chemicals.

Ecologists and biologists now have an unprecedented opportunity to influence and even guide decisions about water resources. How might they contribute? The most critical need is to develop monitoring, assessment, regulatory, and restoration approaches—grounded in rigorous ecological principles—that evaluate the complex dynamics of degradation at local levels and the cumulative regional impacts of human-induced disturbance (Karr and Dudley 1981).

Many indicators (see table 6.4) of the health of biological systems have been tested (Ford 1989; Gray 1989; Levin, Harwell, Kelly, and Kimball 1989; Karr 1990). Each is sensitive at different levels of degradation and to different kinds of human-caused stress. In addition, the indicators vary considerably in ease of measurement. That several biological indicators appear repeatedly in many studies by many biologists, however, suggests an unusual consensus. The complexity of biological systems and the diversity of factors responsible for degradation make it unlikely that any metric will be sensitive enough for all circumstances, and biologists are prone to reject many promising specific approaches because they cannot be generalized. In fact, aspects of those promising indicators should be integrated to create more robust biological monitoring.

The success of the IBI in a wide range of stream sizes and geographic areas comes because it integrates the independent discoveries of many investigators. Like any index, however, the IBI alone is inadequate. The IBI was first restricted to fish communities. Too much time and energy have been expended arguing about which taxon is most appropriate. With robust and general metrics, any major taxon could give beneficial insights. One could develop a fish IBI, a macroinvertebrate IBI, and so on, or suites of metrics that effectively integrate taxa. Whatever the choice, any use of the IBI concept should comprise a broad array of metrics evaluating conditions from individual, population, community, and ecosystem levels.

Ecologists need to support efforts to incorporate biology into evaluations of water resources. Ecologists responsible for program evaluations must strive to overcome the tendency to amass unorganized data. In many respects, the inability or reluctance to distill biological meaning from large quantities of data—by means of rigorous, accurate, yet easily understood analyses—has diminished the role that biology has played in water resource management.

Finally and most important, ecologists should try to develop ways to apply advances in ecological theory to program evaluation. The need for biological input into evaluating water resources is similar in many ways to the need that stimulated conservation biology years ago (Schonewald-Cox, Chambers,

Table 6.4. Biological Indicators Used to Assess the Condition of a Water Resource with the Goal of Protecting Human Health while Maintaining Biotic Integrity of a Specific Resource

Bioassay   Exposing test organisms, in a laboratory setting, to various concentrations of suspected toxicants or dilutions of whole effluent
  Single species test
  Multispecies (microcosm, mesocosm) test
Biosurvey   Collecting a representative portion of the organisms living in a water body to determine the characteristics of the aquatic community
  Individual/species population (may involve selection of indicator species)
    Tissue analysis for bioaccumulation
    Biomarkers (genetics or physiology)
    Biomass/yield
    Growth rates
    Gross morphology (external or internal)
    Behavior
    Abundance/density
    Variation in population size
    Population age structure
    Disease or parasitism frequency
  Community/ecosystem structure (may involve indicator taxa or guilds)
    Species richness/diversity
    Relative abundances of species
    Tolerants/intolerants
    Abundance of opportunists
    Dominant species
    Community trophic structure
    Extinction
  Function
    Production to respiration ratio
    Production biomass ratio
    Biogeochemical cycle nutrient leakage
    Decomposition
  Landscape
    Habitat fragmentation patch geometry
    Linkages among patches
    Cumulative effects across landscapes

MacBryde, and Thomas 1983; Soule 1987). A core issue is how to use ecological knowledge to improve our ability to measure and interpret the effects of pollution or other human impacts on biological assemblages. We must be able to translate this knowledge into statements about the condition (ecological health) of these systems. Ecology as a discipline must contend with questions such as: Given existing spatial and temporal variation, how can we optimize sampling design to detect patterns? How do we identify and define impairment? How can we improve our ability to detect initial impairment (sensitive indicators, early-warning indicators), as opposed to detecting only massive degradation? What should be done to apply integrative, ecological approaches to monitoring in a wide diversity of biological systems?

Ecological research has greatly expanded knowledge of stream ecology and the dynamics of water resource systems, yet unfortunately these insights have not been effective in protecting those systems. The time is ripe to change this trend.

NOTE

An earlier version of this essay appeared as "Biological Integrity: A Long-Neglected Aspect of Water Resource Management" in *Ecological Applications* 1 (1991): 66–84. © by Ecological Society of America. Used by permission of the publisher.

REFERENCES

Angermeier, P. L., and J. R. Karr. 1986. "Applying an Index of Biotic Integrity Based on Stream-Fish Communities: Considerations in Sampling and Interpretation." *North American Journal of Fisheries Management* 6:418–29.

———. 1994. "Biological Integrity versus Biological Diversity as Policy Directives." *BioScience* 44:690–97.

Bain, M. B., and J. M. Boltz. 1989. *Regulated Streamflow and Warmwater Stream Fish: A General Hypothesis and Research Agenda*. U.S. Fish and Wildlife Service Biological Report 89 (18).

Brooks, R. P., and R. M. Hughes. 1988. "Guidelines for Assessing the Biotic Communities of Freshwater Wetlands." In *Proceedings of the National Wetland Symposium: Mitigation of Impacts and Losses*, ed. J. A. Kusler, M. L. Quammen, and G. Brooks. Berne, N.Y.: Association of the State Wetland Managers. 276–82.

Brown, R. E., J. J. Hazdra, L. Keith, I. Greenspan, and J. B. G. Kwapinski. 1973. "Frequency of Fish Tumors in a Polluted Watershed as Compared to Non-Polluted Canadian Waters." *Cancer Research* 33:189–98.

"Changing the Clean Water Act: Reflections on the Long-Term." 1983. *Journal of the Water Pollution Control Federation* 55:123–29.

"A Comparison of the Regional Distribution of Water Supply and Demand with Population Growth." 1990. *Balance Data* (27): 6.

"Congressional Staffers Take a Retrospective Look at PL 92-500: Part 2." 1981. *Journal of the Water Pollution Control Federation* 53:1370–77.

Davis, W. S., and T. P. Simon, eds. 1995. *Biological Assessment and Criteria: Tools for Water Resource Planning and Decision Making*. Boca Raton, Fla.: Lewis Publishers.

Deegan, L. A., J. T. Finn, S. G. Ayvazian, and C. Ryder. 1993. "Feasibility and Application of the Index of Biotic Integrity to Massachusetts Estuaries (EBI)." Final Project Report. Woods Hole, Mass.: Ecosystems Center, Marine Biological Laboratory.

Dudley, D. R., and J. R. Karr. 1979. "Concentration and Sources of Fecal and Organic Pollution in an Agricultural Watershed." *Water Resources Bulletin* 15:911–23.

Dufford, W. 1995. "Washington Water Law: A Primer." *Illahee* 11:29–39.

Fausch, K. D., J. R. Karr, and P. R. Yant. 1984. "Regional Application of an Index of Biotic Integrity Based on Streamfish Communities." *Transactions of the American Fisheries Society* 113:39–55.

Fausch, K. D., J. Lyons, J. R. Karr, and P. L. Angermeier. 1990. "Fish Communities as Indicators of Environmental Degradation." *American Fisheries Society Symposium* 8:123–44.

Ford, J. 1989. "The Effects of Chemical Stress on Aquatic Species Composition and Community Structure." In *Ecotoxicology: Problems and Approaches,* ed. S. A. Levin, M. A. Harwell, J. R. Kelly, and K. D. Kimball. New York: Springer-Verlag. 99–144.

Fore, L. S., J. R. Karr, and R. W. Wisseman. 1996. "Assessing Invertebrate Responses to Human Activities: Evaluating Alternative Approaches." *Journal of the North American Benthological Society* 15:212–31.

Geldreich, E. E. 1970. "Applying Bacteriological Parameters to Recreational Water Quality." *Journal of the American Waterworks Association* 62:113–20.

Goldfarb, W. 1988. *Water Law.* 2d ed. Chelsea, Mich.: Lewis Publishers.

Gray, J. S. 1989. "Effects of Environmental Stress on Species of Rich Assemblages." *Biological Journal of the Linnean Society* 37:19–32.

Greenfield, D. W., F. Abdel-Hameed, G. D. Deckert, and R. R. Flinn. 1973. "Hybridization between *Chrosomus erythrogaster* and *Notropis cornutus* (Pisces: Cyprinidae)." *Copeia* (1): 54–60.

Heiskary, S. A., C. B. Wilson, and D. P. Larsen. 1987. "Analysis of Regional Patterns in Lake Water Quality: Using Ecoregions for Lake Management in Minnesota." *Lake Reservoir Management* 3:337–44.

Hilsenhoff, W. L. 1987. "An Improved Biotic Index of Organic Stream Pollution." *Great Lakes Entomologist* 20:31–39.

Hughes, R. M., and J. R. Gammon. 1987. "Longitudinal Changes in Fish Assemblages and Water Quality in the Willamette River, Oregon." *Transactions of the American Fisheries Society* 116:196–209.

Hughes, R. M., D. P. Larsen, and J. M. Omernik. 1986. "Regional Reference Sites: A Method for Assessing Stream Pollution." *Environmental Management* 10:629–35.

Jennings, M. J., L. S. Fore, and J. R. Karr. 1995. "Ecological Monitoring of Fish Assemblages in Tennessee Valley Reservoirs." *Regulated Rivers: Research and Management* 11:263–74.

Kareiva, P. 1990. "The Fertile Wedding of Ecological Theory and Agricultural Practice." Book review. *Ecology* 71:1221.

Karr, J. R. 1981. "Assessment of Biotic Integrity Using Fish Communities." *Fisheries* 6 (6): 21–27.

———. 1987. "Biological Monitoring and Environmental Assessment: A Conceptual Framework." *Environmental Management* 11:249–56.

———. 1990. "Bioassessment and Non-Point Source Pollution: An Overview." In *Second National Symposium on Water Quality Assessment.* Washington, D.C.: U.S. Environmental Protection Agency. 4-1 to 4-18.

———. 1993. "Protecting Ecological Integrity: An Urgent Societal Goal." *Yale Journal of International Law* 18:297–306.

———. 1995a. "Clean Water Is Not Enough." *Illahee* 11:51–59.

———. 1995b. "Using Biological Criteria to Protect Ecological Health." In *Evaluating and Monitoring the Health of Large Scale Ecosystems*, ed. D. J. Rapport, C. Gaudet, and P. Calow. New York: Springer-Verlag. 137–52.

———. 1997. "Rivers as Sentinels: Using the Biology of Rivers to Guide Landscape Management." In *The Ecology and Management of Streams and Rivers in the Pacific Northwest Coastal Ecoregion*, ed. R. J. Naiman and R. E. Bilby. New York: Springer-Verlag. In press.

Karr, J. R., and D. R. Dudley. 1981. "Ecological Perspective on Water Quality Goals." *Environmental Management* 5:55–68.

Karr, J. R., K. D. Fausch, P. L. Angermeier, P. R. Yant, and I. J. Schlosser. 1986. *Assessing Biological Integrity in Running Waters: A Method and Its Rationale*. Special Publication 5. Champaign: Illinois Natural History Survey.

Karr, J. R., R. C. Heidinger, and E. H. Helmer. 1985. "Sensitivity of the Index of Biotic Integrity to Changes in Chlorine and Ammonia Levels from Wastewater Treatment Facilities." *Journal of the Water Pollution Control Federation* 57:912–15.

Karr, J. R., and I. J. Schlosser. 1978. "Water Resources and the Land-Water Interface." *Science* 201:229–34.

Karr, J. R., P. R. Yant, K. D. Fausch, and I. J. Schlosser. 1987. "Spatial and Temporal Variability of the Index of Biotic Integrity in Three Midwestern Streams." *Transactions of the American Fisheries Society* 116:1–11.

Kerans, B. L., and J. R. Karr. 1994. "A Benthic Index of Biotic Integrity (B-IBI) for Rivers of the Tennessee Valley." *Ecological Applications* 4:768–85.

Kimerle, R. A. 1986. "Has the Water Quality Criteria Concept Outlived Its Usefulness?" *Environmental Toxicology and Chemistry* 5:113–15.

Leonard, P. M., and D. J. Orth. 1986. "Application and Testing of an Index of Biotic Integrity in Small, Coolwater Streams." *Transactions of the American Fisheries Society* 115:401–15.

Levin, S. A., M. A. Harwell, J. R. Kelly, and K. D. Kimball. 1989. *Ecotoxicology: Problems and Approaches*. New York: Springer-Verlag.

Loftis, J. C., R. C. Ward, and G. M. Smillie. 1983. "Statistical Models for Water Quality Regulation." *Journal of the Water Pollution Control Federation* 55:1098–1104.

Lyons, J. 1992. *Using the Index of Biotic Integrity (IBI) to Measure Environmental Quality in Warmwater Streams of Wisconsin*. Forest Service General Technical Report NC-149. St. Paul: U.S. Department of Agriculture.

McDonnell, L. 1990. "Water Use: The Unfinished Business of Water Quality Protection." In *Second National Symposium on Water Quality Assessment*. Washington, D.C.: U.S. Environmental Protection Agency. 6-1 to 6-8.

Meybeck, M., and R. Helmer. 1989. "The Quality of Rivers: From Pristine Stage to Global Pollution." *Palaeogeography, Palaeoclimatology, and Palaeoecology* (Global and Planetary Change Section) 75:283–309.

Miller, D. L., et al. 1988. "Regional Applications of an Index of Biotic Integrity for Use in Water Resource Management." *Fisheries* 13 (5): 12–20.

Mount, D. I. 1985. "Scientific Problems in Using Multispecies Toxicity Tests for Regulatory Purposes." In *Multispecies Toxicity Testing,* ed. J. Cairns Jr. New York: Pergamon. 13–18.

Oberdorff, T., and R. M. Hughes. 1992. "Modification of an Index of Biotic Integrity Based on Fish Assemblages to Characterize Rivers of the Seine Basin, France." *Hydrobiologie* 228:117–30.

Ohio Environmental Protection Agency. 1988. *Biological Criteria for the Protection of Aquatic Life.* Columbus: Ohio Environmental Protection Agency, Division of Water Quality Monitoring and Assessment, Surface Water Section.

Omernik, J. M. 1987. "Ecoregions of the Coterminous United States." *Annals of the Association of American Geographers* 77:118–25.

Plafkin, J. L. 1989. "Water Quality–Based Controls and Ecosystem Recovery." In *Rehabilitating Damaged Ecosystems,* ed. J. Cairns Jr. Boca Raton, Fla.: CRC Press. 2:87–96.

Plafkin, J. L., M. T. Barbour, K. D. Porter, S. K. Gross, and R. M. Hughes. 1989. *Rapid Bioassessment Protocols for Use in Streams and Rivers: Benthic Macroinvertebrates and Fish.* Washington, D.C.: U.S. Environmental Protection Agency. EPA/444/4-89-001.

Ponting, C. 1991. *A Green History of the World: The Environment and the Collapse of Great Civilizations.* New York: St. Martin's Press.

Risser, P. G. 1988. "General Concepts for Measuring Cumulative Impacts on Wetland Ecosystems." *Environmental Management* 12:585–90.

Rossano, E. M. 1996. *Diagnosis of Stream Environments with Index of Biological Integrity.* Kyoto, Japan: Institute of Freshwater Ecology.

Saylor, C., and E. M. Scott Jr. 1987. *Application of the Index of Biotic Integrity to Existing TVA Data.* Chattanooga: Tennessee Valley Authority. TVA/ONRED/AWR 87/32.

Schindler, D. W. 1987. "Detecting Ecosystem Responses to Anthropogenic Stress." *Canadian Journal of Fisheries and Aquatic Sciences* 44:6–25.

Schonewald-Cox, C., S. M. Chambers, B. MacBryde, and L. Thomas, eds. 1983. *Genetics and Conservation.* Menlo Park, Calif.: Benjamin Cummings.

Slàdaček, V. 1973. "System of Water Quality from the Biological Point of View." *Archiv für Hydrobiologie Ergebnisse der Limnologie* 7:1–128.

Soule, M., ed. 1987. *Viable Populations for Conservation.* New York: Cambridge University Press.

Steedman, R. J. 1988. "Modification and Assessment of an Index of Biotic Integrity to Quantify Stream Quality in Southern Ontario." *Canadian Journal of Fisheries and Aquatic Sciences* 45:492–501.

Suter, G. W., II. 1990. "Endpoints for Regional Ecological Risk Assessments." *Environmental Management* 14:9–23.

Taub, F. B. 1987. "Indicators of Change in Natural and Human-Impacted Ecosystems: Status." In *Preserving Ecological Systems: The Agenda for Long-Term Research and Development,* ed. S. Draggan, J. J. Cohrssen, and R. E. Morrison. New York: Praeger. 115–44.

U.S. Environmental Protection Agency. 1987. *Surface Water Monitoring: A Framework for Change.* Washington, D.C.: U.S. Environmental Protection Agency Office of Water, Office of Policy Planning and Evaluation.

———. 1990. *Biological Criteria: National Program Guidance for Surface Waters.* Washington, D.C.: U.S. Environmental Protection Agency Office of Water Regulations and Standards. EPA-440/5-90-004.

———. 1996. *Summary of State Biological Assessment Programs for Streams and Rivers.* Washington, D.C.: U.S. Environmental Protection Agency Office of Policy, Planning, and Evaluation. EPA 230–R-96-007.

Ward, R. C., and J. C. Loftis. 1989. "Monitoring Systems for Water Quality." *Critical Reviews in Environmental Control* 19:101–18.

Wilk, I. J. 1985. "Responsibilities of Scientists: Examination of the Acid Precipitation Problem." *Science of the Total Environment* 44:293–99.

"WPCF Roundtable Discussion—Congressional Staffers Take a Retrospective Look at PL 92-500." 1981. *Journal of the Water Pollution Control Federation* 53:1264–70.

Yoder, C. O., and E. T. Rankin. 1995. "Biological Criteria Program Development and Implementation in Ohio." In *Biological Assessment and Criteria: Tools for Water Resource Planning and Decision Making,* ed. W. S. Davis and T. P. Simon. Boca Raton, Fla.: Lewis Publishers. 109–44.

7

Ode to the Miners' Canary: The Search for Environmental Indicators

HALLETT J. HARRIS AND DENISE SCHEBERLE

The quest by governments, scientists, and academics to find environmental indicators is not unlike the early reliance of coal miners on the canary. Coal miners looked to the health of the canary to decide if the environment of the underground coal shaft threatened their health in ways that the miners could not otherwise detect. So, too, have we begun to look for indicators of ecosystem health to learn about environmental risks that we cannot otherwise discern. At issue is not only the protection of the ecosystem, including the human community, but also a fundamental reorientation of how we solve environmental problems. Equally as important as capturing a sense of the viability of the ecosystem is the potential for improving environmental program evaluation. If properly established, environmental indicators can be used to provide a sense of whether efforts to protect the environment are working and, perhaps, suggest new directions for policy.

In this chapter we discuss the historical, practical, and political dimensions of environmental indicators used to assess ecosystem health. *Ecosystem health* has been defined as "an ecosystem state in which the environment is viable, liveable and sustainable" and as a concept that "encompasses ecosystem integrity as well as recognizing human societal and economic values, and is a social judgement as well as a scientific exercise" (Canadian Council of Ministers of the Environment 1994:v). Thus, our focus is on the ways in which indicators have been used in science and the promise (or pitfalls) of indicator use by communities wishing to assess ecosystem health.

In the first section we outline the evolution of ecological indicators as well as the historical uses of indicator species in science. In the second section we describe some recent attempts in the United States and Canada to include environmental indicators in policy decisions. We specifically examine binational efforts to identify ecological indicators for the Great Lakes. Next, we provide necessary criteria for selecting indicators and a case study of the indicator selection process in Green Bay in Wisconsin. The chapter concludes with a discussion of the political and social consequences of choosing environmental indicators.

The Evolution of Ecological Indicators in Science

Considering the interest in ecological and environmental indicators in the United States and Canada, one might assume that there is an established literature and well-understood set of ecological principles supporting the concept of indicators. However, a search for the term *indicator* in the index of twelve relatively recent (1985–94) college ecology textbooks revealed that only four contained the word *indicator* in the index. In three cases, the reference was to *indicator species* and involved a very brief discussion of indicator species and pollution. Only one textbook (Brewer 1994) presented any broader discussion of the indicator concept. Thus, the process of identifying ecological and environmental indicators is fairly new (Kelly and Harwell 1990; Schindler 1987; Spellerberg 1991). However, several early scientists recognized the concept of indicators.

The origin of the concept of ecological indicators seems to have arisen with early plant ecologists/plant sociologists in the late nineteenth century. Here, the focus was on individual plants as well as groups of plants (associations) as indicative of certain sets of conditions. The main message of indicator species is that what lives in a place is determined quite closely by the conditions of that place. This is surely the message conveyed by A. G. Tansley and contributors (1911) in *Types of British Vegetation*. Here, the authors assert that a "vegetation-unit is always developed in a *habitat* of definite characteristics" "and that certain kinds of plants are always found associated together under definite conditions of life" (1911:2). Another early example of the indicator concept in European plant ecology is found in the work of J. Braun-Blanquet (1932:2).

The promise of indicator species has been used in applied research. According to R. R. Brooks (1972), the Soviet geologist Karpinsky published observations that different plant associations exist on varying geological substrates. His classical work ultimately led to the science of indicator geobotany. Nonetheless, early geobotanists and plant sociologists believed that reliance should be placed on examination of the whole community rather than on one or two characteristic plants within it. Until the 1950s, botanical techniques for prospecting were used almost exclusively in the Soviet Union while they have now been adopted worldwide (Brooks 1972).

In the United States, Frederick E. Clements published a voluminous monograph on plant indicators in 1920. The bulk of the material was oriented toward the use of plants and plant associations as indicators of agricultural conditions. This grew into the well-accepted methodology of estimating range conditions by evaluating the degree of grazing pressure on a particular grassland. Range managers and ranchers classified the various grassland plants (species) into three categories with respect to the dynamics of plant succession: decreasers, increasers, and invaders. Decreasers are identified as extremely nutritious and palatable species of the climax community that decrease under heavy grazing pressure. Increasers are also climax species of high nutrition that generally increase temporarily under heavy grazing conditions. This is believed to be the result of competitive release. The invaders are undesirable species that replace the increasers and decreasers under continued heavy grazing pressure. A grassland (range) in excellent condition was said to have a high percentage of decreasers and no invaders.

This approach to "range" management incorporates an understanding of the ecology of species (autecology) with a recognition of community (synecology) as the management unit. In conjunction with species composition several other criteria were used to assess range conditions, such as plant vigor, residual mulch, soil condition, and the number of animal units that can be supported per acre. This approach represented an early attempt at more holistic management of an agroecosystem.

Yet another thread of scientific development of indicators is seen as a response to legislative mandates in the Water Quality Act of 1965. Under the act, the federal government began to assist states in establishing and enforcing water quality standards for surface waters. The National Technical Advisory Committee on Water Quality Criteria was formed in February 1967 and its report appeared in April 1968. The committee's challenge was to recommend criteria to protect the public health and welfare for five general areas of water use: recreation and aesthetics; drinking water supplies; fish, other aquatic life, and wildlife; agriculture; and industry. This effort established the following list of physical and chemical measures to monitor improvements in water quality, at least for the protection of freshwater organisms (Department of the Interior 1968):

Dissolved materials	Suspended solids
pH, alkalinity, acidity	Color and transparency
Temperature	Floating materials

Dissolved oxygen
Carbon dioxide
Oil
Turbidity
Toxic substances

Tainting substances
Radionuclides
Plant nutrients
Nuisance growths

While this list certainly contains indicators of water quality, it is dominated by indicators of physical/chemical conditions. The criteria recommended were based on what was then known of the "limits of tolerance" of a relatively few species of aquatic organisms to a limited set of physical and chemical environmental variables. For nearly two decades, one indicator, dissolved oxygen, dominated our thinking and efforts to clean up water pollution. This indicator certainly revealed when organic loading (biological oxygen demand) had been significantly reduced (Harris, Sager, Yarbrough, and Day 1987), but revealed nothing about the biological response resulting from the change. This management effort was not wrong, it was just too limited.

Passage of the National Environmental Policy Act in 1969 required environmental impact statements for federally funded projects. The U.S. Fish and Wildlife Service (1980a, 1980b) developed habitat evaluation procedures as the basis for environmental assessment. The assessment procedure is based on "evaluation species," a select group of species characteristic of particular habitats. Baseline conditions of a project area are assessed against optimum preference of the "evaluation species." The impacts of a project are "quantified" by comparing the habitat suitability index (derived by combining the habitat suitability of all evaluation species) against some projected change brought about by the project. A similar approach was developed by the U.S. Forest Service using management indicator species. The mandate for the development of management indicator species can be traced to the National Forest Management Act of 1976. These methods are, of course, not without their critics (Landers 1990).

During the 1980s, awareness of the scope of environmental problems increased. Toxic substances, particularly bioaccumulating substances, became a national and international concern. A variety of tissue, cellular, and subcellular indicators have been developed as diagnostic screening tools, or biological markers, to evaluate the physiological condition of an organism and to detect exposure to contaminants (McCarthy and Shugart 1990; Fox 1993). The need to develop management strategies that could address the intricate network of interactions within ecosystems and the impacts of human activities upon those natural systems became increasingly obvious. Nowhere was this more obvious than in the Great Lakes (Edwards and Ryder 1990; Francis, Magnuson, Regier, and Talhelm 1979; Harris, Talhelm, Magnuson, and Forbes 1982; Lee,

Regier, and Rapport 1982; Ryder and Edwards 1985). The era of the "ecosystem approach" was ushered in with an emphasis on the importance of ecosystem science and stress ecology (Rapport 1990; Harris, Harris, Regier, and Rapport 1988). The ecosystem approach was conceived of as an interdisciplinary, problem-solving process with the goal of restoring, rehabilitating, enhancing, or maintaining the integrity of a particular ecosystem. While grounded in the science of ecology, the process includes community involvement in identifying a desired future state for the ecosystem. (This approach is discussed in the case study and incorporated into figure 7.1.)

Finally, the use of indicators in science was expanded further under the rubric of ecological risk assessment (Harwell, Cooper, and Flaak 1992; U.S. EPA 1990a). The ecological risk assessment framework departs in emphasis from the human health based hazard exposure model in three ways. First, the assessment goes beyond risks to the organism and considers effects at the population, community, and ecosystem levels. Second, there is no single set of assessment endpoints (indicators) that can be generally applied. Third, the approach goes beyond the traditional emphasis on chemical effects to consider the possible effects of nonchemical stressors (U.S. EPA 1992a).

In sum, the search for indicators among scientists has moved from indicators of natural conditions useful in classifying landscape qualities to indicators of ecosystem pathology and health that are meaningful to decision makers and citizens. However, even as the search for indicators has become more sophisticated and closely linked to risk assessment, fundamental issues of science and policy have persisted. Scientists debate appropriate definitions of ecosystem health and ecosystem integrity. Environmental managers consider the appropriate number of environmental indicators. These issues are discussed in the next section.

Defining and Selecting Environmental Indicators

Environmental indicators range across differing scales of ecological organization, space, and time. The challenge, according to P. J. Seidl (1993), is to identify the state of a ecosystem in which there may be an infinite number of types of measurements but only a few of which, yet unknown, are important. Most searches for indicators in the 1990s have been for those indicators that point to the health of the ecosystem (Rapport 1990). Not all researchers in the field of ecology agree that ecosystem "health" can be measured, because the definition of *health* refers to an organism, rather than an ecological system (Suter 1993). For those scientists, the term *ecosystem integrity* has been found acceptable and was used as early as 1972 in the language of the Clean Water Act. The concept of developing measurements and monitoring for ecosystem integrity appears to have been the goal of many of the papers presented at an EPA symposium held in October 1990 (McKenzie, Hyatt, and McDonald 1990). A commonly referred to definition of *ecosystem integrity* is one offered by James R. Karr and D. R. Dudley: "the ability to support and maintain a balanced, integrated, adaptive biological community having a species composition, diversity and functional organization comparable to that of natural habitat in the region" (1981:56).

H. A. Regier (1990), quoting John Neess, identifies a set of system properties that are characteristic of an ecosystem that has integrity. These properties include energetic, natural ecosystemic processes that are strong and not severely constrained; ecosystems that are self-organizing in emerging and evolving ways; ecosystems that are self-defending against invasions by exotic organisms; ecosystems that are robust, with capabilities in reserve so as to survive and recover from occasional crisis; and ecosystems that are productive of goods and opportunities valued by humans.

This concept of ecosystem integrity suggests that the attributes of ecosystem structure and function should be metrics of concern for ecosystem management (Harris et al. 1987). One example of such incorporation is the use of the term *ecosystem integrity* regarding the Great Lakes Basin in the Great Lakes Water Quality Agreement between the United States and Canada. Accordingly, the Council of Great Lakes Research Managers adopted a definition of *indicator* from C. T. Hunsaker and D. E. Carpenter (1992) that focuses on it as a measure of the attributes of the habitat, the degree of stress, and either the level of exposure to the stressor or the response to that exposure.

A second definition, which was developed by the U.S. Intergovernmental Task Force on Monitoring Water Quality (ITFMWQ) encompasses the notion that indicators must be both scientifically and managerially useful. According to the ITFMWQ, an indicator is "a measurable feature which singly or in combination provides managerially and scientifically useful evidence of environmental and ecosystem quality, or reliable evidence of trends in quality" (U.S. ITFMWQ 1994:TF-1). As shown in table 7.1, the definition drives the indicator selection criteria, which includes measures of scientific validity and technical considerations, practical criteria, and programmatic criteria.

Both of these definitions imply that the indicator will reveal meaningful information about how a stressor or set of stressors have impacted the structure and function of the ecosystem under observation. The definitions also suggest that monitoring data collected under regulatory programs (such as effluent and emissions data from companies) will likely need to be combined

Table 7.1. Indicator Selection Criteria

Scientific validity (technical considerations)
    Measurable/quantitative
    Sensitivity
    Resolution/discriminatory power
    Reproducibility
    Representative
    Reference value
    Data comparability
    Integrates effects/exposures
Practical considerations
    Cost/cost-effectiveness
    Ease of collecting data
Programmatic considerations
    Relevance to desired goal, issue, or agency mission
    Program coverage (suite of indicators that
        encompass range of environmental conditions)
    Understandable by public

Source: Adapted from U.S. ITFMWQ 1994:TF-4.

with other kinds of data to more accurately assess ecosystem health. The selection criteria shown in table 7.1 helps to reduce the field of candidate indicators and evaluate existing databases, while at the same time addressing the concern that environmental indicators be relevant to various policy communities, environmental managers, and the public.

This leads to a final observation about the selection of appropriate and relevant indicators. Because ecosystems are complex and subject to multiple anthropogenic stresses (Harris, Talhelm, Magnuson, and Forbes 1982; Rapport and Regier 1992), a number of indicators at different scales might be necessary to adequately reflect the state of integrity. However, since monitoring an ecosystem is time consuming and usually expensive, it is obviously important to select only those indicators that are crucial for an accurate assessment of ecosystem integrity. The next sections consider efforts by governmental organizations in the United States and Canada to develop environmental indicators that could be used to evaluate progress in environmental programs (and thus, serve a policy function) as well as to describe more holistic changes in ecosystem health.

Environmental Indicators in the United States and Canada

The previous section offered definitions of ecosystem health, ecosystem integrity, and environmental indicators. While choosing indicators and evaluating them using criteria such as those developed by the ITFMWQ may seem somewhat obvious, the process is relatively new for governmental agencies. In the next section we provide a brief overview of the recognition by the political and policy communities of the need for environmental indicators and offer a snapshot description of some of the programs currently underway.

The United States

Three factors appear to have prompted the U.S. Environmental Protection Agency (EPA) to begin a search for ecological indicators. The first is a recognition by government officials and policymakers that environmental problems need to be solved holistically (U.S. EPA 1988). The current regulatory structure is primarily media specific: one environmental law regulates air pollution, another regulates water pollution, and so forth. Thus, the movement of pollutants across media is likely, as regulated industry responds to one law or the other. Similarly, enforcement personnel and staff assigned to one programmatic area may fail to see the big picture because they are addressing pollution in primarily one media (see Rosenbaum in this volume).

The unanticipated consequences of single-media pollution control laws prompted the EPA and state officials to call for multimedia enforcement efforts, supplemental environmental projects across media, integrated environmental programs, and interagency task forces to look at industrial pollution into all media (U.S. EPA 1988).

Moreover, as pollution controls were placed on easily identifiable point sources, more insidious and troublesome problems became evident. Nonpoint sources of pollution, bioaccumulative effects of pollutants, and global environmental problems magnified the need to better understand interactive effects of pollutants as well as the biological, physical, and chemical factors that produce ecosystem responses.

The second factor behind the search for ecological indicators is found in the frustration of congressional policymakers, agency officials, and policy analysts when attempting to determine environmental progress. After twenty years of pollution control laws, it was possible to determine that chemical concentrations of pollutants within airsheds or waterbodies had decreased. But did that make a difference? Was the ecosystem better off? How much of a decrease was necessary to protect the resources? Did it matter that the water had 10 parts per million less of a given substance in 1994 than in 1974?

The collective frustration was a result of the mismatch between a policy output (the results of a regulatory environmental program) and an ecosystemic outcome (a measurable improvement in ecosystem health). While environmental regulatory agencies had established databases for measuring results, progress was measured as a function of regulatory compliance. Compliance data, though extensive, were not always useful in determining environmental improvements (U.S. EPA 1994; U.S. ITFMWQ 1994). Policymakers and agency officials were unable to link demonstrable reductions in industrial emissions and effluent to conclusions about ecosystemic integrity. Thus, as regulatory programs evolved, analysts recognized the need for better indicators of progress in restoring and maintaining ecosystem health.

An early attempt to redirect environmental results occurred in 1984, when congressional hearings were held on the proposed environmental monitoring and improvement bill. The bill called for the establishment of a national commission to serve as the prime coordinating body for the nation's environmental monitoring efforts. It also called for the development of monitoring data that would go beyond compliance data. Although the bill did not pass, the hearings provided a forum for discussing the lack of good information to assess environmental results, as evidenced by this comment by Walter Lyon, an adjunct professor of civil engineering at the University of Pennsylvania: "The sad story is, that in spite of all the energy and money that have gone into environmental programs, the Nation does not have an environmental monitoring system that will tell us: whether or not we are losing ground, whether we are making sound decisions about the environment and what pollution problems to anticipate" (U.S. House Committee on Science and Technology 1984:87).

Identifying meaningful measures of progress was not only an EPA concern. In 1988, the Soil Conservation Service developed the *Water Quality Indicators Guide* to aid in finding solutions to surface water contamination from sediments, animal wastes, nutrients, and pesticides (USDA 1988). In 1991, the U.S. Geological Survey, the EPA, and other federal and state agencies voiced mutual concerns about the ability to monitor improvements in water quality. The Intergovernmental Task Force on Monitoring Water Quality (ITFMWQ) was then created to develop an integrated, nationwide strategy for monitoring water quality. The goal of the ITFMWQ was to "enhance the implementation of defensible water quality programs and management decisions" (U.S. ITFMWQ 1994:i). Part of the ITFMWQ strategy included the development of environmental indicators, as previously discussed.

A related final factor providing impetus for the search for environmental indicators was the move to assess and compare environmental risks. In 1987, the EPA released *Unfinished Business,* which represented the agency's first attempt at comparative risk assessment. *Unfinished Business* was troubling because it noted an inverse relationship between environmental risks the agency was paying attention to and the environmental risks that presented the greatest risk to human health. Given the costs associated with environmental protection, the failure of environmental programs to target the greatest risks and reduce those risks first was disturbing. The Science Advisory Board, in a subsequent report, recommended continuing comparative risk assessments and establishing a framework for risk analysis (U.S. EPA 1990b). Such a framework would need appropriate environmental indicators.

In sum, the move toward ecosystem management and integrated programs in the natural resources and pollution control agencies, combined with the need to better measure progress in environmental protection and the development of a risk assessment framework, were the forces that converged in the political and policy arenas to influence the development of measurable ecological indicators. The next sections describe three efforts to identify and use ecosystem indicators.

The Environmental Monitoring and Assessment Program The Environmental Monitoring and Assessment Program (EMAP) was established in 1988 by the EPA Office of Research and Development at the request of the Science Advisory Board (U.S. EPA 1990a). The Science Advisory Board hoped that EMAP would address increasing concerns about the lack of information to use in making decisions about ecosystem health. Most databases were either related to permit compliance or were limited to monitoring data from relatively small parts of the ecosystem (conversation with Glen Warren of the U.S. Environmental Protection Agency Great Lakes National Program Office 1994). EMAP would bridge decision-making and database gaps by providing a statistically based study of various ecosystems and develop indicators by which to measure changes in the environment.

EMAP defines *indicators* as "characteristics of the environment, both abiotic and biotic, that can provide quantitative information on ecological resources" (U.S. EPA 1993:28). EMAP has identified six broad ecological resource categories: near-coastal waters, inland surface waters, wetlands, forests, arid lands, and agroecosystems. Each category is further divided into ecological resource classes (e.g., oak-hickory forests or sagebrush-dominated shrubland), and routine measurements of indicators are made in resource sampling units within each resource class.

Perhaps the best example of indicator development, measurement, and use within the EMAP scientific community is found in the work of the Environmental Research Laboratory in Duluth, Minnesota, which is coordinating the EMAP program in the Great Lakes (EMAP-GL) (Keough and Griffin 1994). The EMAP-GL indicators are chosen based on their ability to relate to desired endpoints and are characterized by four categories: response indicators, which provide evidence of the biological condition of a resource at the organism, population, community, or ecosystem level; exposure indicators, which provide evidence of the occurrence or magnitude of a response indicator's contact with a chemical or biological stressor; habitat indicators, which include physical, chemical, or biological attributes that characterize conditions necessary to support an organism, population, or community in the absence of pollutants; and stressor indicators, which quantify a natural process, an environmental hazard, or a management action that affects changes in exposure and habitat (U.S. EPA 1992b). The indicators being considered for EMAP-GL appear in table 7.2.

The search for appropriate EMAP-GL indicators is a long-term process and follows a number of steps: identification of issues and valued ecosystem attributes (assessment endpoints); development of a set of candidate indicators; screening of indicators based on evaluation criteria; quantitative testing and evaluation of the performance of research indicators; regional scale demonstra-

Table 7.2. A Sample of Candidate Ecological Indicators Selected by EMAP-GL

Response indicators
  Fish pathology
  Aquatic vegetation
  Lake trout recruitment
  Forage fish populations
  Lake trophic status index
  Lake trout/walleye populations
  Diporeia/Hexagenia abundance
  Chlorophyll composition in water
Exposure indicators
  Contaminant residues in fish
  Zebra mussel/exotic species abundance
  Contaminants in sediments from cores and traps
  Water column toxicity to *Ceriodaphnia*
Habitat indicators
  Sediment physical characteristics
  Water column optical characteristics
  Temperature, pH, routine water chemistry
Stress indicators
  Resource management
  Human population densities
  Atmospheric deposition rates
  Land use and land cover surveys
  Agricultural chemical application rates
  Point and nonpoint source pollutant loading

Source: Adapted from U.S. EPA 1992b:4–2.

tion of the sensitivity, reliability, and specificity of response for development indicators; and implementation of a core set of indicators in a full EMAP program (U.S. EPA 1992b).

Binational Search for Indicators in the Great Lakes The development of environmental indicators under EMAP has several similarities with the search for indicators of the health of the Great Lakes ecosystem undertaken by the International Joint Commission (IJC) in 1994. The Science Advisory Board of the IJC requested the search as a result of the frustration felt by board members in assessing ecosystem improvements and charting a future course for improving the Great Lakes water system. Under the Great Lakes Water Quality Agreement, first signed in 1978 and amended in 1987, the governments of the United States and Canada agreed to "restore and maintain the chemical, physical and biological integrity of the waters of the Great Lakes Basin Ecosystem" (IJC 1994c:4). In the same agreement, the IJC was given responsibility for evaluating progress and making recommendations for U.S. and Canadian governmental actions.

In 1993, the IJC established the Indicators for Evaluation Task Force to develop a framework from which to evaluate progress in restoring the health of the Great Lakes. The task force decided to convene a workshop in October 1994 in Windsor, Ontario, to assist with the identification of indicators of physical, chemical, and biological integrity.

Five categories of stressors were identified prior to the meeting: invasion of exotic species, presence of nutrients, presence of persistent toxic substances, physical stresses (such as land-use patterns and shoreline development), and human activities (including population growth, recreation demand, resource use, and agriculture).

Workshop participants, including over forty representatives from Canadian and U.S. governments, universities, and the IJC staff, divided into subgroups according to stressor category. Each group then identified the desired outcomes for ecosystem health and developed a long list of candidate indicators. Potential indicators were then ranked according to nine criteria: necessary and sufficient, data availability, costs, integrative capacity, scientific validity, certainty and quality of results, understandability by technical and lay persons, policy relevance, and ability to establish reference values.

The end product of the collaborative effort in 1994 was a list of possible indicators of Great Lakes Basin ecosystem health. Over seventy environmental indicators were presented by the five workgroups. The task force subsequently developed a list of indicators arranged according to nine desired outcomes for the Great Lakes Basin ecosystem (IJC 1996). Among the desired outcomes are fishability, swimmability, and drinkability of the waters of the Great Lakes. Maintaining the integrity of biological communities and the virtual elimination of inputs of persistent toxic substances are also listed as desired outcomes. Indicators are grouped according to their ability to measure progress toward a desired outcome. For example, the number of fish consumption advisories for each Great Lake is the chosen indicator for measuring progress toward the fishability outcome (IJC 1996:27). This indicator is further defined by the number of sport and commercial species that have advisories and by the total geographic area that is restricted for commercial fishing. Similar indicators, or suites of indicators, are provided to measure each desired outcome.

Canada

Canada, like the United States, is increasing formal efforts to develop ecological indicators as a way of monitoring ecosystem health, charting progress in environmental programs, and improving ecosystem-based management. Environment Canada established a task force to develop a national set of environmental indicators to use for the state of the environment reporting and issued an initial report in 1991. The Canadian Council of Ministers of the Environment (CCME), the major intergovernmental forum in Canada for collective action on environmental issues of national concern, recognized the need for a common environmental information framework in 1990. As in the EMAP and EMAP-GL processes, the CCME produced a framework for evaluating ecosystem health that uses environmental indicators as a tool in September 1994.

The four principal steps identified by the CCME in the ecosystem management framework are collate the existing ecosystem knowledge base, articulate ecosystem goals and objectives, select ecosystem health indicators to gauge progress toward goals and objectives, and conduct targeted research and monitoring (CCME 1994). Thus, the selection of indicators of ecosystem health are central to effective ecosystem management.

This cursory review of U.S. and Canadian attempts to develop selection criteria for environmental indicators, choose those indicators, and incorporate them into ongoing monitoring provides some snapshot examples of how different governmental organizations are approaching ecosystem management. All of these examples are at the national or regional level. The next section, however, offers a more detailed discussion of one particular approach to incorporating environmental indicators into environmental management strategies. The case study also offers a conceptual framework that may be useful for developing environmental indicators within a smaller geographic or political arena.

Green Bay as a Case Study in Ecosystem Management

It is important to understand that the ecosystem approach used in Green Bay did not just happen, it evolved. That evolution has led to the development of a logical, strategic framework for the rehabilitation of large-scale aquatic ecosystems. As early as the 1920s, scientists and the community recognized that Green Bay and its tributaries were degraded (Harris, Sager, Yarbrough, and Day 1987). Little happened to improve the situation until the passage of the Water Quality Act in 1965. From the mid-1960s through the 1980s, the approach to improving water quality focused on the discharge of pollutants into the surface water. The ecosystem approach in the Great Lakes arose as a result of the failure of this conventional management approach to restore the integrity of the ecosystem.

The Green Bay initiative was marked by the promotion of ecosystem science developed primarily through the University of Wisconsin Sea Grant Institute (Smith, Ragotzkie, Andren, and Harris 1988). Most importantly, the process promoted the sharing of new ecosystemic knowledge with scientists, resource managers, and citizens. It also fostered the participation of citizens, who are stakeholders of the resource, and led to the development of citizen action groups (V. A. Harris 1992). In many ways, then, the framework is the result of societal learning and civic science (Lee 1993).

The process was formally recognized in 1985 when the Wisconsin Department of Natural Resources (WDNR) initiated the development of a remedial action plan for Green Bay, which had previously been designated as an area of concern (WDNR 1988). The EPA funded the Green Bay Mass Balance Study in 1987 and engaged the WDNR as a full partner to lead the effort. Since 1987, an update of the Green Bay remedial action plan was completed; a coalition of stakeholders was formed to address sediment contamination; and a small citizens' group (Northeast Wisconsin Waters for Tomorrow) organized an effort to incorporate cost-effective strategies to the ecosystem approach. A basinwide effort to address nonpoint pollution grew out of the expanding influence of the Green Bay remedial action plan. Most importantly for the purpose of this chapter, environmental indicators have been selected that are meaningful to the public, consistent with ecosystem goals, and easily communicated (see figure 7.1). Indicator selection would not have been as promising without a framework for ecosystem management.

The Indicator Selection Process

The framework depicted in figure 7.1 represents the process that evolved from

190   *The Search for Environmental Indicators*

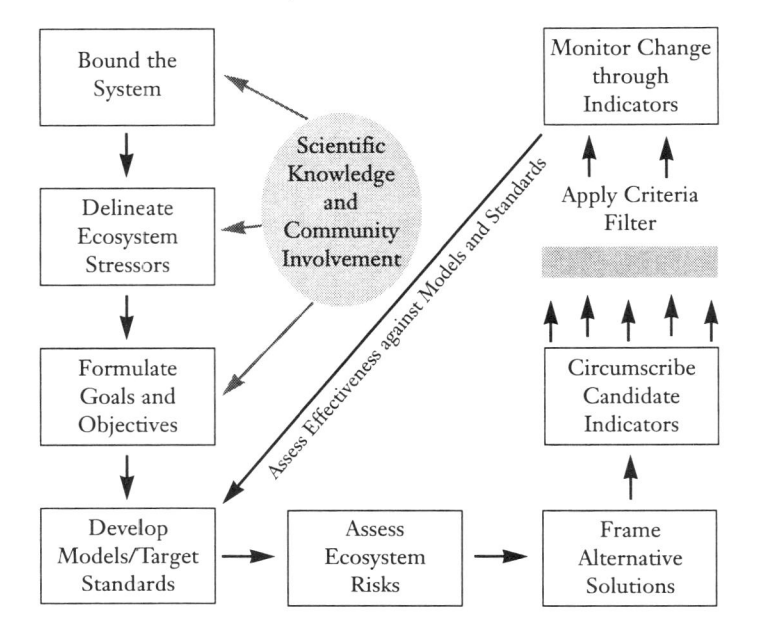

```
┌──────────┐                                    ┌──────────┐
│ Bound the│◄──┐        ╱⎺⎺⎺⎺⎺⎺⎺╲              │ Monitor  │
│  System  │   │       ╱ Scientific╲             │  Change  │
└──────────┘   │      │  Knowledge  │            │ through  │
     │         │      │     and     │            │Indicators│
     ▼         │      │  Community  │            └──────────┘
┌──────────┐   │       ╲Involvement╱                  ▲
│Delineate │◄──┤        ╲⎽⎽⎽⎽⎽⎽⎽╱              ┌──────────┐
│Ecosystem │   │                                 │  Apply   │
│Stressors │   │                                 │ Criteria │
└──────────┘   │                                 │  Filter  │
     │         │                                 └──────────┘
     ▼         │                                      ▲▲▲▲
┌──────────┐   │                                 ┌──────────┐
│Formulate │◄──┘                                 │Circumscribe│
│Goals and │                                     │ Candidate │
│Objectives│                                     │ Indicators│
└──────────┘                                     └──────────┘
     │                                                ▲
     ▼                                                │
┌──────────┐   ┌──────────┐                    ┌──────────┐
│ Develop  │──►│  Assess  │──────────────────► │  Frame   │
│Models/   │   │Ecosystem │                    │Alternative│
│Target    │   │  Risks   │                    │Solutions │
│Standards │   │          │                    │          │
└──────────┘   └──────────┘                    └──────────┘
```

Figure 7.1. Process of Selecting Indicators of Ecosystem Health

the Green Bay experience; it also suggests a process that may be useful in selecting ecological indicators in other geographic areas.

Bound the System The first step in the process is to determine just what area needs to be managed. Ideally, it will be some cohesive ecological unit, such as a watershed or a whole drainage basin. The problem of identifying the area of concern cannot be easily dismissed. Boundaries set too narrowly in the initial phases may force reconsideration of the area of concern later in the process. On the other hand, if the boundaries are too large, policymakers will have trouble focusing public attention on the issues.

In the case of Green Bay, the management agency, WDNR, argued strongly to limit the area of concern to less than 1 percent of the drainage basin. Scientists argued that the stressors impacting the bay were further up in the watershed and that a larger area should be designated. The WDNR was successful in bounding the area of concern. In hindsight, this decision has proved useful. Confining the area of concern to where the problems were manifest (the head of the bay) rather than the entire 6,640 square mile area provided a distinct focal point for policymakers and the public. Although the public advisory committee of the remedial action plan was considering expanding the boundaries as of

1997, the smaller area drew the attention of local citizens and policymakers to the problem initially and allowed them to discover on their own the significance of upstream stressors.

Delineate Ecosystem Stressors Ecosystem stressors cannot be correctly identified without some understanding of how the ecosystem responds to particular stressors. This requires basic and applied research on the ecosystem. State agencies usually do not have the personnel or the money to undertake such research. In the case of Green Bay, the University of Wisconsin Sea Grant Institute developed an ecosystem-based research program funded over a period of fifteen years by the National Oceanic and Atmospheric Administration. Therefore, federal funding that allowed university participation was an integral part of needed research.

After the research base is developed, existing knowledge about the ecosystem is shared through workshops in which scientists and resource managers use stress ecology to analyze the system and identify stressors. The stress-response approach that is part of the workshop links analysis to management because it requires managers and scientists to interact and more effectively transfer information. This phase in the process is important because indicators must be linked to stressors (Harris, Talhelm, Magnuson, and Forbes 1982, Harris et al. 1987; Francis et al. 1979).

Formulate Goals and Objectives Identification of a desired future state is essential and requires public participation guided by scientific understanding of the ecosystem. Presumably, the desired future state is attained by meeting a specific set of objectives to eliminate or reduce existing stressors (WDNR 1993; V. A. Harris 1992). It is in this step that citizens identify what they want from the ecosystem and crystallize their understanding of ecosystem health.

For Green Bay, the desired state includes "a healthy bay environment, a balanced edible sport and commercial fishery, productive wildlife and plant communities, and water-based recreational opportunities, balanced shoreline uses, good water quality that protects human health and wildlife, and an environmentally sound and economical transportation network" (WDNR 1991:5). In this process, the guiding light is still integrity of the ecosystem, but societal values (such as sport fishing and recreation) and economic values (such as shipping) are viewed as legitimate components of the desired future state.

Develop Models and Target Standards Once the goals and objectives are determined, models are developed that predict the response of the system to changes in the magnitude of particular stressors. Indicators are linked to the predictive stress/response models (McAllister 1991; Sager and Richman 1991; Harris,

Erdman, Ankly, and Lodge 1993; WDNR 1993; Millard and Sager 1994). Then standards, such as water quality standards under the Clean Water Act, can be targeted to reflect the predictions of the models. Ideally, as remediation occurs, the result is the desired future state.

Assess Ecosystem Risks The objective of assessing ecosystem risks is to identify opportunities for greatest risk reduction by managers and scientists. The identification of ecological risks upon which environmental protection efforts should be focused requires an ecological risk assessment methodology that is based on anthropogenic stressors affecting an ecosystem and a set of impaired use criteria (for a general discussion, see U.S. EPA 1992a). The outcome permits attention to be drawn to key stressors and associated indicators (Harris, Wenger, Harris, and Devault 1994).

For Green Bay, this step was important. When resource managers and scientists first began to compare risks, the assumption was that the greatest risk to the bay was from polychlorinated biphenyls (PCBs). As the ecological risk assessment process continued, however, risks posed by exotic species and wetland and shoreland filling were found to be potentially more damaging to the health of the ecosystem.

Frame Alternative Solutions In the final step before indicator selection, interdisciplinary task groups consider key stressors and formulate alternative management strategies. Cost-effectiveness and social issues are considered along with scientific and technical information (Mercurio 1995; McIntosh et al. 1993; Northeast Wisconsin Waters for Tomorrow, Inc. 1994). Solutions for addressing risks in the bay have focused on alternative methods to remediate contaminated sediments and alternative best management practices to reduce nutrients and soil erosion in the watershed.

Circumscribe Candidate Indicators At this stage, key stressors have been identified, remedial actions have been devised, and expected responses have been determined. Citizens, public officials, agency staff members, and others have participated in the process and have an understanding of the needs of the ecosystem and, hopefully, a shared vision of the desired outcomes of ecosystem management. The field of candidate indicators has evolved as part of the process and is linked to both community involvement and scientific understanding of ecosystem stressors. The list of candidate indicators, then, has been circumscribed by the process. Still, as depicted in figure 7.1, the array of candidate indicators is much larger than what can be measured. Thus, criteria for ranking candidate indicators and arriving at a suite of indicators must be developed.

Apply Criteria Filter Once the candidate indicators are chosen, they must be evaluated based on selection criteria. Criteria for choosing among candidate indicators may vary according to the area of concern, but usually include common elements such as data availability, reliability, cost considerations, and political feasibility (the ITFMWQ list depicted in table 7.1 is representative of possible selection criteria).

For Green Bay, a long list of candidate indicators was developed (Harris et al. 1987; see table 7.3). The results of the risk assessment analysis (Harris, Wenger, Harris, and Devault 1994) and the application of indicator selection criteria resulted in the identification of nine indicators of ecosystem health (see figure 7.2). The indicators selected are reflective of the key stressors but are certainly less sophisticated than most of the indicators listed in table 7.3. This is the result of applying practical and programmatic considerations to the candidate indicators as well as looking for scientific validity and striving to communicate to the public.

Table 7.3. Ecosystem Indicators for Green Bay

1. Input-output property of the system: nutrients, particulate materials, energy
2. Residence time: water (circulation/flushing rates), toxins, particulates, nutrients (nutrient cycling)
3. Sedimentation rates: organic carbon flux, toxics flux, hypolimnetic O_2 levels
4. Primary production: macrophytes versus phytoplankton, algal biomass and food quality (for zooplankton grazing)
5. Secondary production and biomass: detrital food chain, zooplankton food quality (for planktivore grazing, fish mortality and natality, reproduction and growth rate of preferred fish species)
6. Trophic structure: length of food chain, limitation of energy flow between one level and another, species size distribution and species diversity, P/B level can be specific to trophic levels and turnover rates
7. Ratio of pelagic versus benthic biomass
8. Carbon transfer efficiency: applies at each level of food chain, especially important in detrital food chain
9. Connectivity: within and between trophic level, affects stability of system
10. Predator/prey species abundance: applies at each trophic level
11. Turnover rate of slowest variable in the system: keystone species are at top of food chain and have longest life cycle, physicochemical reactions (such as remineralization of sediment nutrients) are important
12. Conditions of top piscivore: top predator growth rate productivity, body burden of toxic chemicals
13. Key indicator species: representative of most selective species at various trophic levels, integrator of environmental quality
14. Toxic effects: inhibition on productivity (primary and secondary), organismal physiology, bioaccumulation factors, fish body burdens
15. Variability of system parameters over time: flux in physical/chemical conditions, community functions, species functions, pollution levels
16. Uses and perceptions of uses: aesthetics, human values (economic basis)
17. Length of time for a system to return to functional (healthy) state
18. Ratio of fossil to solar fuel in system: needed for an estimate of true or unsubsidized net productivity

Source: Harris et al. 1987.

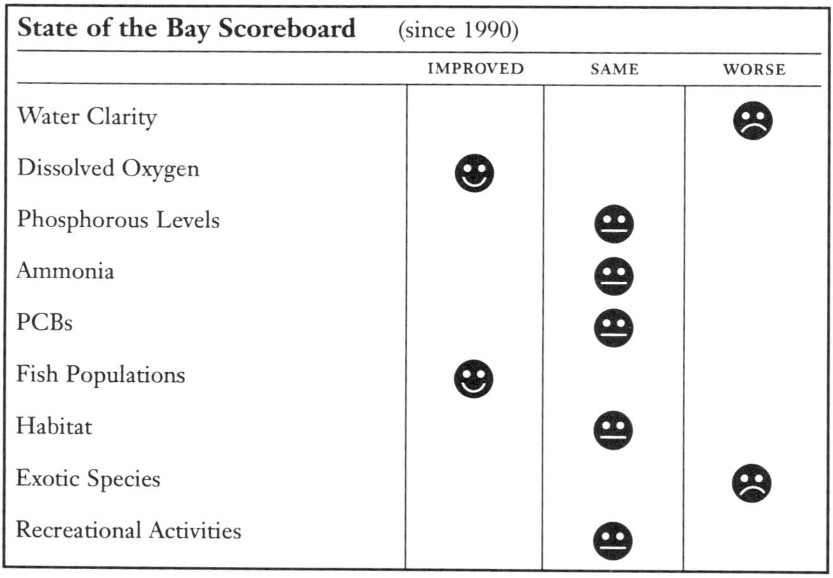

State of the Bay Scoreboard (since 1990)	IMPROVED	SAME	WORSE
Water Clarity			☹
Dissolved Oxygen	☺		
Phosphorous Levels		😐	
Ammonia		😐	
PCBs		😐	
Fish Populations	☺		
Habitat		😐	
Exotic Species			☹
Recreational Activities		😐	

Figure 7.2 State of the Bay Scoreboard

Monitor Change through Indicators Change in the system will undoubtedly occur. However, the question is, Is the ecosystem responding to the remedial actions undertaken as part of the management plan or to other factors? For example, the exponential growth of zebra mussel populations in the bay will likely result in increased water clarity due to the mussels' particle filtering. Water quality goals (and the water quality indicator) may be reached, but the objectives of reducing sediment loading may not be achieved. Thus, the symptom being monitored may improve but this is misleading if considered in isolation from its cause. In short, monitoring changes in carefully selected indicators should provide data from which to assess changes and link remedial actions to improvements in ecosystem integrity.

Finally, the feedback loop in figure 7.1 illustrates the need to periodically assess the effectiveness of the chosen ecological indicators against the models and standards developed in the early stages of the process.

In the end, indicators are ineffective if they cannot be understood by the informed public and decision makers. Two reports titled *State of the Bay* were published in 1990 and 1994 (H. J. Harris 1990, 1994). The State of the Bay Scoreboard (figure 7.2) highlights the ecosystem indicators, summarizes the 1994 report, and evaluates progress toward the desired future state.

Some may argue the scorecard of progress in the bay is oversimplified. Indeed it is, but the data upon which it is based is always available to those with more sophistication and those who can use it. If managers are to bring about change, the public must be informed and ecosystem indicators must convey the message.

The Power of an Indicator

From the foregoing discussion, several conclusions about the search for environmental indicators can be drawn. The first is that the search must be for a set of appropriate indicators, rather than just one or two. Most importantly, the suite of indicators selected must be "necessary and sufficient" to provide an accurate measurement of ecosystem health over time. This will most likely require a composite of physical, biological, and chemical indicators. However, just as crucial is the ability to identify the most appropriate indicators to be used. The problem common to all searches described in this chapter was not identifying relevant indicators but trying to cull those indicators into a manageable number. Thus, the challenge in developing and using indicators comes from both directions. A few indicators probably will not satisfy the "necessary and sufficient" condition, and too many indicators will guarantee that nothing is accomplished because of a lack of resources, data, and time.

Second, it is important to relate indicators to specific goals and objectives for the ecosystem under consideration. For EMAP-GL, indicators are being developed to monitor the health of the Great Lakes; the effort of the IJC is in response to the charge of restoring and maintaining the integrity of the Great Lakes; for CCME, the goal is enhanced ecosystem management. The development of indicators for Green Bay occurred only after consensus was reached about ecosystem objectives. Any attempt to develop indicators must effectively link those indicators to socially relevant goals for the area. Otherwise, the search for indicators will proceed without the backing or understanding of political communities or stakeholders.

In the case of Green Bay, community involvement meant attempting to create meaningful dialogue with citizens, to create a sense of shared ownership and responsibility toward the ecosystem. It also meant developing indicators that could be visualized by lay persons and easily translated into messages about appropriate next steps for rehabilitating the bay. Stakeholder involvement requires considerable planning, communication, and commitment on the part of ecosystem managers. Yet, it seems a necessary precondition to developing indicators that not only measure ecosystem health but also provide an opportunity for public understanding.

Third, the search for ecological indicators raises important questions about the ability to establish universal or regional indicators. Recent searches for ecological indicators have not alleviated issues of appropriate scale. If indicators are used to improve ecosystem management, how large is the area of concern? Even if specific ecosystem categories are developed, such as those under EMAP, are all ecosystems (or subcategories, such as oak-hickory forests) in that category alike? Although attempts are underway in both the United States and Canada, ongoing efforts suggest that we are a long way from a viable set of national environmental indicators, if we ever get them at all. A more promising approach lies in the linkage of indicators to particular ecosystem goals and then the selection of those indicators that will accurately measure progress toward the goals, as recently done by the IJC.

This leads to a fourth conclusion about the efforts to establish environmental indicators. That is, even when indicators are screened using appropriate selection criteria, those indicators should be rigorously tested to evaluate their validity and reliability. The suite of environmental indicators should be evaluated for their ability to produce the measurements that can lead to reasonable assessments about changes in ecosystem integrity. As this fledgling effort continues, agency officials, academics, and ecologists will undoubtedly conclude that some indicators are more susceptible to measurement error, others are simply not amenable to quantitative expression, and others chosen as biological "surrogates" may not measure the biological health of ecosystem communities.

Finally, the search for ecological indicators will be productive only when institutions have the authority to implement decisions and when state, provincial, or federal governments have sufficient resources to operate ecosystem-based programs. The presence of ecological indicators suggests that traditional, single-media programs must expand. The search for indicators places additional burdens on scientists, who must not only fill in the data gaps by increasing their understanding of the ecosystem but also establish credible methodologies and ensure data quality. As described in the CCME report, "an indicator is only as good as the data upon which it is based" (CCME 1994:25).

This prompts a last comment. Our research suggests a warning: while the search for ecological indicators is an important, even crucial, activity in making better decisions about the management of ecosystems or better understanding improvements in or deterioration of ecosystem health, ecological indicators are not perfect. Nor can policymakers rely on ecological indicators any more than the miner relied on the canary. Ecological indicators may be either assets to policy decisions or liabilities, depending on the care given to their selection and the ongoing efforts to evaluate their usefulness.

REFERENCES

Braun-Blanquet, J. 1932. *Plant Sociology: The Study of Plant Communities.* Trans., rev., and ed. G. D. Fuller and H. S. Conard. New York: McGraw-Hill.

Brewer, R. 1994. *The Science of Ecology.* New York: Harcourt Brace College Publishers; New York: Saunders College Publishing.

Brooks, R. R. 1972. *Biological Methods of Prospecting for Minerals.* New York: John Wiley and Sons.

Canadian Council of Ministers of the Environment. Water Quality Guidelines Task Group. 1994. "A Framework for Developing Goals, Objectives, and Indicators of Ecosystem Health: Tools for Ecosystem Based Management." Report.

Clements, F. E. 1920. *Plant Indicators.* Publication 290. Washington, D.C.: Carnegie Institute.

Department of the Interior. National Technical Advisory Committee on Water Quality Criteria. 1968. *Water Quality Criteria.* Washington, D.C.: GPO.

Edwards, C. J., and R. A. Ryder. 1990. *Biological Surrogates of Mesotrophic Ecosystem Health in the Laurentian Great Lakes.* Windsor, Ontario: International Joint Commission.

Fox, G. A. 1993. "What Have Biomarkers Told Us about the Effects of Contaminants on the Health of Fish Eating Birds in the Great Lakes: The Theory and a Literature Review." *Journal of Great Lakes Research* 19 (4): 722–36.

Francis, G. R., J. J. Magnuson, H. A. Regier, and D. R. Talhelm, eds. 1979. *Rehabilitating Great Lakes Ecosystems.* Technical Report 37. Ann Arbor, Mich.: Great Lakes Fishery Commission.

Harris, H. J. 1990. *The State of the Bay: 1990.* Green Bay: University of Wisconsin–Green Bay Institute for Land and Water Studies.

———. 1994. *The State of the Bay: 1993.* Green Bay: University of Wisconsin–Green Bay Institute for Land and Water Studies.

Harris, H. J., T. C. Erdman, G. T. Ankly, and K. B. Lodge. 1993. "Measures of Reproductive Success and PCB Residues in Eggs and Chicks of Forster's Tern on Green Bay, Lake Michigan—1988." *Archives of Environmental Contamination and Toxicology* 25:304–14.

Harris, H. J., V. A. Harris, H. A. Regier, and D. J. Rapport. 1988. "Importance of the Nearshore Area for Sustainable Development in the Great Lakes with Observations on the Baltic Sea." *Ambio* 17 (2): 112–20.

Harris, H. J., P. E. Sager, S. Richman, V. A. Harris, and C. J. Yarbrough. 1987. "Coupling Ecosystem Science with Management: A Great Lakes Perspective from Green Bay, Lake Michigan, U.S.A." *Environmental Management* 11 (5): 619–25.

Harris, H. J., P. E. Sager, C. J. Yarbrough, and H. J. Day. 1987. "Evolution of Water Resource Management: A Laurentian Great Lakes Case Study." *International Journal of Environmental Studies* 29 (1): 53–70.

Harris, H. J., D. R. Talhelm, J. J. Magnuson, and A. M. Forbes, eds. 1982. *Green Bay in the Future: A Rehabilitation Prospectus.* Technical Report 38. Ann Arbor, Mich.: Great Lakes Fishery Commission.

Harris, H. J., R. B. Wenger, V. A. Harris, and D. S. Devault. 1994. "A Method for Assessing Environmental Risk: A Case Study of Green Bay, Lake Michigan, U.S.A." *Environmental Management* 18 (2): 295–306.

Harris, V. A. 1992. "From Plan to Action: The Green Bay Experience." In *Under RAPS: Toward Grassroots Ecological Democracy in the Great Lakes Basin,* ed. J. H. Hartig and M. A. Zarrull. Ann Arbor: University of Michigan Press. 37–58.

Harwell, M. A., W. Cooper, and R. Flaak. 1992. "Prioritizing Ecological and Human Welfare Risks from Environmental Stresses." *Environmental Management* 16 (4): 451–64.

Hunsaker, C. T., and D. E. Carpenter, eds. 1992. *Environmental Monitoring and Assessment Program: Ecological Indicators.* Research Triangle Park, N.C.: U.S. Environmental Protection Agency, Office of Research and Development.

International Joint Commission. 1994a. "Indicators for Evaluation of Progress under the Great Lakes Water Quality Agreement." Draft workshop report.

———. 1994b. "Indicators for Evaluation Workshop: Participant Information Package."

———. 1994c. *Revised Great Lakes Water Quality Agreement of 1978.* Windsor, Ontario: International Joint Commission.

———. Indicators for Evaluation Task Force. 1996. *Indicators to Evaluate Progress under the Great Lakes Water Quality Agreement.* Windsor, Ontario: International Joint Commission.

Karr, J. R., and D. R. Dudley. 1981. "Ecological Perspective on Water Quality Goals." *Environmental Management* 5 (1): 55–68.

Kelly, J. R., and M. A. Harwell. 1990. "Indicators of Ecosystem Recovery." *Environmental Management* 14 (5): 527–45.

Keough, J. R., and J. Griffin. 1994. "Technical Workshop on EMAP Indicators for Great Lakes Coastal Wetlands: Summary Report." Report for the National Biological Survey, Jamestown, N.D.

Landers, P. B. 1990. "Ecological Indicators: Panacea or Liability?" In *Ecological Indicators,* ed. D. H. McKenzie, D. E. Hyatt, and V. J. McDonald. New York: Elsevier Applied Science. 2:1295–1318.

Lee, B. J., H. A. Regier, and D. J. Rapport. 1982. "Ten Ecosystem Approaches to the Planning and Management of the Great Lakes." *Journal of Great Lakes Research* 8 (3): 505–19.

Lee, K. N. 1993. *Compass and Gyroscope: Integrating Science and Politics for the Environment.* Washington, D.C.: Island Press.

McAllister, L. S. 1991. "Factors Influencing the Distribution of Submerged Macrophytes in Green Bay, Lake Michigan: A Focus on Light Attenuation and Vallesnerva Americana." M.S. thesis, University of Wisconsin, Green Bay.

McCarthy, J. G., and L. R. Shugart. 1990. "Biological Markers of Environmental Contamination." In *Biomarkers of Environmental Contamination,* ed. J. G. McCarthy and L. R. Shugart. Boca Raton, Fla.: Lewis Publishers. 3–14.

McIntosh, T. H., R. C. Urig, H. Qiu, T. Sugiharto, and J. J. Lardinais. 1993. "Use of AGNPS, EPICWQ, and SWRRBWQ Computer Models for Water Quality: Land Management Decisions in N.E. Wisconsin." In *Proceedings of the Agricultural Re-*

search to Protect Water Quality Conference, Minneapolis, MN, February 1993, vol. 2. Ankley, Ia.: Soil and Water Conservation Society. 619–21.

McKenzie, D. H., D. E. Hyatt, and V. J. McDonald, eds. 1990. *Ecological Indicators.* 2 vols. New York: Elsevier Applied Science.

Mercurio, J. 1995. *The Fox River Coalition: A Regional Partnership Dedicated to Cleaning Up the Contaminated Sediment and Improving Water Quality on the Fox Valley.* Madison: Policy and Planning Section, Wisconsin Department of Natural Resources, Bureau of Water Resource Management. Pub-WR-382-95.

Millard, E. S., and P. E. Sager. 1994. "A Comparison of Phosphorous, Light Climate, and Photosynthesis between Two Culturally Eutrophied Bays in the Great Lakes: Green Bay, Lake Michigan and the Bay of Quinte, Lake Ontario." *Canadian Journal of Fishery Aquatic Science.* 51 (11): 2579–90.

Northeast Wisconsin Waters for Tomorrow, Inc. 1994. *Toward a Cost-Effective Approach to Water Resource Management in the Fox-Wolf River Basin: A First Cut Analysis.* Appleton, Wisc.: Fox-Wolf Basin 2000.

Rapport, D. J. 1990. "Evolution of Indicators of Ecosystem Health." In *Ecological Indicators,* ed. D. H. McKenzie, D. E. Hyatt, and V. J. McDonald. New York: Elsevier Applied Science. 2:121–34.

Rapport, D. J., and H. A. Regier. 1992. "Disturbance and Stress Effects on Ecological Systems." In *Complex Ecology: The Part-Whole Relation in Ecosystems,* ed. D. C. Patten and S. E. Jorgensen. Englewood Cliffs, N.J.: Prentice-Hall. 107–17.

Regier, H. A. 1990. "Indicators of Ecosystem Integrity." In *Ecological Indicators,* ed. D. H. McKenzie, D. E. Hyatt, and V. J. McDonald. New York: Elsevier Applied Science. 1:183–200.

Ryder, R. A., and C. J. Edwards, eds. 1985. *A Conceptual Approach for the Application of Biological Indicators of Ecosystem Quality in the Great Lakes Basin.* Windsor, Ontario: International Joint Commission.

Sager, P. E., and S. Richman. 1991. "Functional Interactions of Phytoplankton and Zooplankton along the Trophic Gradient in Green Bay, Lake Michigan." *Canadian Journal of Fishery and Aquatic Science* 48 (1): 116–22.

Schindler, D. W. 1987. "Detecting Ecosystem Responses to Anthropogenic Stress." *Canadian Journal of Fishery and Aquatic Science* 44 (supp. 1): 6–25.

Seidl, P. J., ed. 1993. *Toward a State of the Great Lakes Basin Ecosystem.* Windsor, Ontario: International Joint Commission.

Smith, P. L., R. A. Ragotzkie, A. W. Andren, and H. J. Harris. 1988. "Estuary Rehabilitation: The Green Bay Story." *Oceanus* 31 (3): 12–28.

Spellerberg, I. 1991. *Monitoring Ecological Change.* New York: Chapman and Hall.

Suter, G. W. 1993. *Ecological Risk Assessment.* Ann Arbor, Mich.: Lewis Publishers.

Tansley, A. G., ed. 1911. *Types of British Vegetation.* London: Cambridge University Press.

U.S. Department of Agriculture. Soil Conservation Service. 1988. *Water Quality Indicators Guide: Surface Waters.* Washington, D.C.: GPO. SCS-TP-161.

U.S. Environmental Protection Agency. Office of Policy, Planning, and Evaluation. 1987. *Unfinished Business: A Comparative Assessment of Environmental Problems.* Washington, D.C.: GPO.

———. Research Strategies Committee. Science Advisory Board. 1988. *Future Risk: Research Strategies for the 1990s.* Washington, D.C.: GPO. SAB-EC-88-040.

———. Office of Research and Development. 1990a. *Environmental Monitoring and Assessment: Ecological Indicators.* Washington, D.C.: GPO. EPA/600/3-90-060.

———. Science Advisory Board. 1990b. *Reducing Risk: Setting Priorities and Strategies for Environmental Protection.* Washington, D.C.: GPO. SAB-EC-90-021.

———. Risk Assessment Forum. 1992a. *Framework for Ecological Risk Assessment.* Washington, D.C.: GPO. EPA/630/R-92-001.

———. Office of Research and Development. 1992b. *Great Lakes Monitoring and Research Strategy.* Washington, D.C.: GPO. EPA/620/R-92-001.

———. Office of Research and Development. 1993. *Program Guide: Environmental Monitoring and Assessment Program.* Washington, D.C.: GPO. EPA/620/R-93/012.

———. 1994. "State of the Lakes Ecosystem Conference: Integration Paper." Draft for discussion purposes. EPA 905-D-94-0002.

U.S. Fish and Wildlife Service. Division of Ecological Services. 1980a. *Habitat as the Basis for Environmental Assessment.* Ecological Services Manual 101. Washington, D.C.: GPO.

———. 1980b. *Habitat Evaluation Procedures (HEP).* Ecological Services Manual 102. Washington, D.C.: GPO.

U.S. House Committee on Science and Technology. 1984. *Environmental Monitoring Improvement Act Hearings.* 98th Congr., 2d sess. Mar. 28.

U.S. Intergovernmental Task Force on Monitoring Water Quality. 1994. *Water Quality Monitoring in the U.S.: 1993 Report.* Washington, D.C.: U.S. Geological Survey.

Wisconsin Department of Natural Resources. 1988. *Lower Green Bay Remedial Action Plan.* Madison: Wisconsin Department of Natural Resources. PUBL-WR-175-87.

———. 1991. *The Green Bay Remedial Action Plan: A Summary.* Madison: Wisconsin Department of Natural Resources. PUBL-WR-243.

———. 1993. *Lower Green Bay Remedial Action Plan: 1993 Update.* Madison: Wisconsin Department of Natural Resources.

PART 4

ECONOMICS AND EVALUATION METHODOLOGY

8

Economic Approaches to Evaluating Environmental Programs

JOHN B. BRADEN AND CHANG-GIL KIM

In the evaluation of environmental programs, economic analysis is becoming increasingly important. Economics provides not only a set of analytical tools for estimating the benefits and costs of environmental programs but also quantitative criteria for program performance, insight into the likely responses of those affected by the programs, and a logical framework for integrating diverse information requirements.

While economic analysis has much to offer as an evaluation tool, the U.S. Congress has resisted economic scrutiny of the environmental programs it has created. When the present generation of environmental laws were being enacted in the 1970s, economic perspectives were essentially absent from the political debate (Luken and Fraas 1993). The laws tended more toward notions of acceptable risks to human health and technologically achievable solutions than toward economic benefits and costs. Some of the statutes were written broadly enough to enable administrative agencies to use economic criteria in decisions about implementation. Other statutes, however, carefully circumscribed the consideration of economics. Table 8.1 reveals how some parts of U.S. environmental laws allow the consideration of benefits and costs and others do not.

The statutes that place some categories of benefits or costs off-limits prevent the determination of whether a program is worth more, in economic terms, than it costs. As is apparent from table 8.1, the only regulations for which the statutes allow full economic evaluations are in the areas of new source and motor vehicle regulations under the Clean Air Act, hazardous chemicals under the Toxic Substances Control Act, and pesticide regulation under the Federal Insecticide, Fungicide, and Rodenticide Act.

Although Congress has been reluctant to use economic criteria to evaluate

Table 8.1. Analyses Allowable under Environmental Statutes

	Benefits			Costs			
Act/Regulation	Pollution Reduction	Health	Welfare/ Environment	Compliance Cost	Cost-Effectiveness	Economic Impacts	Benefits/ Costs
Clean Air Act							
NAAQS—primary[a]	X	X					
NAAQS—secondary[a]			X				
Hazardous air pollutants		X					
Motor vehicle standards[b]	X	X	X	X	X	X	X
New source standards	X	X	X	X	X	X	X
Fuel standards[b]	X	X	X	X			X
Clean Water Act							
Private treatment works	X		[c]	X	X	X	
Public treatment works	X						
Safe Drinking Water Act							
Maximum contamination levels		X	X	X		X	
Toxic Substances Control Act							
Hazardous Chemicals	X	X	X	X	X	X	X
Resource Conservation and Recovery Act		X	X				
Federal Insecticide, Fungicide, and Rodenticide Act							
Pesticide Regulation	X	X	X	X	X	X	X

Source: Luken and Fraas 1993:101.
a. NAAQS = National Ambient Air Quality Standards.
b. Type of analysis depends on grounds for control.
c. Includes non-water-quality environmental impacts only.

environmental programs, the executive branch has been much more enthusiastic, at least at the presidential and budgeting levels. As early as 1971, the Office of Management and Budget established the "quality of life review" process for proposed regulations, including environmental regulations (Ash 1975). In 1974, President Ford became the first of three successive chief executives to order executive branch agencies to scrutinize the economic consequences of all major regulatory proposals. Ford's main concern was inflationary potential, since prices were rising rapidly at the time, but his order also called for a review of the probable costs, benefits, and risks (Executive Order 11821, 39 Fed. Reg. 41501.41502, Nov. 29, 1974). President Carter's version (Executive Order 12044, 43 Fed. Reg. 12661.12665, Mar. 24, 1978) called for a presentation of regulatory alternatives and an analysis of their respective economic consequences. But Carter's order did not call for the agencies to base their decisions on net benefit calculations. This additional step was taken later in the most well-known and enduring of the executive orders, President Reagan's Executive Order 12291 (46 Fed. Reg. 13193.13198, Feb. 19, 1981; Smith 1984). Reagan's order states, in part:

a. Administrative decisions shall be based on adequate information concerning the need for and consequences of proposed government action;
b. Regulatory action shall not be undertaken unless the potential benefits to society for the regulation outweigh the potential costs to society;
c. Regulatory objectives shall be chosen to maximize the net benefits to society;
d. Among alternative approaches to any given regulatory objective, the alternative involving the least net cost to society shall be chosen.
e. Agencies shall set regulatory priorities with the aim of maximizing the aggregate net benefits to society, taking into account the condition of the particular industries affected by regulations, the condition of the national economy, and other regulatory actions contemplated for the future.

Especially significant in this passage are the references to the "net benefits to society" (Luken and Fraas 1993:98). By stating that "regulatory action shall not be undertaken unless the potential benefits to society for the regulation outweigh the potential costs to society," this order established economic performance as a criterion for choosing not only the means of accomplishing a specific regulatory goal but also the goal to be pursued. Of course, the provisions of the executive order are subsidiary to the provisions of the statutes that authorize regulations; elsewhere in the order appears the qualification "to the extent permitted by law." So, in cases where statutes explicitly rule out economic balancing, the hands of the executive branch are tied.

While presidents and the Office of Management and Budget have placed confidence in economic evaluation, the program agencies often do not share this view. Economic studies are a way for the president and the Office of Management and Budget to extract information from the program agencies and to gain more control over budgets. Moreover, most agency personnel are specialists in the technologies or social pathologies that the agencies were created to address: engineers, chemists, and biologists in the case of the U.S. Environmental Protection Agency. Many are skeptical of economics. Accordingly, the agency cultures typically place physical or demographic measures of achievement above economic measures.

Presidential enthusiasm for economic evaluation during the 1970s and 1980s caused the economic evaluation of environmental programs to grow. Methods that have been developed are firmly grounded in the well-established procedures of benefit-cost analysis (Gramlich 1981; Schmid 1989). They build on decades of experience with economic analysis of major governmental investments (Stokey and Zeckhauser 1978).[1] But, environmental programs present special challenges. The greatest challenge arises because environmental goods and services do not pass through markets. Since the prices arising in markets are the usual metrics of economic value, economists have been forced to find surrogates. The 1970s and 1980s were a time of intense developmental work on the estimation of nonmarket values (Cummings, Brookshire, and Schulze 1986; Braden and Kolstad 1991; Freeman 1993). By the 1990s, this work matured sufficiently to be accepted in administrative proceedings on monetary liability for pollution damages (Kopp and Smith 1993).

While the new techniques for monetary valuation have been important contributions, economics has much more to say about the assessment of environmental programs. Another important area of insight concerns the types of programs that are used. A distinction is often made between "command-and-control" programs and "incentive-based" alternatives. The command-and-control approach sets quantitative performance standards. Many of those standards are based on specific technologies that polluters, in turn, are encouraged to use. Thus, these regulations end up being highly prescriptive because they effectively lead polluters to adopt the same abatement strategies irrespective of their own unique circumstances. Incentive-based approaches, on the other hand, provide financial incentives for pollution abatement but leave it up to each polluter to decide how and how much to abate. The incentives can include levying fees on emissions or issuing allowances for a target level of pollution and permitting sources to buy and sell those allowances so as to minimize their cost of compliance. The scholarly literature on these options has grown voluminous (e.g., Anderson et al. 1977; Schelling 1983; Tietenberg 1985) and the impetus for their use has been growing, not only in the United States (e.g., *Project 88* 1991) but also in other nations (e.g., Braden, Folmer, and Ulen 1996).

The presidential executive orders calling for economic evaluation provided a forum for the consideration of incentive-based policies just as they motivated progress on nonmarket valuation techniques. In particular, those orders called on the agencies to identify and explore a range of regulatory alternatives. The growing scholarly consensus for incentive-based policies eventually induced the U.S. Environmental Protection Agency (EPA) to consider them. Starting around 1980, for example, based on provisions in the 1977 amendments to the Clean Air Act, the EPA approved limited versions of emissions allowance trading. The same idea was also applied in several other environmental programs during the course of the 1980s, most notably the phasing out of leaded gasoline (Hahn 1989). After years of trial and refinement of tradeable allowances in these limited applications, a full-fledged market for pollution allowances was authorized in the 1990 Clean Air Act Amendments as part of a national strategy to reduce emissions of sulfur dioxide (a precursor of acid rain) into the atmosphere. Although some of these markets have proven to be considerable successes, notably the market for lead additive allowances that was used to expedite the transition to unleaded fuel in the early 1980s, there are also many cases in which they have had little direct impact (Hahn 1989). In addition, compromises in the goals of environmental policy sometimes must be accepted to make markets work. For example, to ensure a large market with many potential traders for sulfur dioxide allowances, sources in the eastern United States may trade on a one-to-one basis irrespective of their specific location even though it is generally recognized that emissions from the Ohio River Basin are more problematic than emissions from the Eastern Seaboard or New England. In principle, the allowance market could actually increase the concentration of emissions from these more consequential sources.

The increasing prominence of incentive-based approaches represents a significantly new role for economics in environmental policy. Instead of serving only as a tool to analyze those programs, economics is being used through markets as a key part of the implementation strategy.

In the following chapter we describe more specifically the topics introduced above. A large share of the chapter is a nontechnical review of the basic economic theory of social program evaluation—welfare economics and its application through benefit-cost analysis. In the next section we describe the techniques that economists have developed for estimating nonmarket values. In the next section we turn to the special problems of evaluating programs that attempt to reduce risks. Many environmental programs are of this kind; their consequences for individuals are uncertain, so they are more difficult for individuals to evaluate. In the final section we return to the use of economic incentives as an implementation option for environmental programs.

Before going further, we should acknowledge two limitations in our approach. The first is that in focusing on economic aspects of environmental programs, we will have little to say about physical and biological dimensions of pollution problems. This lack of attention should not be interpreted as a lack of appreciation. For economics to be useful in this area, it must be solidly connected to the physical and biological processes linking the generation of wastes to the impacts on humans and ecosystems. The second limitation is that economics embodies certain assumptions about preferences, behavior, and interpersonal comparison. For example, it establishes individuals rather than communities as the unit of analysis, and it essentially weights individuals' preferences for environmental protection according to the amount of money each person would spend to accomplish it. These assumptions enable economists to derive succinct, aggregate monetary measures of impact—measures that translate readily into policy considerations. On the other hand, some critics disagree with these assumptions, and they discredit economic evaluation. Other chapters in this book may be examined for approaches that embrace a wider range of behavioral postulates.

Economic Models for Environmental Program Evaluation

This section addresses, first, the criteria by which net impacts are ordinarily judged and, second, different analytical models for aggregating the benefits and costs. We present in summary form the basic principles of benefit-cost analysis.

Criteria for Assessing Economic Performance

Economic impacts are usually discussed in two dimensions: amount and distribution. The amount is measured in dollars (or some other currency unit) of costs and benefits. The difference between costs and benefits—net benefits—is a measure of whether a given set of expenditures produces outputs that are more or less valuable than the resources they use up. In this sense, net benefits are a measure of economic efficiency—whether resources are being used to produce goods and services of the highest possible value.

The economist Wilfredo Pareto (1848–1923) suggested an economic-efficiency criterion that has long been used as the standard for program evaluation. The Pareto criterion states that a program leads to an improvement in economic welfare if some or all of the individuals in the economy are made better off and none is made worse off. A situation in which no one can be made better off without someone else suffering a loss is said to be a "Pareto optimum." A Pareto optimum represents a maximally efficient use of resources. In addition, a Pareto optimum has an implicit distributional dimension. It protects the status quo in the sense that no one can be made worse off than they were in the absence of the proposed program. Some people may be made much better off by a program, but as long as no one is hurt, the criterion is satisfied.

The Pareto criterion cannot evaluate a change that makes some individuals better off and others worse off. Since most environmental programs involve changes of this kind, the strict Pareto criterion is of limited usefulness. Efforts to find a more useful criterion (e.g., Bergson 1938; Kaldor 1939; Hicks 1939; Scitovszky 1941; Rawls 1971) stumble against the need to count some individuals' preferences more than others', to have actual payment of compensation for losses, or to return to a naive state in which individuals do not know their social and economic positions. As a practical matter, the economic approach to program evaluation accepts a version of the criterion called "potential Pareto improvement." This version sums up the monetary gains and losses, to whomever they accrue, and determines if the sum is positive or negative. If the benefits exceed the costs, then the program is deemed to be economically justified even though the losers may never actually receive compensation.

Conventional economic analyses, then, emphasize efficiency and appear to say nothing about distribution. But, the efficiency calculations are themselves affected by the initial distribution of wealth. This is because the efficiency calculations take stated monetary values at face value. Since the marginal value of money is generally held to diminish with one's economic status, wealthier people can and will bid more than poor people for a program that satisfies the same basic human needs. Thus, wealthier people have more "votes" in efficiency calculations.

To address the fairness or equity of public programs, including environmental programs, the analyst must generally go beyond the efficiency calculations and describe the demographic and regional distribution of benefits and costs.[2] Analytical tools are available for making these calculations, as we indicate below when discussing input-output and general equilibrium analysis.

Economic Models for Program Evaluation

In this section we consider two prominent economic approaches to program evaluation: benefit-cost analysis (BCA) and cost-effectiveness analysis (CEA). The former is used where beneficial impacts can be valued in monetary terms. The latter is pertinent where the the "best" (i.e., lowest-cost) method of satisfying a given noneconomic performance objective is selected. CEA is essentially a limited application of the same techniques used in BCA.

In both approaches, one of the greatest challenges is to determine how program impacts ripple through the economy, affecting different industries, groups, and regions. Following the description of BCA and CEA is a discussion of methods for assessing those ripple effects, specifically input-output analysis and computable general equilibrium analysis.

Benefit-Cost Analysis BCA requires an estimation of the direct and indirect effects of the program, translation of the effects into common units, and a determination of the net impact on social welfare. BCA applies the criterion of potential Pareto improvement. Typically, the benefits in the BCA are based on the willingness of potentially affected individuals to pay for environmental regulatory programs. In practice, the BCA in the regulatory impacts analysis under any executive order, such as EO 12291, should include not only estimates of benefits and costs that can be quantified but also a description of health and unmeasurable benefits (U.S. EPA 1987). A project is deemed economically worthwhile if the total benefits generated over time are greater than the total social costs incurred over the same time.[3] To compare outcomes realized at different times, the stream of future benefits and costs are discounted by an appropriate rate of interest to determine present value equivalents.[4] The results are usually expressed as a net present value (*NPV*) number (the difference between the present value of benefits and costs), a benefit-cost ratio (*B/C*), or an internal rate of return (*IRR*) (the rate of discount that causes the net present value to be zero). The corresponding decision rules of BCA are positive *NPV*, *B/C* > 1, and *IRR* > *i*, where *i* is some benchmark rate of interest (see table 8.2).[5] All of these rules signify economically efficient programs.

In addition to providing insight into the merit of individual projects or programs, the preceding decision rules are sometimes adapted to rank a number of proposals that are competing for finite budget resources. The higher its *NPV*, *B/C* ratio, or *IRR*, the more highly ranked is a proposal. This application of the decision criteria raises what is known as the capital rationing problem. In simple terms, the problem is that the projects are not perfectly divisible and the ones that are ranked highest according to the decision rules may not exactly fit within the overall budget. Maximizing returns to the budget may require selecting projects out of order. Imagine, for example, four projects in which the estimated net present benefits and budget outlays are (budget outlays in parentheses): A: 35 (25); B: 20 (15); C: 13 (10); and D: 6 (5). If the budget is limited to less than 55, then choices will have to be made. If it is 30, for example, the best strategy is to accept proposals A and D.

Table 8.2. Decision Rules in Benefit-Cost Analysis

Criterion	Decision Rule
Net present value (*NPV*) $$\sum_{t=0}^{T} \frac{(B_t - C_t)}{(1+r)^t}$$	A program is desirable if *NPV* > 0 Choosing the program with the largest *NPV*
Benefit-cost ratio (*B/C*) $$\frac{\sum_{t=0}^{T} \frac{B_t}{(1+r)^t}}{\sum_{t=0}^{T} \frac{C_t}{(1+r)^t}}$$	A program is desirable if *B/C* > 1 Choosing the program with the largest *B/C* ratio among mutually exclusive programs
Internal rate of return (*IRR*) $$\sum_{t=0}^{T} \frac{(B_t - C_t)}{(1+r^*)^t} = 0$$	A program is desirable if $r^* > i$ (market rate of interest) Choosing the program with the largest *IRR*

Note: The variables are B_t = benefit in period *t*; C_t = costs in period *t*; *r* = discount rate; *T* = program time horizon; r^* = internal rate of return.

The study by Alan J. Krupnick and Paul R. Portney (1991) of a proposal to strengthen the enforcement of ambient ozone standards under the Clean Air Act illustrates the application of benefit-cost analysis to an environmental program. As a respiratory irritant, ground-level ozone is a threat to asthmatics and others with impaired respiratory function. It is a major component of smog and can impede plant growth. Ozone forms from chemical reactions of nitrous oxides (NO_x) with volatile organic compounds (VOCs) in the presence of sunlight. It has been regulated in the United States since 1971, but many cities remain in violation of the federal standards.[6] In anticipation of amending the ozone abatement requirements, several groups studied the costs and benefits of further abatement in the late 1980s. Subsequently, in 1990, the Clean Air Act was amended to require tougher abatement measures in ozone nonattainment areas.

Krupnick and Portney evaluated the proposals to reduce emissions of VOCs. Specifically, they looked at technologies identified in earlier studies that would have the combined effect of reducing discharges in nonattainment areas by 35 percent (individual cities would realize reductions from 25 percent to 50 percent depending on the types of sources). These reductions are predicted to bring about one-third of the nonattainment areas into compliance.

The costs of abatement were assumed to equal the expenditures on new technologies—things such as less volatile fuels and printing inks. (Expenditures are likely to overestimate true economic costs, but often they must be used because more accurate data are not available.) The cheaper options were assumed to be adopted first and the more expensive options adopted only as required to achieve the 35 percent goal. The capital and operating costs of the new technologies were combined into a present value, and the present value was apportioned out over the life of the equipment as an annualized value. Depending on different assumptions about the technologies, the annualized cost of compliance in nonattainment areas was estimated to be $6.6 billion to $10.0 billion, or $1,800 to $2,700 per ton of VOCs avoided. Associated costs in attainment areas would increase the estimated total cost to between $8.8 billion and $12.8 billion per year.

With respect to the benefits of VOC reduction, Krupnick and Portney concentrated on acute (short-term) health impacts. They used dispersion models to estimate the consequences for ambient ozone levels, then used epidemiological and clinical studies respectively to develop two different estimates of the health effects of lower ozone. Finally, they used monetary estimates from survey data of the value of avoiding asthma attacks ($25 each), restricted activity days ($20 each), and days of mild coughing ($5 each). Combining these steps, they found that the health improvements in nonattainment areas are valued at $250 to $800 million per year depending primarily on whether clinical (higher) or epidemiological (lower) studies were used to estimate the health impacts. They concluded that the huge costs of VOC abatement will produce only meager benefits and that a less ambitious plan for VOC abatement would stand a better chance of balancing the benefits and costs.[7]

The study by Krupnick and Portney is one of many benefit-cost analyses that have been performed for environmental programs. A summary of benefit-cost studies of environmental programs prepared within the federal government is presented in U.S. EPA (1987), and an example covering proposals for change in the Clean Water Act is U.S. EPA (1994). In addition, the EPA periodically issues comprehensive economic assessments of its programs, for example, U.S. EPA (1990).

As the study by Krupnick and Portney indicates, benefit-cost analysis is usually applied to very specific proposals. It organizes evidence in response to the question, Will proposal X produce positive net benefits? But BCA does not address the question, What proposal would produce positive net benefits? The difference between these questions is whether the policy objectives are allowed to vary. A methodology that could answer the second question would provide more guidance in the design of environmental programs.

Cost-Effectiveness Analysis In many circumstances it is difficult to estimate the economic benefits of environmental protection. A useful alternative to BCA under these circumstances is CEA. In CEA, costs are measured in monetary terms, but benefits are not. CEA is used to select among policy measures the one that will minimize the costs of realizing a specific policy objective, typically a quantitative performance standard or tolerance level for humans, plants, animals, and property.

Environmental programs are particularly prone to difficult problems of benefit measurement because many of the consequences have no associated prices. A good example is the case of programs that limit exposure to toxic or hazardous chemicals. Attaching dollar values to human health and life is a complex and controversial matter (Cropper and Freeman 1991). CEA circumvents these difficulties by focusing on an approach that will minimize the cost of a particular reduction in mortality or morbidity.[8]

CEA uses the same evaluation methodologies as BCA, except that the benefit stream is not evaluated. Alternative policies or programs are compared on the basis of the present value of costs or the annualized equivalent. The benefit-cost ratio and the internal rate of return have no analogues in cost-effectiveness analysis.

As an illustration of the use of CEA, Scott E. Atkinson and Donald H. Lewis (1974) compared two approaches to achieving a given ambient air quality standard. One approach would impose uniform emission standards and the other would target sources that could abate most cheaply. The criterion was minimum cost. Not surprisingly, they found that cost-based targeting would achieve the ambient standard much more cheaply. Similarly, the Office of Technology Assessment (1988) analyzed different strategies for reducing mobile-source emissions of VOCs. The results showed that the most cost-effective approach is to reduce the volatility of gasoline.

Input-Output Analysis The discussion of benefit-cost analysis to this point glosses over many issues that arise in its application. Some of the most compelling issues have to do with establishing monetary values of outcomes that do not ordinarily have prices. We will return to these in a later section. Another important issue is how to trace the ripple effects of a program through the economy. Most public programs set in motion a series of economic adjustments: people changing jobs, products gaining and losing value, technologies changing, and goods and services mixing. It is not enough to consider only the primary impacts. A full analysis will also consider the secondary effects.

The leading methodology for analyzing the ripple effects is input-output (I-O) analysis (Leontief 1970). I-O models are tools for tracing the secondary impacts by way of a complex system of interindustry and interregional economic linkages.[9] The heart of I-O analysis consists of so-called I-O tables, or matrixes. These tables describe the relationship between resource use and final outputs in the economy. The economic system is divided into separate producing sectors. Each sector sells its output to other sectors and to final buyers, such as consumers, governments, and export buyers. Such demands are final in the sense that goods and services supplied to these categories pass out of the production system into a consumption system. Simultaneously, each sector buys goods and services from other producing sectors, as well as labor, land, and capital inputs, to carry out production. The models can be subdivided regionally as well, in order to identify the geographical incidence of the impacts as well as the sectoral consequences.

The technical interdependence between undesirable (such as pollution) and desirable outputs can be described by technical coefficients (I-O coefficients) added to the structural matrix of the economy. By linking environmental damages to inputs or outputs of the market system, any change in the output level of pollutants can be traced to either changes in the final demand for specific goods and services or changes in the technical structure of one or more sectors of the economy.

I-O models have been customized in several ways to address environmental programs. John H. Cumberland and Bruce N. Stram (1976), for example, show how an I-O model can be extended to estimate the future impacts of emission of pollutants such as radioactive wastes, sulfur oxides, and pesticides. Jeong J. Rhee and John A. Miranowski (1984) show how the effects on industry of pollution control regulations can be captured by two multipliers: "the pollution control multiplier" and "a modified Keynesian multiplier." Using the annual time series (1947–85) of I-O tables for the U.S. economy, Dale Jorgenson and Peter Wilcoxen (1990) analyze the macroeconomic impact of environmental programs by simulating the long-term growth of the U.S. economy with and without regulation. John L. R. Proops, Malte Faber, and Gerhard Wagenhals (1993) formulate an I-O model to estimate the economic ripple effects of CO_2 reductions in Germany and the United Kingdom. Their result showed that the electricity industry in both countries would be of central importance in any attempt to reduce CO_2 emissions. Abdul Qayum (1994) reformulates Leontief's model to treat environmental goods, such as clean air and fresh water, in the same way that regular market goods are treated.

While it can yield important insights into the full direct and indirect effects of environmental programs, I-O analysis has some important limitations. One difficulty is that it assumes fixed I-O coefficients, which implies simple linear relationships. In reality, however, economic relationships are frequently nonlinear and their relationships change over time. I-O models also preclude substitution between inputs in production processes, including the substitution of less polluting inputs and processes, and this is highly unrealistic. Over time, particularly as technologies change, substitution does occur. Finally, I-O models are infrequently updated due to data and technical limitations, so they are commonly out of date.

Computable General Equilibrium Analysis A more flexible type of sectoral adjustment framework is provided by computable general equilibrium (CGE) models. Like I-O analysis, the aim of CGE analysis is to track the economic ripple effects of a change in policy or technology. In CGE models, however, not only can output and input quantities change but prices also can rise or fall as required to reconcile changes in supply and demand.[10] The only fixed parameters are technologies, behavioral assumptions, some factor supplies, and policy instruments. Given this structure, the model simulates the working of a market economy in which prices and quantities adjust to clear markets for products and factors.

The economic analysis of environmental programs using CGE models is a relatively recent development. Lars Bergman (1990, 1996) formulates a CGE model with markets for emission permits and emission control technologies and uses the model to estimate the impacts on factor prices and resource allocation of pollution reductions in Sweden. The results suggest that reducing the emission of major pollutants such as sulfur oxides (SO_x), nitrogen oxides (NO_x), and carbon dioxide (CO_2) will have widespread effects on prices and consumption. Also, in this line of research, using a static CGE model, Gunter Stephan, Renger V. Nieuwkoop, and Thomas Wiedmer (1992) evaluate the distributional and allocational effects of a CO_2 control program in Switzerland under several compensation policies. Their results show that public acceptance of greenhouse gas reduction programs could be expanded and the economic cost of these programs reduced if carbon taxes were accompanied by a redistribution of the tax revenues.

CGE models are particularly appropriate for quantifying the economic effects of environmental regulatory programs at the national or international level. The CGE approach is more realistic than the I-O approach because it allows for more things to change in response to policy or program initiatives. But, CGE models are also more difficult to implement computationally and cannot offer nearly as much detail on regions or sectors as I-O models.

In this section, we have introduced criteria and models used by economists to evaluate environmental policies. These criteria and models apply to aggregate effects and provide summary indicators of the economic performances of a particular program or proposal. We have said too little to this point, however, about the basic building blocks of economic analysis—the monetary measures of the primary impacts on individuals or businesses. It is only after these data are developed that the analyst can proceed to estimate the aggregate economic effects. In the next section, we examine the techniques used by economists to measure those primary impacts.

Economic Valuation of Program Impacts

Economic Benefits and Costs

An important challenge in the economic evaluation of programs is to calculate the monetary value of benefits and costs. As is evident from the preceding section, the economic approach presumes that program impacts can be measured in monetary terms.

A particular challenge arises in estimating monetary measures of the benefits of environmental programs. By "benefits" we refer to the health, aesthetic, or recreational improvements accomplished through environmental programs. Many of the benefits of environmental protection do not pass through markets; some do not even involve consumption as we usually think of it; and others, such as national defense and public radio, are jointly consumed. These features of environmental goods pose large obstacles to the measurement of monetary value.

Various categories of environmental values are summarized in table 8.3. Since market prices are commonly used as estimates of monetary value in economic studies, their absence for environmental benefits is a major handicap. It is impossible to determine the economic value of an environmental program without monetary measures of the primary nonmarket impacts.[11]

Economics holds that the willingness to pay for something is the measure of its value. In principle, whether or not the "thing" is actually consumed or merely experienced is irrelevant. All that matters is that people are willing to pay to defend or improve the environment. Nevertheless, the inclusion of intrinsic values (values not based on use, such as existence value as defined in table 8.3) is a matter of intense debate in benefit-cost analysis of environmental programs. For example, Donald H. Rosenthal and Robert H. Nelson (1992) argue that existence value should not be included in benefit-cost analysis since the concept is poorly defined and very difficult to measure.[12] Emery N. Castle, Robert P. Berrens, and Richard M. Adams (1994) add that existence values may be associated with market goods as well as nonmarket goods, so attempts to

Table 8.3. Classification of Total Value

Categories	Features	Example
Use values		
Consumptive use value	Arises from the consumptive use of a wildlife resource	Hunting, fishing, or trapping
Nonconsumptive use value	Arises from contact with nature	Visiting national parks
Indirect use value	Arises from indirect contact with nature that does not change nature or reduce its availability to others	Reading about, viewing pictures of, or watching TV programs about nature
Intrinsic values		
Existence value	Arises from altruistic motives, including bequests	Contributions to environmental preservation programs that will never result in direct or indirect use of the protected resources
Option value	Arises from uncertainty about the future	Support for environmental programs that may someday be sought for direct or indirect use

include them for environmental commodities alone may introduce bias. On the other hand, Raymond J. Kopp (1992) argues that existence value is well-defined in theory and measurable in principle using contingent valuation techniques (discussed later in this section) and thus should be included in economic evaluations of environmental programs.

Theoretical Framework for Economic Valuation

In the absence of actual markets for environmental goods and services, the underlying problem for measuring the economic impacts of environmental programs is how to define theoretically appropriate and empirically observable monetary measures of changes in individual well-being. The theoretically correct measure is the maximum amount of money an individual would be willing to pay to secure a change in environmental quality or the minimum amount one would be willing to accept to forgo it.[13] These amounts would leave the individual indifferent between having the money and having the environmental change.

These simple concepts are captured in the expressions willingness to pay (WTP) and willingness to accept (WTA) payment. The WTP concept presumes that the affected people are entitled to the level of welfare in the absence of change. WTA, on the other hand, presumes they are entitled to the level of welfare in the presence of the change. As a practical illustration of the difference, consider a proposal to add acreage to a wilderness area. If people were asked what would be the maximum they would pay to accomplish this change, then we are measuring WTP. Such a payment would leave them indifferent between two scenarios: A, keeping the money and doing without the extra acreage, and B, giving up the money but gaining the additional wilderness. For those who do not favor wilderness, the payment could be negative; that is, they might have to be compensated to be as well off with the extra acreage. In the case of WTA, on the other hand, economists would try to determine the minimum compensation people would require to do without the added wilderness. Those who favor wilderness would require positive compensation while those who would be made worse off by more wilderness should accept negative compensation (that is, they should be willing to make a payment).

Both of these value measures are based on the assumption of substitutability in preferences between environmental quality and money. If the effect of a change in environmental quality translates into a small change in income, economic theory suggests that the WTP and WTA measures of value should be approximately the same (Willig 1976; Randall and Stoll 1980). The main source of divergence should be due to differences in income implied by the different points of reference, but since the difference between the points of reference is "small," the monetary measures should be nearly the same.

The hypothesis that WTP and WTA are approximately equal in most circumstances has not been confirmed in many empirical studies of environmental quality changes. The values elicited in these studies often reveal large differences. Ann Fisher, Gary H. McClelland, and William D. Schulze (1988) review a number of comparative studies and find that the ratio of WTA to WTP ranges from 1.6:1 to 16.6:1. This is disconcerting because it suggests that the monetary values for environmental quality are fragile and, hence, unreliable guides for environmental program evaluation. W. Michael Hanemann (1991) provides an explanation.[14] His idea is that the degree of substitutability between environmental and nonenvironmental goods plays a role along with income effects. In his scheme, for unique goods, WTP and WTA measures are likely to diverge because money is a better substitute for the state without those goods than for the state with them. More money will be required to give up unique goods (WTA) than to remain in the deprived state (WTP). But, the empirical evidence concerning Hanemann's explanation is mixed. Table 8.4 summarizes two studies that set out to test whether the ease of substitution makes a difference in monetary valuation. Wiktor L. Adamowicz, Vinay Bhardwaj, and Bruce Macnab (1993) found no substantial evidence for the substitution hypothesis. For both close substitutes and poor substitutes, the WTA:WTP ratio approached 2:1. Jason F. Shogren, Seung Y. Shin, Dermot J. Hayes, and James B. Kliebenstein (1994), however, found ratios of 3:1 to 8:1 for goods with no close substitutes but only 1:1 to 3:1 for goods with close substitutes. Their results suggest that substitutability does make a difference, although the discrepancies remain potentially large even for close substitutes.

Table 8.4. Comparison of WTP and WTA

Study		WTP	WTA	WTA/WTP	Features
Adamowicz, Bhardwaj, and Macnab 1993		$28.5	$48.5	1.70	With substitute
		36.6	70.0	1.91	Without substitute
Shogren, Shin, Hayes, and Kliebenstein 1994[a]	[I]	0.60	5.06	8.43	Without substitute
	[N]	0.71	2.36	3.32	Without substitute
	[F]	0.86	3.03	3.52	Without substitute
	[I]	2.37	6.55	2.76	With substitute
	[N]	2.90	3.06	1.06	With substitute
	[F]	2.62	3.25	1.24	With substitute

Note: All values are in year-of-study dollars.
a. The bracketed characters I, N, and F indicate inexperience, naive levels, and informed levels, respectively, according to the number of experimental trials conducted per person.

Whatever the explanation, that WTP and WTA measures often produce very different estimates of monetary value for environmental programs is discouraging to the cause of economic evaluation. Without a clear consensus on the basic property rights from which to evaluate change, it is possible to justify significantly different value estimates and, hence, very different conclusions about the economic desirability of a particular program. In practice, the developing consensus is to take the most conservative approach. This involves using WTP measures of value. With WTP measures, the values expressed are constrained (at least in concept) by status quo wealth positions and are less likely to represent wishful efforts to capture realized windfalls in the form of large compensation payments from supposed "deep pockets."

The Techniques of Economic Valuation

It is one thing to define the characteristics of an ideal money measure of welfare change and quite another to implement such an ideal measure. As we have seen, economic theory does not speak with one voice on the matter of valuation. The ambiguity is compounded when it comes to actual empirical measurement of values. There are several techniques, none of them perfect, and the different techniques can produce different answers to the same valuation question.

Developing practical measurement techniques for nonmarket values has been a major concern for environmental economists since the early 1970s. The techniques can be broadly classified as indirect (based on market data for related goods) and direct (based on synthetic data for the good itself).[15] Table 8.5 summarizes the applicability of these techniques to various types of environmental impact categories.

Indirect Methods Indirect methods are based on the behavior of households in markets related to environmental goods and services. They are called "indirect" because the methods involve inferring economic values from relationships between environmental quality and various market goods.

As a representative example of indirect methods, the hedonic price method (HPM) involves inputting the value of environmental quality from observed market prices for goods that are linked to the environment.[16] The most common linked goods are real estate (i.e., housing, land, or both) and labor. Property values, for example, are probably influenced by proximity to attractive water bodies or the presence of scenic vistas as well as by size, proximity to schools and commercial centers, and other determinants. Likewise, wage rates are probably affected by known exposure to hazardous chemicals as well as by skill lev-

Table 8.5. Economic Valuation Methods for Environmental Impacts

Environmental Impact Category	Hedonic Property	Hedonic Wages	Travel Cost	Contingent Valuation
Air pollution				
Morbidity	NU	L	NU	X
Mortality	L	X	NU	NU
Aesthetics				
Visual and sensory	X	L	NU	X
Recreation	NU	NU	X	X
Water pollution				
Recreation	L	NU	X	X
Aesthetics	X	NU	L	X
Ecosystems	NU	NU	NU	X
Drinking water	NU	NU	NU	?
Toxic substances				
Morbidity and mortality	X	NU	X	X
Noise				
Nuisance	X	NU	NU	X

Key: X indicates a used technique; L indicates very limited application; NU indicates nonusable technique; and ? indicates possible but not yet utilized
Source: Adapted from Pearce and Markandya 1989:64.

el, region, and unionization. The HPM attempts to extract the component value of the environmental quality characteristics from the overall prices of real estate parcels or the wage rates for jobs.

As developed by Sherwin Rosen (1974), the HPM consists of a two-step procedure. First, a hedonic price equation is estimated. This equation decomposes the price of a composite good into marginal implicit prices of characteristics. Second, the implicit prices for an individual characteristic, such as the environmental variable, are used to estimate a relationship between the implicit price and the quantity demanded of the characteristic. This second stage produces a marginal WTP function for the characteristic.

Indirect methods generally impose large data requirements and sophisticated statistical demands. In many cases, the same data set can produce a wide range of value estimates depending on the assumptions made about the structure of preferences. Since we really do not know exactly how to characterize preferences, we have little basis on which to choose among competing estimates.

As an example of the HPM, David Harrison and Daniel A. Rubinfeld (1978) use housing market and census tract data for Boston to estimate the economic benefits of air pollutant concentration with and without emission control programs. Their premise is that households consider air pollution levels as well as the quantity and quality of housing services and other neighborhood characteristics in making their housing choices. Using census data, Harrison and Rubinfeld specify a hedonic value function, differentiate the hedonic equation with respect to air pollution attributes (using nitrogen oxide levels as a proxy variable), and regress the resulting implicit price against observed air pollution attributes and other variables (such as income levels). The total benefit of cleaner air equals the difference in area under the resulting marginal WTP function between air quality levels with and without the programs. Harrison and Rubinfeld's results show that the valuation placed on a marginal improvement in air quality is quite sensitive to the specification of the hedonic housing value equation.

Another indirect method, the travel cost model (TCM), is widely used for valuing recreation sites. This model measures the WTP for a site by relating the cost of traveling to a recreation site to the attributes of the site. The premise is that the site cannot be used without travel and, further, that a more attractive site will evoke more expenditure on travel by consumers. The extra expenditures provide an estimate of the monetary value of the added quality of the site. Data requirements for this approach include market prices for travel, data on alternative recreation sites, and survey data on the numbers, origins, and demographics of travelers. The use of TCMs is obviously limited to a particular type of nonmarket good—visited sites. The model works best when applied to the valuation of a single site and conditions at alternative sites can be expected to remain constant.

As an example of the TCM, V. Kerry Smith, Raymond B. Palmquist, and Paul Jakus (1991) apply the methodology to evaluate quality improvements of sport fishing in the Albermarle-Pamlico Estuary on the coast of North Carolina. They use survey data on fishing trips, the fishing equipment used, fish caught, characteristics of the sites visited, and the respondents' socioeconomic attributes (income, age, horsepower of fishing boat, and experience). The site characteristics consist of an estimate of fish availability and measures of the effluent loadings (nitrogen and phosphorus) from each coastal area as proxies for water quality in each location. Travel costs include the vehicle operating costs and the opportunity cost of travel time. Using the travel cost data to determine WTP for each site, Smith and his colleagues then embed the travel cost relationships in a hedonic pricing framework to identify the effects of site quality on the WTP functions. In the first stage, the hedonic travel cost functions are estimated for each site characteristic. In the second stage, demand for catch, the key quality variable, is estimated with both marginal price and the quantity as dependent variables. The results are used to calculate the benefit per person per trip for improvements in the catch rate. This information could be used to determine the fishing benefits of programs to enhance fish stocks or improve water quality.

In addition to the HPM and the TCM, there are a variety of other indirect approaches to evaluating environmental quality (Braden and Kolstad 1991; Freeman 1993). The methods are tailored to particular applications. In the case of the health impacts of pollution, for example, some analysts have estimated values by measuring associated expenditures on medical care and wages lost through illness or death. A similar approach has been taken in valuing air pollution damages to building materials by totaling up the costs of preventive and repair measures.

What all of the indirect methods have in common is the notion that nonmarket effects influence the way people behave in markets and those changes in behavior reveal the value placed on nonmarket outcomes. There are three important limitations to indirect approaches. First, as we have noted, the estimation of the nonmarket values is often very difficult. The analyst must work backward from observed consumption in the market through a preference structure and then forward from the preference structure to draw inferences about the unobserved demand for environmental quality. Any one of these steps is challenging in itself, and the process is ultimately ambiguous because we do not know the true structure of preferences. Second, market-based measures are rarely complete, even for use-based values. The premise is that nonmarket values can be realized only through the consumption of an identifiable set of market goods.[17] But, many environmental goods and services do not readily fit this description. The consumption of dirty air or loud noise, for example, cannot be avoided completely unless one remains confined by filters, but then there is the loss of personal freedom to consider. The loss of freedom surely has an effect on welfare, but this effect is not captured in the expenditures on the defensive equipment. Third, as was suggested in table 8.5, not all environmental values stem from actual consumption, either direct or indirect. Nonuse values are essentially impossible to estimate from market data.

These three limitations present tall obstacles to program evaluation. Because these indirect methods are cumbersome, complex, and piecemeal, they have stimulated economists to pursue monetary valuation techniques that do not depend on market behavior but, instead, elicit values for environmental quality directly—direct methods.

Direct Methods Direct methods use surveys or experimental markets to elicit individuals' monetary values for changes in the environment. They are called "direct" because they present choices about the specific environmental variables of interest rather than the surrogates on which indirect methods depend. Instead of using travel costs to value improved lake quality, for example, one could simply ask a sample of people what the improved quality is worth to each of them. The responses could then be aggregated up to a value for the full population.

The most widely used of the direct approaches, and the one most suited to field use, is the contingent valuation method (CVM). CVM uses surveys to ask people how much they would be willing to pay to change some aspect of environmental quality.[18] The basic assumptions are that people have coherent preferences and will truthfully express those preferences in responding to a hypothetical questionnaire. As a branch of survey research, CVM is subject to many of the limitations of that field, including the potential for sample bias and framing bias.[19]

Many analysts have used the survey technique to estimate the monetary value of programs to improve air and water quality, protect endangered species, and achieve other environmental goals.[20] Alan Randall, Berry Ives, and Clyde Eastman (1974), for example, used surveys to estimate the benefits of reducing the visual impairment caused by emissions and transmission lines of the Four-Corners power plant in the southwestern United States. Respondents were shown photographs representing the current situation, a scenario of modest changes, and another case representing high levels of abatement and relocation of transmission lines. They were then asked how much they would pay to achieve an improvement from one level to another. Other examples include the work of David Brookshire, Mark A. Thayer, William D. Schulze, and Ralph C. D'Arge (1982), who used survey methods to measure the benefits of improving air quality in the Los Angeles metropolitan area, and that of Kevin J. Boyle and Richard C. Bishop (1987), who employed the survey method to investigate the benefits of preservation programs for the bald eagle.

The reliability of CVM responses depends critically on the design of the survey instrument. How information is presented and how valuation questions are asked can make significant differences in the responses (Mitchell and Carson 1989; Hausman 1993). In this respect, CVM is subject to criticisms analogous to those raised about indirect methods: the judgment of the analyst can affect the results.

Another type of direct approach actually tries to simulate markets for environmental commodities. Much research on valuation questions has been done in experimental settings like those more typically associated with psychological research. For example, Shogren, Shin, Hayes, and Kliebenstein (1994) conducted experimental auctions for different types of commodities to explore the importance of product substitution. One of their "commodities" was risk reduction for food-borne pathogens. Subjects were asked to declare their maximum WTP for safer food or their minimum WTA compensation for taking a test product food instead of safer food. Subjects were required to eat the food to receive a reward. The laboratory experiments take place in controlled settings, so certain stimuli can be precisely controlled and their influence on observed behaviors tested.

A smaller body of research has been able to create real (although not self-sustaining) markets for environmental commodities instead of relying on hypothetical questionnaires or laboratory markets. A leading example is the work of Richard C. Bishop and Thomas A. Heberlein (1979), who orchestrated a market in permits for goose hunting in Wisconsin. They made monetary offers to purchase permits already held by hunters. The researchers estimated the demand for duck hunting by varying the amounts offered and seeing how many permits were relinquished.[21] This is precisely the type of information needed to establish monetary values, and because real money was changing hands, the results are equivalent to regular market-based estimates. Unfortunately, however, it is rare to have the resources available to implement such a study or to be able to package environmental goods and services in ways that are amenable to market transactions.

To summarize, direct methods attempt to elicit monetary values directly rather than through choices about other, related goods and services. Because they are direct, it is much easier to extract exactly the value information that program analysts need. Direct methods also have the advantage of being able to capture intrinsic values that do not relate to use, and they can be structured to capture in a single exercise the composite economic value of all types of benefits (use as well as intrinsic) stemming from a particular program. Market-based measures cannot compare in the breadth of values that can be investigated through direct methods. To their detriment, however, direct methods do not perfectly emulate the process of market valuation. CVM, in particular, elicits expressed preferences as opposed to revealed preferences. Many economists are reluctant to place confidence in values that do not require the commitment of resources. Even the results from laboratory or simulation studies in which money is used may not be generalizable because of the uniqueness of the laboratory setting. Other concerns with indirect methods are that respondents may not be as careful with their choices as they would be in the real world, and their responses may be unduly influenced by the selective information that the administrator provides.

Total Economic Value The preceding section examined a fundamental problem in the economic evaluation of environmental programs—the problem of attaching monetary values to goods and services that are not priced in markets. Economic evaluation begins with monetization. The second step in economic evaluation is to aggregate the monetized benefits and costs and draw inferences about the net impact on society.

Aggregation raises complex problems, especially where the monetary values of various categories of impacts must be combined. As John P. Hoehn and Alan Randall (1989) have pointed out, the order in which the categories are valued can change the respective monetary values and, hence, the sum of the values. The only way to avoid this problem in most cases is to elicit a single value from each consumer for the entire complex of impacts, and the only practical (albeit controversial) way to do this is through survey methods.

Risk Analysis

Objectives of Risk Analysis

Environmental impacts may be divided into two groups: predictable and unpredictable. By predictable, we mean that a certain set of consequences always follows a release of pollution. So, for example, a release of chlorofluorocarbons into the atmosphere always degrades the stock of stratospheric ozone and a release of unoxidized organic material into a water body always reduces the supply of dissolved oxygen. Unpredictable impacts are those that do not necessarily follow from the simple act of discharging pollutants but instead depend on a confluence of several unpredictable factors to have serious consequences. Tropospheric ozone formation is a prime example. Tropospheric ozone forms from volatile hydrocarbons reacting with nitrous oxides in the presence of sunlight. The effect of a release of VOCs depends on the rates at which other chemicals are being released and on climatic factors, so it is not fully predictable. Similarly, the effects of many hazardous chemicals are not predictable at the level of individual organisms, although they may be predictable at the level of a population.

From the point of view of individual dischargers, potential victims, or regulators, unpredictable environmental impacts can be said to be sources of risk. Risk refers to outcomes that are probabilistic rather than deterministic. These outcomes may be managed in two dimensions: the probabilities of occurrence and the magnitude of occurrence. Thus, for example, the handling and transporting of hazardous chemicals might be regulated to reduce the probability of a release, and the chemical properties might be regulated to control the severity of a spill if it does occur.

A special set of tools is needed to characterize risky environmental impacts. These tools are known collectively as risk analysis. Risk analysis is a systematic process for describing and quantifying variable outcomes. It draws on knowledge from a variety of disciplines, such as toxicology, radiology, engineering, economics, and statistics. Risk analysis has the objective of describing the distribution of potential outcomes and the costs and benefits of changing those outcomes through the intervention of public programs.

Methodology of Risk Analysis

Risk analysis consists of three interrelated steps: hazard identification, risk assessment, and risk evaluation (Covello and Merkhofer 1993).

Hazard Identification Hazard identification involves identifying the risk agents and the conditions under which they can produce adverse health effects. In reality, there exist many risk agents, such as toxic metals, solvents, pesticides, electromagnetic radiation, radioactivity, and so forth. For most hazardous substances such as these, we need sophisticated techniques to identify adverse health effects since those effects are often subtle and latent. There are many methods for assessing hazards, as shown in table 8.6, and among them are econometric methods designed to draw inferences about the factors that cause hazards. To provide the basic information on potential causes and effects in such cases, risk analysts use a variety of techniques, including epidemiological studies, in vivo animal bioassays, short-term in vitro cell and tissue culture tests, and structure-activity analyses (Cohrssen and Covello 1989).

Risk Assessment Once the hazard has been characterized, the analysis enters its second step—risk assessment. Here, the unique challenges of unpredictability take on prominence. This step is where the outcomes must be described statistically, that is, a probability must be assigned to each level in the range of possible outcomes.

Risk assessment involves describing and quantifying the impacts and their probabilities. According to the National Research Council (1983), the analytical procedures of risk assessment consist of four steps: source/release assessment, exposure assessment, dose-response assessment, and risk characterization.[22] The first step, source/release assessment, involves quantifying the extent to which a risk source releases or otherwise introduces risk agents into the human environment. Quantitative techniques for release assessment are monitoring, performance testing and accident investigation, statistical methods, and modeling. The second step, exposure assessment, is the process of measuring the intensity, frequency, and duration of human or other population exposures to risk agents. This technique provides quantitative data on individuals, populations, or ecosystems that are, or may be, exposed to a risk agent; the concentrations

Table 8.6. Techniques for Hazard Identification

Category	Techniques
Analogies	Barometric analysis Case studies Historical comparisons
Trend analysis	Simple extrapolation Trend projection Time series analysis Curve fitting Statistical modeling
Probabilistic approaches	Bayesian statistics Markov analysis Monte Carlo simulation Parametric analysis
Simulation methods	Panel testing Physical modeling Structural modeling Surrogate testing Toxicological testing
Survey methods	Authority opinion Brainstorming Checklists Delphi technique Monitoring Panel survey Public preferences
Tracing methods	Econometric modeling Event tree analysis Feedback loop analysis Input-output analysis Relevance tree analysis

Source: Adapted from Conservation Foundation 1985:23.

of the risk agent; and the duration and other characteristics of exposures. The third step, dose-response assessment, provides quantitative data on the specific amounts of a risk agent that may reach the organs or tissues of people, other living creatures, and physical assets. This information provides a basis to estimate the percentage of these populations that might be harmed or injured and, where relevant, the characteristics of such populations. Finally, the fourth step, risk characterization, integrates the results of the previous steps. The outcome is a risk statement that includes one or more quantitative estimates of risk.

Risk Evaluation Risk evaluation entails comparisons and judgments about the significance of risks. The evaluations may consider the effects of different policy and program configurations. The evaluation of the cost and benefits of reduced risk is a formidable and often controversial task. While the cost of regulating hazardous substances, for example, can be measured with some degree of accuracy, the benefits are elusive and difficult to quantify. For example, valuing risks to human health implicitly requires valuing human morbidity or mortality. Furthermore, people seem to value risky propositions differently from certain ones of equivalent expected value, and they may even be inconsistent in valuing different risk scenarios that have equivalent consequences and probability distributions.

The economic problem of determining an optimal level of risk and safety in society has been approached in two ways. One relies on the measurement of WTP and the other takes a human capital approach.

The WTP approach has been advocated by economists because it has a firm grounding in traditional economic theory. The WTP approach estimates the value of the welfare change that would result from changing the environmental risks. As we have discussed, the welfare change equals the maximum amount an individual would be willing to pay to reduce risk or the minimum compensation an individual would be willing to accept for an increase in risk. According to the literature, there are three techniques for applying the WTP approach: hedonic wage-risk studies, defensive expenditures assessment, and contingent valuation.[23] The hedonic wage-risk method was already described briefly. It attempts to measure a wage premium associated with more risky occupations. The defensive expenditure method estimates WTP for risk reductions based on an individual's actual revealed preferences for durable and nondurable self-protection mechanisms, such as smoke detectors, seat belts, medicine, and behavioral reactions to smoking. The contingent valuation method uses surveys or experiments to evaluate risks and can be applied to environmental risks that are not readily measured in job injury data.

Although the WTP approach is generally accepted by economists, the human capital approach is more common in actual use. The human capital approach values risk reduction by examining actual financial consequences over the lifetime of an injured party. In the case of a health impact, for example, it would include medical costs, the present value of lost earnings, and perhaps a value for pain and suffering, rather than the minimum compensation that the victim would require. This approach has an advantage in that it uses reasonably well-defined values to evaluate risk reductions. The process can be readily understood by noneconomists in legal proceedings to determine compensation for injuries. When combined with mortality tables and individual earnings data, the benefits of health risk reduction can readily be computed. However, the approach has some important drawbacks. One is that it reflects any distortions that may exist in real markets, with the consequence, for example, that it may generically assign a lower value to the lives of women and minorities because they tend to be paid less and a zero value to retired individuals because they are earning nothing.

In summary, the techniques of risk evaluation are essentially specialized applications of general economic valuation techniques. The special character of risk evaluation lies in that outcomes are not certain but are instead only possibilities. This quality can be confusing to humans, and there are competing theories about how we perceive and respond to risk. Most of the evaluation approaches, be they hedonic analysis, defensive expenditure estimation, or the human capital approach, use observable market data and assume humans behave rationally and consistently. Unfortunately, however, there is much evidence to suggest that we do not really fully understand how humans respond to uncertain events. Furthermore, as is also true for certain events, it can be difficult to isolate the WTP values that we really seek from market data. Contingent valuation can more readily be focused on WTP values, but they are subject to disturbing incentive and perception problems that are further confounded by the probabilistic nature of the outcomes. We must conclude, then, that the economic evaluation of risky environmental phenomena is a long way from an exact science. Perhaps the best strategy for evaluation under these circumstances is triangulation—trying several different approaches simultaneously to explore the range of possible values. A narrow range of values generated by different techniques provides confidence that the results are good approximations of the "true" monetary value.

Economic Incentives for Pollution Prevention

In the main body of this chapter we have focused on economic methods for evaluating the substantive objectives of environmental programs, such as how much abatement to require or how aggressive to be in protecting endangered species. This is the kind of question for which evaluation is most often sought. In this brief section, however, we return to a theme raised in the introductory section: evaluating different strategies for pursuing those objectives. Economics offers much insight into questions of strategy, just as it helps in evaluating objectives.

As we noted earlier, economists make a broad distinction between command-and-control policies and incentive-based policies. In the context of environmental programs, quantitative restrictions on pollution fit in the former category while charges for pollution fit in the latter. In principle, with full information about the benefits and costs of pollution control, both approaches could yield the same, socially efficient outcome. A system of perfectly selective pollution restrictions would accomplish the same efficient assignment of abatement as would a charge system. In practice, however, pollution restrictions are rarely perfectly selective. More usually, they only loosely respect differences in benefits and costs. A charge system, on the other hand, is highly selective, with individual polluters able to make their own decisions about abatement based on their specific costs and benefits.

Incentive-based approaches are intended to work in the long run as well as the short run. That is, if it is clear that present decisions will have financial consequences in the future, this realization should affect present decisions. For instance, if today's producers of toxic compounds are liable for illnesses their chemicals will cause twenty years from now, they will more carefully screen the products they sell. In this way, incentive-based approaches encourage damage prevention. Rather than the government looking ahead and tell polluters what they must do now to prevent future damages, incentives (especially those created by liability) make it in the polluter's own interest to look ahead and take precautionary action. It is this decentralization of responsibility, together with the greater latitude in choices and the motivation to look ahead, that so enamors economists of incentive-based strategies (Portney 1990).

Incentive strategies are not always available, nor are they always preferable. Sometimes, as with transferable emissions allowances, they must be used within a regulatory framework. The point is that program analyses should not focus on program goals to the exclusion of program design. An evaluation of environmental programs should consider both their objectives and the strategies they use. The two dimensions are interconnected. A more efficient strategy will improve the economic appearance of a particular goal. More ambitious goals may be worthwhile if program costs can be held down. Thus, a complete economic evaluation should consider strategies and goals in combination.

Conclusion

In this chapter we have outlined economic methods of environmental program evaluation. To a large degree, the economic approach makes use of well-established principles of benefit-cost analysis. Rather than covering those principles in detail, we have emphasized the unique issues that arise in environmental programs—especially the problems of assigning monetary values to environmental goods and services and evaluating risk. These are areas in which the practical needs of program evaluation have led to new advances in economic scholarship.

In addition to techniques of evaluating environmental objectives, we have suggested that economics provides much broader insights into environmental programs. Economic thinking, for example, leads to the conclusion that financial incentives in some cases may accomplish more innovative abatement and pollution prevention than would quantitative restrictions. Thus, an important message to take from this chapter is that economics provides not only tools for environmental program evaluation but also guidance about the design of environmental programs.

NOTES

1. Current guidelines for benefit-cost analysis within the executive branch are given in Darman (1992).

2. For an example of a study that primarily addresses the distributional consequences of environmental programs, see Been (1994).

3. For more complete expositions of BCA, see Gramlich (1981); Schmid (1989); and Oh (1992).

4. Discounting adjusts for the fact that a dollar today is worth more than a dollar at some future date because the present dollar can be invested in the interim. The discount rate represents the productive value of money. Which rate of interest to use is a matter of debate (Lind et al. 1982). Among the candidates are loan rates for private corporations, borrowing rates for individual consumers, borrowing rates for the government, and a special "social discount rate" that would take into account the interests of future generations.

5. Since the internal rate of return is a solution to a polynomial expression, it can take on multiple values. Generally, only one of these values is relevant to the decision criteria.

6. The U.S. standard is 0.12 parts per million measured over a one-hour interval. To be in compliance, the ozone concentration must not be exceeded more than three days in any three-year period.

7. The study by Jane V. Hall et al. (1992) provides different, somewhat higher estimates of the damages from ozone pollution. If these estimates were accepted, the benefit-cost ratio for VOC abatement would be more favorable.

8. A thorough exposition of the CEA in health and medical practices is presented in Weinstein and Stason (1993:49–55).

9. Further details on environmental I-O models are given in Førsund (1985) and Proops, Faber, and Wagenhals (1993).

10. A general equilibrium model is one in which all markets clear in equilibrium. CGE models incorporate equilibrium links among the production structure, the pattern of demand, and incomes of various groups. For a general discussion of the CGE model, see Bergman (1990, 1991).

11. A thorough discussion of the total economic value of an environmental resource is presented in Per-Olav Johansson (1990). The calculation of the total economic value

requires that values be combined for several different classes of consequences. For more detailed expositions of the theoretical basis for the total economic value, see Randall and Stoll (1983); Hoehn and Randall (1989); and Randall (1991).

12. Even though we do have the tools for measuring the existence value (such as the survey instrument–contingent valuation method), Rosenthal and Nelson argue that outcomes of the survey method would be significantly changed since estimates of the existence value depend heavily on a hypothetical situation in which respondents are asked to give their evaluation of nonuse-related objects. So, they consider measuring the existence value as unmeasurable (1992:117–21).

13. More technically, WTP represents the maximum sum of money an individual would be willing to pay rather than do without an increase in some good, such as an environmental amenity. Similarly, WTA is the minimum payment an individual would require to voluntarily forgo an improvement that otherwise would be experienced. Both WTP and WTA can apply to negative changes as well as to positive changes (Freeman 1993).

14. Another line of explanation comes from the psychology literature. Individuals appear to view losses differently from gains, a phenomenon referred to as cognitive dissonance. Prospect theory argues that cognitive dissonance is at work in the kinds of valuation exercises posed by environmental programs, in which consumers perceive that they will either gain or lose relative to the status quo. For more detailed exposition, see Kahneman and Tversky (1979); Fisher, McClelland, and Schulze (1988).

15. An extensive review of economic valuation is provided by Cummings, Cox, and Freeman (1985); Johansson (1987); Braden and Kolstad (1991); Freeman (1993); and James (1994).

16. Indirect methods for valuing nonmarket goods consist of hedonic property values, avoidance expenditures, referendum voting, and travel cost (Freeman 1993:23–26). Sherwin Rosen (1974) developed a general theoretical framework of HPM using hedonic prices to analyze the demand for and supply of attributes for differentiated products. The term *hedonic prices* is defined as "the implicit prices of attributes and are revealed to economic agents from observed prices of differentiated products and the specific amounts of characteristics associated with them" (Rosen 1974:34).

17. This is known as the weak complementarity assumption. See Freeman (1993:104–13).

18. Numerous authors of articles and books have discussed the theoretical validity and application of CVM. A thorough exposition of CVM is presented in Mead (1993) and Hanemann (1994).

19. *Sample bias* refers to the selection of respondents who are unrepresentative of the broader population. *Framing bias* refers to the selective provision of information in a way that leads subjects to certain conclusions.

20. Richard T. Carson and colleagues (1994) list nearly 1,700 applications of contingent valuation methods.

21. Bishop and Heberlein (1979) also were able to compare the demand estimates from the simulated market to monetary values elicited through CVM studies. They concluded that WTP measures exhibit a tendency to be more or less accurate, while

WTA measures are liable to be substantially overestimated. This result implies that CVM should be considered a lower bound in WTP and an upper bound in WTA.

22. In the context of environmental program evaluation, there are several models of risk assessment. The model developed by the National Research Council of the National Academy of Sciences has been widely used by the EPA for assessing the risks of cancer and other health risks that result from exposure to chemicals. The National Research Council model consists of four categories: hazard identification, dose-response assessment, exposure assessment, and risk characterization. The Covello-Merkhofer (1993) model consists of four interrelated but conceptually distinct steps: release assessment, exposure assessment, consequence assessment, and risk estimation. In this chapter we briefly review the NRC model since it is widely accepted by risk analysts.

23. For more detailed explanations of WTP estimation methods, see Viscusi (1985); Fisher, Violette, and Chestnut (1989); and Shogren (1990).

REFERENCES

Adamowicz, W. L., V. Bhardwaj, and B. Macnab. 1993. "Experiments on the Difference between Willingness to Pay and Willingness to Accept." *Land Economics* 69 (4): 416–27.

Anderson, F. R., A. V. Kneese, P. D. Reed, S. Taylor, and R. B. Stevenson. 1977. *Environmental Improvement through Economic Incentives.* Baltimore: Johns Hopkins University Press.

Ash, R. L. 1975. "Evaluation of the Inflationary Impact of Major Proposals for Legislation and for the Promulgation of Regulations or Rules." Circular A-107 to the Heads of Executive Departments and Establishments, Office of Management and Budget.

Atkinson, S. E., and D. H. Lewis. 1974. "A Cost-Effectiveness Analysis of Alternative Air Quality Control Strategies." *Journal of Environmental Economics and Management* 1 (3): 237–50.

Been, V. 1994. "Locally Undesirable Land Uses in Minority Neighborhoods: Disproportionate Sitting or Market Dynamics?" *Yale Law Journal* 103 (6): 1383–1422.

Bergman, L. 1990. "The Development of Computable General Equilibrium Modeling." In *General Equilibrium Modeling and Economic Policy Analysis,* ed. L. Bergman, D. W. Jorgenson, and E. Zalai. Cambridge: Basil Blackwell. 3–30.

———. 1991. "General Equilibrium Effects of Environmental Policy: A CGE-Modeling Approach." *Environmental and Resource Economics* 1 (1): 43–61.

———. 1996. "Sectoral Differentiation as a Substitute for International Coordination of Carbon Taxes: A Case Study of Sweden." In *Environmental Policy with Political and Economic Integration: The Europe Union and the United States,* ed. J. B. Braden, H. Folmer, and T. S. Ulen. Cheltenham: Edward Elgar Publishing. 329–48.

Bergson, A. 1938. "A Reformulation of Certain Aspects of Welfare Economics." *Quarterly Journal of Economics* 52 (Feb.): 310–34.

Bishop, R. C., and T. A. Heberlein. 1979. "Measuring Values of Extra Market Goods:

Are Indirect Measures Biased?" *American Journal of Agricultural Economics* 61 (5): 926–30.

Boyle, K. J., and R. C. Bishop. 1987. "Valuing Wildlife in Benefit-Cost Analysis: A Case Study Involving Endangered Species." *Water Resource Research* 23 (5): 943–50.

Braden, J. B., H. Folmer, and T. S. Ulen, eds. 1996. *Environmental Policy with Political and Economic Integration: The Europe Union and the United States.* Cheltenham: Edward Elgar Publishing.

Braden, J. B., and C. D. Kolstad, eds. 1991. *Measuring the Demand for Environmental Quality.* Amsterdam: North-Holland.

Brookshire, D., M. A. Thayer, W. D. Schulze, and R. C. D'Arge. 1982. "Valuing Public Goods: A Comparison of Survey and Hedonic Approaches." *American Economic Review* 72 (1): 165–71.

Carson, R. T., J. Wright, A. Alberini, N. Carson, and N. Flores. 1994. *A Bibliography of Contingent Valuation Studies and Papers.* La Jolla, Calif.: Natural Resource Damage Assessment.

Castle, E. N., R. P. Berrens, and R. M. Adams. 1994. "Natural Resource Damage Assessment: Speculations about a Missing Perspective." *Land Economics* 70 (3): 378–85.

Cohrssen, J. J., and V. T. Covello. 1989. *Risk Analysis: A Guide to Principles and Methods for Analyzing Earth and Environmental Risks.* Washington, D.C.: U.S. Council on Environmental Quality.

Conservation Foundation. 1985. *Risk Assessment and Risk Control.* Washington, D.C.: Conservation Foundation.

Covello, V. T., and M. W. Merkhofer. 1993. *Risk Assessment Methods: Approaches for Assessing Health and Environmental Risks.* New York: Plenum Press.

Cropper, M. L., and A. M. Freeman III. 1991. "Environmental Health Effects." In *Measuring the Demand for Environmental Quality,* ed. J. B. Braden and C. D. Kolstad. Amsterdam: North-Holland. 165–211.

Cumberland, J. H., and B. N. Stram. 1976. "Empirical Application of Input-Output Models to Environmental Problems." In *Advances in Input-Output Analysis,* ed. K. R. Polenske and J. V. Skolka. Cambridge: Ballinger Publishing. 365–88.

Cummings, R. G., D. S. Brookshire, and W. D. Schulze. 1986. *Valuing Environmental Goods: An Assessment of the Contingent Valuation Method.* Totowa, N.J.: Rowman and Allanheld.

Cummings, R. G., L. A. Cox Jr., and A. M. Freeman III. 1985. "General Methods for Benefit Assessment." In *Benefit Assessment: The State of the Art,* ed. J. D. Bentkover, V. T. Covello, and J. Mumpower. Dordrecht: D. Reidel Publishing. 161–91.

Darman, R. 1992. "Guidelines and Discount Rates for Benefit-Cost Analysis of Federal Programs." Circular No. A-94 rev., Transmittal Memorandum 64 to the Heads of Executive Departments and Establishments, Office of Management and Budget.

Fisher, A., G. H. McClelland, and W. D. Schulze. 1988. "Measures of Willingness to Pay versus Willingness to Accept: Evidence, Explanations, and Potential Reconciliation." In *Amenity Resource Valuation: Integrating Economics with Other Disciplines,* ed. G. L. Peterson, B. L. Driver, and R. Gregory. State College, Pa.: Venture Publishing. 127–34.

Fisher, A., D. Violette, and L. Chestnut. 1989. "The Value of Reducing Risks of Death: A Note on New Evidence." *Journal of Policy Analysis and Management* 8 (1): 88–100.

Førsund, F. R. 1985. "Input-Output Models, National Economic Models, and the Environment." In *Handbook of Natural Resource and Energy Economics,* vol. 1, ed. A. V. Kneese and J. L. Sweeney. Amsterdam: North-Holland. 325–41.

Freeman, A. M., III. 1993. *The Measurement of Environmental and Resource Values: Theory and Methods.* Washington, D.C.: Resources for the Future.

Gramlich, E. 1981. *Benefit-Cost Analysis of Government Programs.* Englewood Cliffs, N.J.: Prentice-Hall.

Hahn, R. W. 1989. "Economic Prescriptions for Environmental Problems: How the Patient Followed the Doctor's Orders." *Journal of Economic Perspectives* 3 (2): 95–114.

Hall, J. V., A. M. Winer, M. T. Kleinmann, F. W. Lurmann, V. Brajer, and S. D. Colome. 1992. "Valuing the Health Benefits of Clean Air." *Science* 255 (Feb. 14): 812–17.

Hanemann, W. M. 1991. "Willingness to Pay and Willingness to Accept: How Much Can They Differ." *American Economic Review* 81 (3): 635–47.

———. 1994. "Valuing the Environment through Contingent Valuation." *Journal of Economic Perspectives* 8 (4): 19–43.

Harrison, D., and D. A. Rubinfeld. 1978. "Hedonic Housing Prices and Demand for Clean Air." *Journal of Environmental Economics and Management* 5:81–102.

Hausman, J. A., ed. 1993. *Contingent Valuation: A Critical Assessment.* Amsterdam: North-Holland.

Hicks, J. 1939. "The Foundations of Welfare Economics." *Economic Journal* 49 (Dec.): 696–712.

Hoehn, J. P., and A. Randall. 1989. "Too Many Proposals Pass the Benefit Cost Test." *American Economic Review* 79 (3): 544–51.

James, D. 1994. *The Application of Economic Techniques in Environmental Impact Assessment.* Dordrecht: Kluwer Academic Publishers.

Johansson, P.-O. 1987. *The Economic Theory and Measurement of Environmental Benefits.* New York: Cambridge University Press.

———. 1990. "Valuing Environmental Damage." *Oxford Review of Economic Policy* 6 (Spring): 34–50.

Jorgenson, D. W., and P. J. Wilcoxen. 1990. "Environmental Regulation and United States Economic Growth." *Rand Journal of Economics* 21 (2): 314–40.

Kahneman, D., and A. Tversky. 1979. "Prospect Theory: An Analysis of Decision under Risk." *Econometrica* 47 (Mar.): 263–91.

Kaldor, N. 1939. "Welfare Propositions in Economics and Interpersonal Comparisons of Utility." *Economic Journal* 49 (Sept.): 549–52.

Kopp, R. J. 1992. "Why Existence Should Be Used in Cost-Benefit Analysis." *Journal of Policy Analysis and Management* 11 (1): 123–30.

Kopp, R. J., and V. K. Smith, eds. 1993. *Valuing Natural Assets.* Washington, D.C.: Resources for the Future.

Krupnick, A. J., and P. R. Portney. 1991. "Controlling Urban Air Pollution: A Benefit-Cost Assessment." *Science* 252 (Apr. 26): 522–28.

Leontief, W. 1970. "Environmental Repercussions and the Economics Structure: An Input-Output Approach: A Comment." *Review of Economics and Statistics* 52 (3): 262–71.

Lind, R. C., K. J. Arrow, G. R. Corey, P. Dasgupta, A. K. Sen, T. Stauffer, J. E. Stiglitz, J. A. Stockfisch, and R. Wilson. 1982. *Discounting for Time and Risk in Energy Policy.* Washington, D.C.: Resources for the Future.

Luken, R. A., and A. G. Fraas. 1993. "The U.S. Regulatory Analysis Framework: A Review." *Oxford Review of Economic Policy* 9 (Winter): 96–111.

Mead, W. J. 1993. "Review and Analysis of State-of-the Art Contingent Valuation Studies." In *Contingent Valuation: A Critical Assessment,* ed. J. A. Hausman. Amsterdam: North-Holland. 306–37.

Mitchell, R. C., and R. T. Carson. 1989. *Using Surveys to Value Public Goods.* Washington, D.C.: Resources for the Future.

National Research Council. 1983. *Risk Assessment in the Federal Government: Managing the Process.* Washington, D.C.: National Academy Press.

Oh, H.-S. 1992. *Economics of Natural Resources and the Environment.* Seoul, Korea: Bummoonsa.

Portney, P. R., ed. 1990. *Public Policies for Environmental Protection.* Washington, D.C.: Resources for the Future.

Project 88—Round II Incentives for Action: Designing Market-Based Environmental Strategies. 1991. A public policy study sponsored by Senators T. E. Wirth and J. Heinz. Washington, D.C.

Proops, J. L. R., M. Faber, and G. Wagenhals. 1993. *Reducing CO_2 Emissions: A Comparative Input-Output Study for Germany and the UK.* Berlin: Springer-Verlag.

Qayum, A. 1994. "Inclusion of Environmental Good in National Accounting." *Economic System Research* 6 (2): 159–66.

Randall, A. 1991. "Total and Nonuse Values." In *Measuring the Demand for Environmental Quality,* ed. J. B. Braden and C. D. Kolstad. Amsterdam: North-Holland. 303–21.

Randall, A., B. Ives, and C. Eastman. 1974. "Bidding Games for Valuation of Aesthetic Environmental Improvements." *Journal of Environmental Economics and Management* 1 (1): 132–49.

Randall, A., and J. R. Stoll. 1980. "Consumer's Surplus in Commodity Space." *American Economic Review* 70 (3): 449–55.

———. 1983. "Existence Value in a Total Valuation Framework." In *Managing Air Quality and Scenic Resources at National Parks and Wilderness Areas,* ed. R. D. Rowe and L. G. Chestnut. Boulder: Westview Press. 265–74.

Rawls, J. A. 1971. *Theory of Justice.* Cambridge, Mass.: Harvard University Press.

Rhee, J. J., and J. A. Miranowski. 1984. "Determination of Income, Production, and Employment under Pollution Control: An Input-Output Approach." *Review of Economics and Statistics* 66 (1): 146–50.

Rosen, S. 1974. "Hedonic Prices and Implicit Markets: Product Differentiation in Pure Competition." *Journal of Political Economy* 82 (1): 34–55.

Rosenthal, D. H., and R. H. Nelson. 1992. "Why Existence Value Should Not Be Used in Cost-Benefit Analysis." *Journal of Policy Analysis and Management* 11 (1): 116–22.

Schelling, T., ed. 1983. *Incentives for Environmental Protection.* Cambridge, Mass.: MIT Press.

Schmid, A. A. 1989. *Benefit-Cost Analysis: A Political Economy Approach.* Boulder: Westview Press.

Scitovszky, T. 1941. "A Note on Welfare Propositions in Economics." *Review of Economic Studies* 9 (1): 77–88.

Shogren, Jason F. 1990. *A Primer on Environmental Risk Analysis.* Staff Report 90–SR 46. Ames: Center for Agricultural and Rural Development, Iowa State University.

Shogren, Jason F., S. Y. Shin, D. J. Hayes, and J. B. Kliebenstein. 1994. "Resolving Differences in Willingness to Pay and Willingness to Accept." *American Economic Review* 84 (Mar.): 255–70.

Smith, V. K., ed. 1984. *Environmental Policy under Reagan's Executive Order: The Role of Benefit-Cost Analysis.* Chapel Hill: University of North Carolina Press.

Smith, V. K., R. B. Palmquist, and P. Jakus. "Combining Farrell Frontier and Hedonic Travel Cost Models for Valuing Estuarine Quality." *Reviews of Economics and Statistics* 73 (4): 694–99.

Stephan, G., R. V. Nieuwkoop, and T. Wiedmer. 1992. "Social Incidence and Economic Costs of Carbon Limits: A Computable General Equilibrium Analysis for Switzerland." *Environmental and Resource Economics* 2 (6): 569–91.

Stokey, E., and R. Zeckhauser. 1978. *A Primer for Policy Analysis.* New York: W. W. Norton.

Tietenberg, T. H. 1985. *Emissions Trading: An Exercise in Reforming Pollution Policy.* Washington, D.C.: Resources for the Future.

U.S. Environmental Protection Agency. Office of Policy Analysis. 1987. *EPA's Use Benefit-Cost Analysis: 1981–1986.* Washington, D.C.: GPO. EPA 230-05-87-028.

———. Office of Policy Analysis. 1990. *Environmental Investments: The Cost of a Clean Environment.* Washington, D.C.: Office of Policy Analysis. EPA 230-11-90-083.

———. Office of Water. 1994. *President Clinton's Clean Water Initiative: Analysis of Benefits and Costs.* Washington, D.C.: GPO. EPA 800-R-94-002.

U.S. Office of Technology Assessment. 1988. *Urban Ozone and the Clean Air Act: Problems and Proposals for Change.* Washington, D.C.: GPO.

Viscusi, W. K. 1985. "The Valuation of Risks to Life and Health: Guidelines for Policy Analysis." In *Benefit Assessment: The State of the Art,* ed. J. D. Bentkover, V. T. Covello, and J. Mumpower. Dordrecht: D. Reidel Publishing. 193–210.

Weinstein, M. C., and W. B. Stason. 1993. "Foundation of Cost-Effectiveness Analysis for Health and Medical Practices." In *Cost-Effectiveness in the Nonprofit Sector: Methods and Examples from Leading Organizations,* ed. G. L. Schmaedick. Westport, Conn.: Quorum Books. 45–57.

Willig, R. D. 1976. "Consumers' Surplus without Apology." *American Economic Review* 66 (4): 589–97.

9

Evaluation of State Environmental Programs: Observations on Selected Cases

DAVID H. MOREAU

The Sedimentation Pollution Control Act and the Water Supply Watershed Protection Act are two of North Carolina's most significant initiatives to protect the environment and public health that were taken outside the framework of federal legislation. The first of these programs is now over twenty years old; the second is relatively recent in origin and is still undergoing changes based on feedback from early efforts to implement the act. Both have been subjected to evaluation, and I will review those efforts by comparing what was done to evaluate these two programs with what would have been required if more formal methods of risk and benefit-cost analysis had been used. The primary question of interest here is not how well evaluators performed benefit-cost or risk analysis. Rather the primary questions of interest here are How closely does experience with these two programs conform with more formal models? What additional steps would be necessary to bring this experience into greater conformity with more formal methods? and What practical limitations were or could have been encountered in bringing about that conformity? I will focus on the estimation of costs of resources required to implement the programs; efforts to link physical, chemical, and biological outcomes of programs to resource inputs; and efforts to estimate economic benefits of those outcomes.

Environmental Values in Benefit-Cost Analysis: A Historical View

Because benefit-cost analysis (BCA) had its origins in water resource planning and has been adapted to account for environmental values not established by market processes, it is quite natural to place the evaluation of environmental programs related to water resources in that context. In the 1920s and early 1930s, the Army Corps of Engineers was using a crude form of BCA to evaluate multiple purpose water projects. More formal adoption of BCA is generally traced to language in the Flood Control Act of 1936 (PL 74-738, 49 Stat. 1570), which declared: "The Federal Government should improve . . . navigable waters . . . for flood control purposes if the benefits to whomsoever they may accrue are in excess of the estimated costs." Application of this principle became widespread over the next two decades. In 1946 the Federal Interagency River Basin Committee initiated a process to establish principles and practices for estimating benefits and costs for water projects. Results of that process were published as *Proposed Practices for Economic Analysis of River Basin Projects* in May 1950, a document that became known as the "green book." The committee acknowledged that public policy may be influenced by factors other than economic considerations, but maximization of benefits less cost is a "fundamental requirement for the formulation and economic justification of projects and programs" (1950:5). Among other distinctions, the committee differentiated between "tangible" and "intangible" effects, that is, those that are potentially measurable in monetary terms and those that are not, such as improvements in public health, protection of scenic values, and prevention of loss of life. These effects were to be described and not "overlooked," but according to Beatrice H. Holmes (1972:32) one of the key issues in the intense debates over appropriate uses and misuses of BCA in the 1950s and 1960s was the adequacy with which intangible benefits and costs were being treated in project evaluation.

Among the first to adapt BCA to incorporate environmental values was a panel of consultants appointed by the Bureau of the Budget to develop a broader set of principles to guide water resource planning. In their report (Bureau of the Budget 1961), the panel members stated that their all-embracing objective is to maximize national welfare, but that national welfare consists of multiple components that cannot be captured by a single measure. Among the components that they identified were the increase of national income, equitable distribution of income, and preservation of aesthetic and cultural values. Their view of the multiple-dimensional nature of national welfare led them to recommend that standards and criteria be framed in terms of multiple objectives rather than the single objective of national income.

Views of the panel members were reflected in a revised set of standards and procedures for water resource planning published as Senate Document 97 of the second session of the 87th Congress in 1962 (U.S. Water Resources Council 1962). The multiple objective framework was further refined by the Water Resources Council (WRC) established by the Water Resources Planning Act of 1965. When the WRC published its *Proposed Principles and Standards for*

Planning Water and Related Land Resources in December 1971, it had been strongly influenced by the enactment of the National Environmental Policy Act in 1969. The WRC recognized two equally weighted objectives, namely, national economic development and environmental quality. These two objectives were to guide the formulation of alternative plans, but beneficial and adverse effects of each plan were to be evaluated in four separate accounts, one for each of the objectives, one for regional development, and one for social well-being. The final version of these regulations, referred to commonly as "Principles and Standards," was adopted by the WRC in 1973.

President Jimmy Carter used analyses performed under these regulations to strongly oppose a large number of water projects that had been previously authorized by Congress. Carter went on to direct the WRC to make "Principles and Standards" more inclusive of environmental values by requiring that for every project designed to develop water resources an alternative would have to be developed that maximized environmental values. The Reagan administration reversed directions. Secretary of the Interior James Watt was a strong proponent of water projects, and the WRC and its regulations were prime targets for change. The WRC was "zeroed" in the budget, and "Principles and Standards" was replaced by *Economic and Environmental Principles and Guidelines for Water and Related Land Resources Implementation Studies* (U.S. WRC 1983). These nonbinding guidelines no longer had the force of binding rules, but they shifted federal administrative policy to place greater emphasis on economic aspects of BCA.

Despite these multiple objective adaptations of BCA, economists have continued to develop concepts and methods for measuring benefits to environmental quality under the economic efficiency objective. Much of that work since the early 1970s is reviewed by Nick Hanley and Clive L. Spash (1993). John Braden and Chang-Gil Kim summarize important developments in the field in chapter 8 of this volume, in which they trace the earliest initiatives to the Office of Management and Budget's quality of life review in 1971. They argue that the Office of the President and the Office of Management and Budget have been less reticent to embrace economic evaluation of environmental programs than either Congress or program agencies. Evidence cited above suggests that, although relative weights given to the two objectives have been shifted from one administration to the next, water resource development agencies have tended to keep separate accounts for valuing economic and environmental effects on plans and projects.

State governments have been more reticent than federal agencies to adopt formal methods for evaluating programs. That reticence results in part from the lack of expertise and experience in the field. The two cases I examine in this chapter illustrate some of the obstacles that are encountered when evaluations using more formal methods are undertaken.

Sedimentation and Erosion Control Program in North Carolina

North Carolina's sedimentation and erosion control program, established by the Sedimentation Pollution Control Act of 1973, has as its goal the minimization of damage to streams caused by erosion and sedimentation from land-disturbing activities. Construction projects that will disturb more than one acre of land are required to have an approved sedimentation control plan before work can be initiated. Residential and commercial developments and large public projects such as highways, airports, reservoirs, and pipelines are primary targets of the program. Agriculture is exempt, and forestry is partially exempt. Any local government may operate its own program if its ordinance is approved by the state and if it passes biennial performance evaluations. In all areas of the state not covered by local ordinances, the program is administered by the Division of Land Resources (DLR) of the Department of Environment, Health, and Natural Resources (DEHNR) operating through seven regional offices.

The workload on DLR grew steadily from a level of about 800 plans a year in 1980 to about 3,000 plans a year in 1990. About 25 percent of plans that are submitted are disapproved, necessitating revisions and resubmittal by applicants. DLR's rate of field inspections also grew to over 11,000 a year by 1990.

A benefit-cost framework for evaluating this program is sketched in figure 9.1. A framework similar to this was envisioned by the North Carolina Sedimentation Control Commission, the appointed body that establishes the formal rules under which DLR administers the program, when it requested and funded an evaluation of the program in 1989.

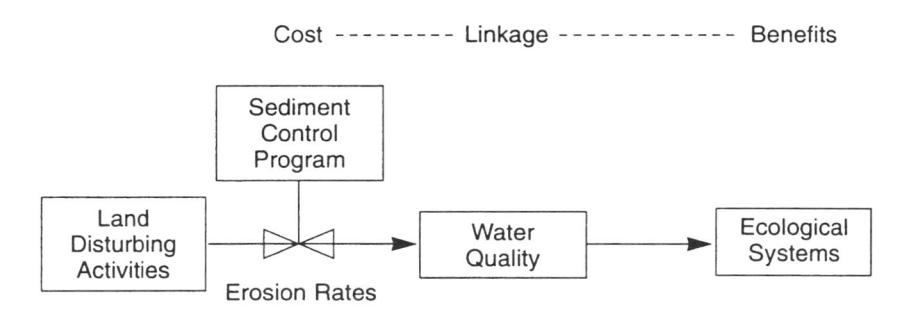

Figure 9.1. Framework for Benefit-Cost Analysis of Sediment Control Program

The request for proposals written by the members of the commission called for estimation of costs—including legal and administrative costs—costs of plan preparation and implementation, and other costs to the development community; an assessment of program effectiveness, including adequacy of coverage, technical guidance and training, administrative efficiency, enforcement, and control measures; and a description of benefits. The request for proposals stated that it was unlikely that benefits could be measured in simple economic terms (Gardner 1988).

Estimation of Cost

Costs of the program were relatively easy to estimate. Administrative costs were estimated from direct expenditures by DLR and nineteen local programs, and although those expenditure data do not include some items that would be in a full cost accounting, they provide a reasonable first approximation. Staff time and expenditures were allocated to program administration, plan review, inspection and monitoring, general surveillance, clerical, and miscellaneous. Costs were derived by activity levels to get the average cost for each plan reviewed ($99 for state offices; $196 for local program offices) and the cost per inspection ($31 for state offices; $32 for local program offices).

Compliance costs were a bit more difficult to obtain. Plans were obtained for 127 randomly selected construction projects in nine local areas. Quantities of materials, labor, and other cost centers were taken from the plans for each of these projects. Unit prices for these quantities were obtained from local contractors and engineering firms. The total cost was obtained by multiplying quantities by appropriate unit prices. Total expenditures were then reduced to the average cost per acre of the project ($1,500; Malcolm et al. 1990:V-6). Variations in cost across the three geographic provinces of the state and between state and local programs were also estimated.

Linking Inputs to Outputs

Outputs of the program could be measured at several different end points, namely, reduction in rates of erosion leaving project sites, improvements to the quality of water (lower turbidity) in receiving streams, and improvements to aquatic ecosystems that would be adversely affected by sediment. The evaluation was only modestly successful in linking program activities to outputs, and most of the assessment of program effectiveness was based on attitudinal surveys, not on erosion and sedimentation surveys.

Any attempt to link program activities to the goal of reducing sediment load is severely handicapped by the absence of any monitoring of the rates at which sediment is transported from project sites. If it was required on all sites, such monitoring would add considerable cost to the program, but that cost could be reduced significantly through selection of a random sample of projects for which monitoring is required. The lack of monitoring for erosion from non-project areas is also a limiting factor.

Enforcement actions are based not on water quality monitoring but on the presence of off-site sediment that can be obviously traced to a particular project. Program evaluators used that same type of indicator to estimate program effectiveness, classifying each of the 128 field sites as having either no failure (no apparent off-site sedimentation), a minor failure, or a major failure, in which the dividing line between a major and minor failure was based on a somewhat arbitrary amount of sediment deposited off-site but near the project. One of the primary measures of program effectiveness was the proportion of projects that fell into each of these categories.

If the data were available, a better measure of effectiveness would have been rates of erosion leaving project sites with and without the program. Instead, while reporting that one-third of the sites they visited had experienced major failures (VI-46), the evaluators observed that "even on sites judged to have incurred a major loss of sediment, the sedimentation rate and total accumulation can be inferred to be much less than was observed at the beginning of the program.... Clearly, the sediment control program in North Carolina is having a profound effect in reducing off-site sedimentation" (VI-70). Indicators used for evaluation, however, would not lead one to that conclusion.

Some attempts were made to link program activities to other measures of outputs, but they met with very little success. In one exploratory exercise, an attempt was made to relate fishing yield by weight to existence of the sediment control program. For each of seven locations in the state, yields were regressed on time and a dummy variable representing the existence of the program. Given the simplicity of the models, it is not surprising that they performed poorly in predicting yields. These failures can be attributed to the masking effect of other sources of sediment (primarily from agriculture), failure to account for the fate of sediment in streams, an incomplete understanding of the relationship between fish populations and sediment, and exclusion of any variable to account for fishing pressure.

An effort was also made to estimate the loss of storage due to sedimentation in water supply reservoirs. That work met with difficulties due to limited information about several factors—the attribution of sediment loads to sources, rates of loss of storage, and reliable costs for replacing or dredging lost storage. The same impediments were encountered when an effort was made to estimate loss of storage in flood control reservoirs.

No effort has been made by DEHNR or program evaluators to estimate improvements in water quality or improvements to aquatic ecosystems that can be attributed to the program. Although sediment is consistently cited by DEHNR as the leading cause of degraded water quality in North Carolina, monitoring of stream quality for sediment is quite limited. As of 1990 DEHNR operated 378 water quality stations, one for approximately every 100 miles of stream. Quality was sampled quarterly, and sediment was not measured directly in those surveys. The monitoring program is clearly not designed to detect trends, and DEHNR has warned that because of shifting guidelines for biennial water quality assessments, it is scientifically inappropriate to compare past reports to determine if water quality is getting better or worse (DEHNR 1990:14).

Estimation of Benefits

When the beneficial effects of a program measured in physical, chemical, and biological terms cannot be closely linked to program initiatives, as is the case with the sediment control program in North Carolina, the estimation of economic benefits becomes even more problematic. Furthermore, if economic benefits are estimated, the validity of those estimates is brought into question.

Difficulties encountered in efforts to link the sediment control program to fish yields and loss of storage in reservoirs have been discussed previously. It could be argued that poor results obtained from these efforts could be overcome by better designed models and additional data collection. That might be true if the evaluation process is given sufficient financial resources and is extended over a sufficient length of time in which meaningful water quality and ecosystem responses can be measured.

Despite these limitations, two attempts were made to gain information about benefits of the program in North Carolina, namely a contingent valuation survey and an attitudinal survey. In the contingent valuation survey, 1,200 persons were asked how much they would pay for the state to administer, monitor, and enforce sediment control measures. They were given no information about what beneficial outcomes would follow or about what levels of uncertainty might be associated with estimates of those outcomes. Results of the survey led evaluators to conclude that North Carolinians would be willing to pay substantially more for the program than they actually pay, resulting in a very favorable conclusion about the economic worth of the program. These results are subject to question, however. Would responses to the survey have been different had the respondents been given additional information about the outcomes, including uncertainty about those outcomes? Would responses to the survey have been different if respondents had been told that even if the program was well administered, improvements in water quality could not be detected?

The attitudinal survey was conducted to solicit information from a variety of organizations, including those with regulatory authority, professional organizations, trade associations, and environmental groups. A prepared questionnaire was used to conduct telephone and personal interviews with fifty-seven representatives from thirty-seven organizations.

Among several other questions, each of these representatives was asked to judge the balance between benefits and costs of the program. Most specifically, each was asked to rate the balance as being 1: very beneficial; 2: somewhat beneficial; 3: balanced; 4: somewhat costly; or 5: very costly. When asked this question, representatives were given neither definitions of benefits and costs nor factual information about how the program was performing. Without more structured questions, respondents were given considerable latitude in forming their responses. Given the nature of these questions, it is not surprising that responses reflected advocacy positions of groups represented by respondents. Representatives of several groups voted either 1 or 2 as follows: soil conservation professionals, 100 percent; environmental organizations, 95 percent; Commission members, 76 percent; the regulated community, 33 percent. While opinion surveys about costs and benefits are important, they can hardly be treated as substitutes for estimates of actual costs and benefits.

Water Supply Protection Program

Several efforts at the state level in the 1980s to get local governments to adopt voluntary programs to protect water supply watersheds from development met with limited success. A bill to establish the Legislative Watershed Protection Study Committee to examine the need for a statewide watershed protection program passed in 1988, and that committee held hearings and drafted the bill that in 1989 became House Bill 156, the Water Supply Watershed Classification and Protection Act (WSWCPA). The act establishes a mandatory local program of watershed protection that is consistent with statewide minimum performance standards to be set by the Environmental Management Commission (EMC). The EMC was directed by the act to adopt watershed classifications and to assign to each water supply watershed in the state an appropriate classification.

The EMC held public hearings on proposed classifications and standards in 1990. After a series of lightly attended educational meetings and public hearings at which most participants expressed support for the standards, the EMC voted to adopt a slightly modified version of these rules. In May 1991, however, developers of one large project that would be affected by the rules requested that the EMC invalidate the rules because the EMC failed to adhere to required administrative procedures. While the EMC refused to invalidate any portion of its standards, it did agree to place the entire set of classifications and standards before public hearings again in August 1991. The second set of hearings was heavily attended, with environmentalists accusing developers of packing the hearings. The final version of the standards was adopted by the EMC in February 1992.

Classifications and Standards

As implemented, the WSWCPA might be characterized as a nondegradation policy similar to those in the federal Clean Air Act and the Clean Water Act. The classifications adopted by the EMC are based on existing levels of development in watersheds, and nothing in the regulations is designed to mitigate existing conditions. The same water quality standards must be met in all classes, but the performance-based standards vary with existing levels of development. Uninhabited watersheds in public ownership were assigned to Class WS-I. Watersheds not subject to much urban development and without known discharges were classified WS-II. Standards for WS-III were designed to hold the line in moderately developed watersheds with only domestic and nonprocess industrial discharges, and WS-IV watersheds were those in which substantial urban development had already occurred. In addition to restrictions on wastewater discharges, nonpoint source control measures include vegetative buffer areas along streams and reservoirs, restrictions on activities and hazardous material use, and development density and impervious surface area limitations either without engineered stormwater control devices (low-density option) or with engineered devices (high-density option). Each watershed includes two areas: a critical area within a half-mile of the water supply and a noncritical area, where controls can be less stringent.

A benefit-cost framework for evaluating this program, as sketched in figure 9.2, could be quite similar to that for the erosion and sedimentation control program. Reductions in intensities of development, preservation of natural buffers, stormwater controls, and resources to manage the program may be viewed as the inputs, each of which has associated costs. Outputs of the program include reductions in rates at which pollutant loads are generated from urban activities, reductions in the flux of pollutants entering streams in affected watersheds, improved quality of receiving streams and reservoirs, and reduced risks to the health of benefited customers.

```
Cost ---------- Linkage ------------ Benefits
```

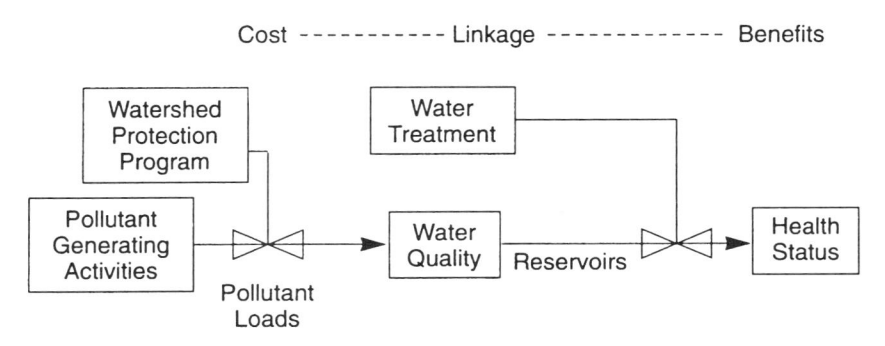

[Watershed Protection Program] → [Water Treatment]
[Pollutant Generating Activities] —Pollutant Loads→ [Water Quality] [Reservoirs] → [Health Status]

Figure 9.2. Framework for Benefit-Cost Analysis of Evaluation of Water Supply Watersheds Protection Program

Estimation of Cost

At least two questions about economic impact arose during debate on these regulations: Would they pose a significant constraint on the supply of land that is available for new development? and, How would they affect the cost of development and, ultimately, the price of housing? Debate on these issues led to an analysis of selected economic aspects of the regulations, published as part of the public information packet distributed in advance of the hearings in 1991. That analysis included results of an assessment by DEHNR's Division of Planning and Assessment, based largely on a review of the literature; an analysis by the Water Resources Research Institute (WRRI); and an analysis by the staff of the Division of Environmental Management, based largely on a prior study of beachfront property in North Carolina and a study by the Chesapeake Bay Critical Area Commission.

Land Availability The WRRI report (Moreau, Watts, Purdy, and Grey 1992) showed that in very few locations would the regulations have a significant effect on land supply for residential housing. Gross development densities were estimated using a geographic information system to capture 1990 U.S. census counts of housing within each of the 359 watersheds in classes WS-II, WS-III, and WS-IV (WS-I watersheds are virtually uninhabited). Only 9 of the 359 watersheds in these classes had gross densities in 1990 as high as one unit per acre, and those watersheds covered only 30.4 square miles, less than 0.3 of 1 percent of land in classified watersheds and less than 0.1 of 1 percent of land in the state. Even with generous allowances for publicly owned land and other unbuildable areas, the supply of land for residential development is hardly affected.

Cost of Development The WRRI report also made an assessment of the impact of the rules on land requirements by comparing the densities at which residential subdivisions were developed in the five years before the rules with the densities specified in the rules. At least two indicators of impact are readily measurable: the percentage of developments that would not be affected by the rules and the average percentage increase in land requirements to make development practices consistent with the rules.

Estimates of these quantities were made possible from an analysis of the land consumption relative frequency curve constructed from a sample of 65 subdivisions, all developed since 1985 within water supply watersheds. No development in the sample had a density of 0.25 acres per housing unit (a/hu), 10 percent of the subdivisions consumed less than 0.43 a/hu, and 25 percent consumed less than 0.53 a/hu. The median consumption in these developments was 0.82 a/hu. Under the assumption that the sample is representative of development practices in unregulated watersheds, the curve was used to estimate the percentage of developments that would satisfy the rules. Results are given in table 9.1. That same relative frequency curve of land consumption was used to estimate the average increase in land requirements for subdivisions under the new regulations in comparison with pre-rule development practices. Results of these computations are also given in table 9.1.

Under the assumption that regulations would not affect prices of undevel-

Table 9.1. Compliance Rates and Additional Land Required for Compliance by a Sample of Subdivisions Built Less than Five Years before the Adoption of Regulations

Class	Percentage of Subdivisions That Would Satisfy Rules		Percentage Increase in Land Requirements	
	Without Stormwater Control	With Stormwater Control	Without Stormwater Control	With Stormwater Control
WS-II—Critical	9.0	81.0	183.0	8.0
WS-II	37.0	89.0	52.0	3.0
WS-III—Critical	37.0	89.0	52.0	3.0
WS-III	81.0	99.0	8.0	0.0
WS-IV—Critical	81.0	99.0	8.0	0.0
WS-IV—Protected	81.0	100.0	8.0	0.0
All	72.0	97.5	18.0	1.0

Source: Moreau, Watts, Purdy, and Grey 1992.

Note: Percentages of subdivisions that would satisfy rules for the All class were calculated by weighting the percentages within each category by the relative sizes (land area) of the categories. Percentage increases in land requirements were calculated for each classification, and those values were weighted by relative sizes of the categories (as measured by land areas) to calculate the All class percentages.

oped land, a rough approximation of effects of the regulations on housing costs can be obtained by changing raw land requirements and holding all other factors constant. Tax assessment data for subdivisions in the sample indicated that the value of developed lots accounts for 10 to 20 percent of the total housing value. If undeveloped land accounts for no more than 50 percent of developed land areas, undeveloped land would account for somewhere in the range of 5 to 10 percent of housing costs. Under those conditions, the rules would cause a rise of 0.5 to 0.9 percent in affected housing units without stormwater controls. There would be very little effect of density regulations on cost in areas where stormwater regulations are adopted. There, installing stormwater controls would be the only cost.

Linking Inputs and Outputs

Relating these inputs and their costs to program outputs is conceptually straightforward—lower development densities generate smaller pollution loads, stormwater controls remove pollutants that would otherwise enter streams, lower pollution loads mean better water quality and less sediment in reservoirs, better water quality means less exposure to harmful substances, and less exposure means better health. Quantification of these linkages is far more difficult, however, because there are even more uncertainties than in the sediment control program. To uncertainties about rates of pollutant generation, behavior of stormwater controls, and fate and transport in streams and reservoirs are added the difficulties of identifying all potentially harmful substances and organisms, accurately portraying the extent to which consumers would be exposed to those substances, and estimating the incidence of diseases from known exposures.

Although not undertaken in direct response to the WSWCPA, a series of watershed management studies by Camp, Dresser, and McKee, Inc. (1988, 1989) have attempted to quantify the relationship between inputs for similar locally instituted programs and reservoir quality. Even though these efforts represent reasonable attempts to predict reservoir quality under a variety of scenarios of future development, they are admittedly simplistic representations of complex processes. They are based on average export coefficients, simple assumptions about delivery and transport in streams, and very simple models of reservoir behavior. The availability of local data also limited the level of certainty with which these calculations were made. Only modest efforts were made to present levels of uncertainty surrounding these calculations.

Results of these calculations showed some impacts on water quality in reservoirs, but they failed to capture many of the kinds of impacts that are known to have occurred in watersheds that are subjected to substantial urbanization and animal agriculture. In particular, they fail to address any risks of pathogenic organisms, particularly those associated with giardia and cryptosporidium that gained such widespread attention in the early 1990s. The mass balance approach taken in these studies does not address potential spills of hazardous materials and animal wastes that pose threats that may be located on or transported across these watersheds. No attempt was made to estimate changes in health status of exposed populations to changes in water quality.

Estimation of Benefits

Estimation of economic values associated with water quality improvements or reduced health risks resulting from the WSWCPA were not possible because estimates of those outputs were never made. At least two factors contributed to the failure to extend the analysis far enough to make such estimates. One is the lack of government staff members' experience with making estimates of economic benefits for any kind of regulatory program, especially one involving life and health. The second factor is related to the first. Many elected and appointed decision makers and a large segment of the public would be unwilling to accept BCA as the primary basis for evaluating environmental, health, and safety programs.

Even if evaluators had attempted to extend the evaluation to include estimation of economic benefits, they would have had to overcome other obstacles. Prediction of water quality improvements resulting from the program would have to go beyond even that which was provided in these studies. The WSWCPA is primarily intended to protect public health, and to calculate economic benefits associated with the program, specific kinds of health risks (particular heavy metals, carcinogens, and pathogens) would have had to be identified. Lead and zinc were the only specific threats to health included. Additional steps would have been required to estimate concentrations of specific substances and organisms at points of use. Effects of water treatment processes and duration in water distribution systems would also have to be taken into account.

If the best available information had been used to calculate changes in concentrations of specific contaminants at points of use and if that information had been used to calculate changes in morbidity and mortality, the assignment of benefits using currently available literature would have been even more difficult. Currently available estimates of economic benefits for improving water quality have not addressed the specific question of willingness to pay for improving the quality of drinking water. In one of the best-known studies, Richard T. Carson and Robert C. Mitchell (1993) estimated individuals' willingness to pay for improving water quality from nonboatable to boatable, from boatable to fishable, and from fishable to swimmable, but they did not address the question of drinking water. W. Kip Viscusi's (1993) review of the literature on estimation of the value of risks to life and health did not include any studies of willingness to pay to improve the quality of drinking water.

Further, predictions of water quality improvements are highly uncertain. Predictions of morbidity and mortality when water quality is known are also highly uncertain. Even when actual water quality conditions are known and health risks have been estimated, estimates of willingness to pay to reduce those risks are subject to considerable uncertainty.

If uncertainties about economic benefits are combined with those about the health risks on which benefits are based and, in turn, if those uncertainties are combined with those about water quality improvements on which health risks are based, error bounds on economic benefits will be quite large indeed. Although the question of how much uncertainty is too much is not easy to answer, the utility of benefit estimation for making decisions about the worthiness of programs to protect drinking water quality surely declines as the level of uncertainty rises.

Valuation of willingness to pay for improved drinking water quality could face special difficulties because of these uncertainties. W. Kip Viscusi, Wesley A. Magat, and Joel Huber (1991) pointed out that in many important instances even the best informed experts do not have perfect information about risks. Exposure levels may be uncertain, individuals have different levels of sensitivity to known exposures, and people are often uncertain about the underlying science. When individuals are asked to make choices based on ambiguous estimates of risks, their decisions are quite often different from choices they make when risks are known. Viscusi's own surveys suggested that individuals are averse to ambiguous information about risk, and the manner in which ambiguous information is presented to survey participants can affect how they value that information.

Conclusions

Unfortunately, the two cases I have reviewed in this chapter are marked by significant ambiguities about physical, chemical, and biological outputs and the risks they pose to environmental resources and human health. Anyone should be cautious about drawing general conclusions from limited experience with only two cases, but the evidence found from these investigations is at least suggestive of what one might expect to find in other cases in which efforts are made to use formal BCA models to evaluate environmental programs.

First, estimating costs of environmental and health programs is no trivial task, but reasonable estimates of costs can be made at relatively modest expense. Expenditure data for program administration is generally readily available. Reasonable estimates of direct effects of regulations on compliance costs were possible in these two instances, one by a statistical sample of actual projects in which regulations were applied and the other by an analysis of how regulations would affect the cost of similar developments not affected by the regulations. Estimation of indirect affects of regulations on selling prices of new and existing residences were not made and would have required substantially greater effort.

Second, in both cases estimating improvements to water quality was difficult. For the sediment control program, prediction of sediment runoff from construction sites with and without regulations was relatively straightforward, but predicting the transport and fate of sediments in streams and its impact on ecosystems was much more problematic. The domination of sediment loads from agriculture would tend to mask effects of controls on construction sites when sediment loads are measured at downstream locations. Predicting effects of watershed regulations on water quality in downstream reservoirs was hampered by the absence of reliable relationships between pollutant loads and particular types of residential and agricultural operations on watersheds, the probabilistic nature of many contaminating events, and reliance on overly simplistic water quality models.

Finally, estimation of economic benefits was stymied, in large part, by the absence of reliable estimates of environmental and health risks. For the sediment control case, evaluators attempted to get around that problem by surveying experts about their opinions of the relative magnitudes of benefits and costs. Because these judgments were elicited from people who were given very little information about actual costs and accomplishments, it was not surprising to find that respondents reflected advocacy positions of the groups they represented. No efforts were made to estimate economic benefits for the water supply watershed protection program. If efforts had been made to do so, evaluators would have been confronted with considerable ambiguity about reductions in risk to which individuals would be exposed.

The primary conclusion to be drawn from these cases is that use of formal BCA and risk analysis should not be attempted unless reasonably reliable predictions of changes in physical, chemical, and biological variables are possible (for evaluability assessment, see Rich in this volume). For programs in which costs are acceptably small and public perception of benefits are sufficient to justify expenditures, formal analysis may be unnecessary. The sediment control program appears to fall in that category. On the other hand, if costs are sufficiently large to justify a more formal analysis, evaluators must be provided sufficient funds to predict outputs with a degree of certainty that is sufficient to make meaningful estimates of risks and economic benefits. Without a level of certainty sufficient to support reliable estimates of risks and associated benefits, multiple objective evaluation similar to that developed for the WRC's "Principles and Standards" and largely preserved under *Principles and Guidelines* remain an appropriate framework for analysis. In that framework, beneficial and adverse effects on the two objectives, economic development and environmental quality, are entered under separate accounts, thereby avoiding the necessity to reduce all relevant information to a common measure when it is not possible to do so with confidence.

REFERENCES

Bureau of the Budget. 1961. *Standards and Criteria for Formulating and Evaluating Federal Water Resources Development.* Washington, D.C.

Camp, Dresser, and McKee, Inc. 1988. *Watershed Management Study: Lake Michie and Little River Watersheds.* Prepared for Durham County, N.C.

———. 1989. *University Lake Watershed Study.* Prepared for Orange Water and Sewer Authority, Carrboro, N.C.

Carson, R. T., and R. C. Mitchell. 1993. "The Value of Clean Water: The Public's Willingness to Pay for Boatable, Fishable, and Swimmable Quality Water." *Water Resources Research* 29 (7): 2445–54.

Federal Interagency River Basin Committee. Subcommittee on Benefits and Costs. 1950. *Proposed Practices for Economic Evaluation of River Basin Projects.* Washington, D.C.: GPO.

Gardner, C. H. 1988. "Evaluation of the North Carolina Erosion and Sedimentation Control Program." A request for proposals. North Carolina Department of Environment, Health, and Natural Resources. Raleigh.

Hanley, N., and C. L. Spash. 1993. *Cost-Benefit Analysis and the Environment.* Brookfield, Vt.: Edward Elgar Publishing.

Holmes, B. H. 1972. *History of Federal Water Resources Programs and Policies, 1961–70.* U.S. Department of Agriculture Miscellaneous Publication 1379. Washington, D.C.: GPO.

Interagency Committee on Water Resources. Subcommittee on Evaluation Standards. 1958. *Proposed Practices for Economic Analysis of River Basin Projects.* Washington, D.C.: GPO.

Malcolm, H. R., A. C. Beard, R. J. Burby, E. J. Kaiser, M. I. Luger, and R. G. Paterson. 1990. *Evaluation of the North Carolina Erosion and Sedimentation Control Program.* Raleigh: Division of Land Resources, North Carolina Department of Environment, Health, and Natural Resources.

Moreau, D. H., K. Watts, R. Burdy, and J. Grey. 1992. *North Carolina's Water Supply*

Watershed Protection Act: History and Economic and Land Use Implications. Report 271. Raleigh: Water Resources Research Institute of the University of North Carolina.

North Carolina Department of Health, Environment, and Natural Resources. 1990. *Water Quality Progress in North Carolina.* Raleigh, N.C.

U.S. Water Resources Council. 1962. *Policies, Standards, and Procedures in the Formulation, Evaluation, and Review of Plans for Use and Development of Water and Related Land Resources.* 87th Cong., 2d sess. S. Doc. 97.

———. 1971. *Proposed Principles and Standards for Planning Water and Related Land Resources. Federal Register* 36 (245) (Dec. 21): 24144–24194.

———. 1973. "Establishment of Principles and Standards." *Federal Register* 38 (174) (Sept. 10): 24778–24869.

———. 1983. *Economic and Environmental Principles and Guidelines for Water and Related Land Resources Implementation Studies.* Washington, D.C.: GPO.

Viscusi, W. K. 1993. "The Value of Risks to Life and Health." *Journal of Economic Literature* 31 (Dec.): 1912–46.

Viscusi, W. K., W. A. Magat, and J. Huber. 1991. "Communication of Ambiguous Risk Information." *Theory and Decision* 31 (1): 158–73.

10

Endogenous Risk and Environmental Program Evaluation

THOMAS D. CROCKER AND JASON F. SHOGREN

Risk is exogenous, beyond the control of people. This view dominates the risk assessment–risk management program evaluation studies used to brace environmental policy. The view is inherent in the assessment-management bifurcation sanctified by the National Academy of Sciences (1983) and is common in scientific and policy discussions about environmental risk (Crocker 1984; Lichtenberg and Zilberman 1988; North and Yosie 1987; Ruckelshaus 1984). It has had a profound influence on how natural scientists, among others, describe risk and on how economists evaluate environmental programs. Since the mid-1980s, the U.S. Environmental Protection Agency has used this bifurcation to try to establish a risk-based approach to environmental management. In this chapter we argue that although this management effort is commendable, the bifurcation on which it is based can seriously mislead policymakers.

Intuition and everyday evidence affirm that people do try to influence the likelihood or the severity of an undesirable event. People invest significant resources in attempts to increase the probability that good things will happen and that bad things will not. If the unfavorable is likely to occur, they try to lessen its impact. Consider your own behavior toward risk. You may invest in a water filter, buy a membership to a health club, jog, eat food low in fat and high in fiber, or apply sunscreen. Each choice alters the risk you face to your health and welfare. How people actively alter their everyday risks and construct opportunities to do so depends on both their attitudes toward risk and their technology to reduce risk. Risk, in fact, is endogenous.

Programs to evaluate environmental risk reductions must systematically incorporate the implications of endogenous risk. Otherwise, Richard Zeckhauser and W. Kip Viscusi's plea for more "systematic strategies for assessing and responding to risk" will fall short (1990:563). Recognition of endogenous risk opens the door to addressing interdependencies in the physical and economic world that have not previously been treated in a consistent, effective manner. In this chapter we explore several key interdependencies in the theory of endogenous risk. We suggest potential avenues to incorporate the interdependencies into evaluations of risk reduction programs.

There are at least three critical interdependencies that are desirable to incorporate into future evaluations of environmental programs. First, endogenous risk implies that the current risk assessment–risk management bifurcation has promoted misleading research exercises in the natural sciences and the risk management disciplines. Risk assessment exercises often will be misspecified unless they simultaneously include both physical and economic manifestations of environmental change. Dismissal of the economic decision-making processes that influence risk exposures will result in underestimating benefits of risk reduction. Second, endogenous risk makes vivid the interdependence between private and collective risk-reduction mechanisms. The current singular focus on collective reduction will result in the undervaluation of societal preference for risk reduction, thereby biasing downward the risk-benefit evaluations central to implementing environmental programs. Third, endogenous risk implies an interdependence among private parties or governments who self-protect. Self-protection may not resolve risk; it may simply transfer the risk through space to another locale or through time to another generation. Evaluators who treat risk as exogenous will promote excessive expenditures on self-protection.

Self-Protection and Environmental Risk

Since the early 1970s an increasing number of scientists have recognized that risk is endogenous. Psychologists concede that individuals perceive substantial control over uncertain events. P. Stallen and A. Tomas, for example, write that "the individual is not so much concerned with estimating uncertain parameters of a physical or material system as he is with estimating the uncertainty involved in his exposure to the threatening event and in opportunities to *influence* or *control* his exposure" (1984:284; emphasis added). Chauncy Starr (1972), an engineer, makes much of the difference between voluntary (endogenous) and involuntary (exogenous) exposures to risk. Indeed, outside the field of economics, discussions of risk typically consider "measures that modify events or reduce the vulnerability to loss" (Kates 1978:38). Examples of private self-protection abound. People move or reduce physical activities when air pollution becomes intolerable. They buy bottled water if they suspect their primary water sources are polluted. They chelate children who have high blood lead concentrations. They apply sunscreen to protect their skin from UV radiation. Other organisms also self-protect. Plants employ variations, premature abscissions, resource sinks, inhibitor proteins, and immune bodies such as granules, fibers, membrane fragments, and viruses. Similarly, animals self-protect by withdrawing to cover, using distractions and diversions, feigning death, releasing chemical repellents, and sending warning signals (Shogren and Crocker 1990).

At the policy level, the success of collective mandates to promote safety often depends upon individual choices. Use of auto seat belts reduces both the probability and the severity of injury, but their mandatory installation cannot guarantee that passengers will choose to wear them. Highway speed limits are similarly effective at reducing fatalities, but only when drivers observe them. In the workplace, initiatives involving personal protective gear (e.g., hard hats) have the same problem: they protect only those workers who wear them. In each case, individual decisions influence both the probability and the magnitude of harm.

Individuals often substitute self-protection for the protection of collectively supplied safety programs. Ian Burton, Robert Kates, and Gilbert White (1978) enumerate diverse examples, including the use of higher strength building materials in response to prospective tornado, storm surge, and earthquake hazards; more thorough weeding and crop storage in response to the prospect of drought; sand-bagging and evacuation in anticipation of floods; and improved nutrition and exercise regimens to cope with health threats. These and similar private coping strategies reduce both the individual's chance of having a threat realized and its magnitude if realized.

Jack Hirshleifer (1970) and other economists have argued that it is always possible to redefine a problem such that the state of nature is independent of human action. This position allows one, as Jean-Jacques Laffont (1980) noted, to continue working within the highly tractable framework of exogenous risk. Consider, however, a situation in which bacterial groundwater contamination threatens a household's drinking water. The probability of illness among household members can be altered if they boil the water. An analyst might define the situation as independent of the household's actions by focusing solely on groundwater contamination, over which the members of the household likely have no control. But this definition is economically irrelevant if the question is the household members' response to and damages from groundwater contamination. Those in the household are concerned about the probability of being made ill and the severity of any realized illness, and they are able to exercise some control over those events. The household members' risk is endogenous because by expending their valuable resources the people can influence probability and severity.

Isaac Ehrlich and Gary Becker (1972) define ex ante efforts to reduce probability as self-protection, s, and ex ante efforts to reduce prospective severity as self-insurance, x. The individual selects s and x to maximize the von Neumann–Morgenstern (von Neumann and Morgenstern 1944) expected utility index, EU; that is,

$$Max_{s,x} EU = [p(s)U(M - s - x) + (1 - p(s))U(M - L(x) - s - x)] \quad (10.1)$$

where p is probability, M is wealth, L is the money equivalent of realized severity, and s and x are expenditures on self-protection and self-insurance against the realization of an undesirable state. Assume $p' > 0$, $p'' < 0$, $L' < 0$, $L'' > 0$. Primes denote the relevant derivatives. The necessary conditions for the individual's optimal levels of self-protection and self-insurance are then

$$s: p'V - pU'(M - s - x) - (1 - p)U'(M - L(x) - s - x) = 0 \quad (10.2)$$

$$x: p'U(M - s - x) - (1 - p)U'(M - L(x) - x - s)(1 + L') = 0 \quad (10.3)$$

where $V = U(M - s - x) - U(M - L(x) - s - x) > 0$, $U' > 0$, and $|L'| > 1$.

Equations 10.2 and 10.3 state the standard result that an individual maximizes expected utility by equating the marginal cost of influencing probability or severity to the marginal wealth acquired. Self-protection and self-insurance activities will not be undertaken if doing so is not expected to increase net wealth. Within this framework a few researchers have explored the theoretical underpinnings and the behavioral implications of endogenous risk (Bradley and Lehman 1986; Viscusi 1979). We now discuss the key interdependencies resulting from endogenous risk.

Risk Assessment and Risk Management

In studies of risk, it has come to be understood that risk assessment quantifies environmental risk; risk management regulates environmental risk. This is explicitly stated by the National Academy of Sciences (1983) and by the U.S. Environmental Protection Agency (U.S. EPA 1984). First, the level of any given risk is quantified by the natural and biomedical sciences. The findings are then made available to the fields of law, politics, philosophy, economics, and the natural sciences so that they can be applied to the risk management process. Assessment occurs on one side, management on the other. But when endogenous risk is at issue, this strict bifurcation warrants considerable skepticism.

The term *endogenous risk* implies that observed risks are functions of natural science parameters *and* an individual's self-protection decisions. The basis by which people make decisions about risk will differ across individuals and across situations with the relative marginal productivities of their self-protection efforts, even though the properties of the natural phenomena that trigger these efforts may apply equally to everyone. It follows that attempts to assess risk levels solely in terms of natural science may be highly misleading—costly self-protection is endogenous and may, therefore, vary systematically in the observed risk data. The sources of the systematic variation are relative prices, incomes, and other economic (and social) parameters that influence any individual's self-protection decisions.

Risk management is usually interpreted to mean the *collective* provision of a more desirable state (Smith 1985). But when individuals privately self-protect, risk assessment and risk management become inseparable. From the individual's perspective, private and collective provisions are substitutes. Individual willingness to substitute one for the other will be influenced by relative productivities and relative prices. For example, a low or zero price for collective provision will lessen the proclivity to self-protect, thus increasing the level of observed risk. People buy less safe vehicles when collective law enforcement assures them that everyone drives cautiously. Similarly, if the individual's cost of access to collective provision is high, the demand for self-protection will increase, thus reducing the observed risk levels that are the stuff of risk assessment. When traveling abroad, many North Americans, for example, drink beer or wine exclusively if they are unsure about the sanitary quality of local public water supplies. To the outside observer who does not account for their self-protection, these people appear to have great intestinal fortitude. Setting the price of, the productivity of, and the terms of access to collective provision is the stuff of risk management.

In general, accurate risk assessment and effective risk management in the presence of endogenous risk require a full accounting of the input and output substitutions of producers and consumers when responding to environmental changes. The biological and physical phenomena that the risk assessor chooses to investigate define the set of opportunities relevant to the risk manager's representation of the individual's economic decision problem. An unrealistically restricted opportunity set limits the description of the producer's and consumer's alternatives for maximizing gains or minimizing losses from an environmental change. This restricted set causes economic estimates of the gains from reducing environmental risks to be understated. The set also implies that the biological and physical manifestations of environmental change depend as much upon producer and consumer decision processes as upon the biological and physical phenomena that trigger the economic reactions (Adams, Crocker, and Katz 1984). Endogenous risk implies that risk assessors must explicitly address the simultaneous nature of how economic decisions affect observed risk and how the natural science features affect economic decisions. Otherwise, evaluations of programs to alter environmental risk will be misdirected.

Incorporating the simultaneous environmental and economic system implied by endogenous risk into environmental program evaluations is achievable today. Recent advances in computer technology now allow the computer to be used as a creative research tool to integrate physical and economic processes into a unified system. An environmental economic modeling system that incorporates reciprocities between the physical and the economic can be used to simulate the impacts of alternative policy scenarios and patterns of activity. One such system is being developed for the EPA to evaluate the risk-benefit trade-offs of alternative pesticide policies. This integrated system, known as the Comprehensive Environmental and Economic Planning and Evaluation System (CEEPES), incorporates reciprocities between economic and environmental indicators of social welfare (Bouzaher et al. 1995). Although there are several steps remaining before the system is completely specified, preliminary results suggest that the approach is feasible. It has already influenced pesticide policy in the EPA by redirecting attention away from restrictions on single chemicals to policies that consider a systems approach to weed control. But to construct such systems, incentives must be provided to reward those who dare to work within the multidisciplinary environment. More effective risk evaluation programs require more interaction among disciplines whose members often regard intellectual traditions and paradigms different from their own with considerable skepticism. In a time of declining research budgets, skepticism readily shades into animosity.

Valuation of Environmental Risk

Restructuring the valuation exercises central to the benefit-cost analysis of environmental hazards is critical to a better accounting of endogenous risk in environmental program evaluations. Restructuring cuts deeper than problems of measurement. It involves acknowledging fundamental deficiencies in perspective that inhibit development of the measures needed for reliable policy guidance. Assumptions of exogenous risk displace the search for the most pertinent data, bias interpretations of available data, and create uncertainty about the effectiveness of program initiatives.

Specifically, the assumption of exogenous risk in benefit-cost analysis can lead to the undervaluation of reduced risk and the misidentification of those who value risk reductions most highly. There are several reasons for undervaluation, all involving the inability of an exogenous risk perspective to disentangle the relative values of private and collective contributions to risk reductions. When risk is considered exogenous to the individual, protection must be supplied collectively. Nonetheless, self-protection is often a viable substitute for collectively supplied protection; it can also expand an individual's opportunities to exploit personal gains from collective provision. Exercising to stay in good physical shape reduces one's susceptibility to the dangers posed by an unsanitary physical environment. This same exercise enhances one's expected longevity when collective efforts have sanitized the environment.

The valuation literature for exogenous risk characteristically assumes that the value of risk reductions declines as risk decreases (Jones-Lee, Hammerton, and Philips 1985). Empirical evidence that this marginal value actually increases is held to be a lapse from rational economic behavior. Jason Shogren and Thomas Crocker (1991b) show, however, that endogenous risk within the traditional expected utility framework can generate behavior consistent with increasing marginal valuations of risk reductions. In particular, if the marginal productivity effects of self-protection on probability differ from the effects on severity, increasing marginal valuations can occur. This result challenges the standard view that those who are at greater risk and who have greater wealth must value a given risk reduction more highly. The poor may have fewer opportunities to self-protect and their marginal productivity in those opportunities they do have may be less than for the wealthy. Calls for environmental "justice" for the poor and uneducated can thus be shown to have an analytical basis that has empirical plausibility. The result also implies that the undervaluations caused by a singular focus on collective risk reductions could increase with the degree of success gained by these collective efforts. As the marginal effectiveness of successive collective provisions declines, the relative effectiveness and therefore the value of private provision increases.

When self-protection and collective protection are perfect substitutes—equally effective in producing risk reductions—the upper bounds on the values participants attached to risk reductions are consistently associated with self-protection, as Jason Shogren (1990) observed in a series of controlled experiments; collective protection always represented the lower bounds. This implies, all else being equal, that individuals prefer self- rather than collective provisions. In these experiments, participants could substitute a single mechanism for self-protection for a single collective protection mechanism.

Individuals have numerous ways in which they can alter risk privately. In contrast, when researchers employ the concept of the value of a statistical life or limb, benefit-cost analyses of reduced environmental risks do not acknowledge the existence of multiple or even single private risk reduction mechanisms. The value of a statistical life is defined as the cost of an unidentified single death weighted by a probability of death that is uniform across individuals. But even if individuals have identical preferences, substantial differences exist in their opportunities for or costs of altering risk. The statistical life or limb approach fails to address the differences in individual risks induced by self-protection. An individual who has ready access to private risk reduction mechanisms will value collective mechanisms less than otherwise. A complete assessment of this individual's value for a given risk reduction thus requires considering willingness to pay for self-provision as well as for collective provision. In essence, by virtue of its exclusive focus on collective provision, the statistical life or limb approach undervalues environmental threats to human health and endorses economically excessive levels of environmental degradation.

The undervaluation of risk reductions induced by inattention to private provision extends as well to risks that are distributed over time. Charles Blackorby, David Donaldson, and David Maloney (1984) have demonstrated that unless one incorporates individuals' preferences toward the availability of goods and services in time into traditional benefit-cost analysis, program evaluations based on an independently derived social time discount rate will be biased (see also Keen 1990). To first calculate the time stream of net benefits and only after that to apply a discount rate to this stream disregards that this application changes the relative values the individual attaches to elements of the stream. The individual will thus reallocate consumption and investment over time so as to smooth satisfaction.

The undervaluation problem can be resolved by assessing the individual's preference for alternative risk reduction strategies. By allowing the individual to reveal whether he or she would prefer to reduce risk privately or collectively or both, or by reducing the probability or severity or both, and by evaluating how the individual times his or her mind (time preference) and minds his or her time, evaluators will have better measures of the value of risk reduction (Crocker and Shogren 1993). Nonmarket valuation techniques such as the contingent valuation method or laboratory experiments can be used further to elicit preferences for alternative risk reduction strategies (Cummings, Brookshire, and Schulze 1986). But without first understanding how people prefer to reduce risk, program recommendations will be based on potentially precise but inaccurate, incomplete information. This produces an unnecessarily restrictive policy environment in which a decision maker's prediction of the consequences of his or her programs may well be undone by individuals' actions. The decision maker's provision of a healthier environment or safer highways may simply cause people to drive faster or be less cautious about personal exposures to illness-producing agents.

Risk Abatement and Transferable Risks

Risks that are endogenous may also be transferable. The concept of transferable risk implies that the individuals protect themselves by simply transferring the risk through space to another location or through time to another generation. The physical or the utilitarian consequences of self-protection from environmental hazards are not limited to the self-protected. We are not putting old wine into new bottles. Transferable risk differs in spirit and in form from the traditional view of pollution externalities. Transferable risk is motivated by intentional behaviors, not by the simple, unintentional residuals of production. Agents select a technology that can transfer a risk; transferable risks create conflict; conflict induces strategic behavior; and well-intended program prescriptions can yield unintended consequences.

Indeed, from the materials balance perspective of Allen Kneese, Robert Ayres, and Ralph d'Arge (1970), most environmental programs do not reduce environmental problems. They do not reduce the mass of materials used or cause them to accumulate in the economy. While continuing to allow waste masses to flow into the environment, the programs simply transfer these masses through time and across space. Future generations and other jurisdictions then suffer the damages.

For example, the midwestern industrial states have reduced regional air pollution problems by building tall stacks at emitter sites. Prevailing weather patterns then transport increased proportions of emissions to the northeastern states and eastern Canada. Clearly the midwestern states have reduced their damages by adopting abatement technologies that increase air pollution damages elsewhere. After twenty years of studies, charges, and countercharges, the acid deposition issue continues to be a point of contention between the highest levels of the U.S. and Canadian governments. Other examples abound. In agriculture, pollution from other sources encourages land, fertilizer, and pesticide substitutions, which produce pollution that affects others. Large present-day use of pesticides accelerates the development of immune insect strains with which future human generations must contend. Some governments forbid the storage of toxins within their jurisdictions, thereby causing the toxins to be stored (or dumped) elsewhere.

The Des Moines Water Works in Iowa provides an especially lucid example. The Water Works built a $3.8 million nitrates removal facility to clean nitrates from the city's Des Moines River drinking water supply. The facility is the world's largest. In 1991, nitrates at the Water Works intake exceeded ten parts per million for twenty-nine days, prompting a legally imposed nitrate alert. Nitrate pollution, it is suspected, promotes stomach cancer and methemoglobinemia (the blue baby syndrome). The removal facility simply transfers this risk, however, in that once removed, the nitrates are dumped back in the Des Moines River to pose threats of stomach cancer and blue babies in downstream towns. L. D. McMullen, manager of the Water Works, notes that "the nitrate is not salable so we will just take it out of the water temporarily. We put it back into the water and someone has to worry about it downstream" (Swoboda 1992:11). McMullen thereby saves Des Moines $200,000 in nitrate disposal costs annually.

Conflict is the inevitable consequence of McMullen's approach. The set of mutually satisfactory agreements is empty under existing environmental programs and individuals purposely try to make others worse off in order to make themselves better off. Pollution recipients are made into pollution perpetrators and natural allies become adversaries. The noncooperative, unilateral use of self-protecting technologies creates environmental conflicts.

Within the simple framework of this chapter, transferable risks imply that the individual's expected utility problem in equation 10.1 must be modified to include the self-protection expenditures, S, of another individual. The first individual's maximization problem then becomes

$$Max_{s,x} EU = [p(s)U(M - s - x) + (1 - p(s, S))U(M - L(x) - s - x)] \quad (10.4)$$

while the other individual's problem is

$$Max_{s,x} EU = [p(S, s)U(M - S - X) + (1 - p(S, s))U(M - L(X) - S - X)] \quad (10.5)$$

where X is the other individual's self-insurance expenditures.

Shogren and Crocker (1991a) use the system in equations 10.4 and 10.5 to show that environmental polices that allow unilateral transfers of risk instead of encouraging cooperative resolutions will result in excessive expenditures on self-protection. In the absence of public limits to individuals' noncooperative self-protection activities, environmental risk reductions can be made prohibitively expensive. Collective strategies that encourage self-protection through risk transfers need to be reconsidered. When transfers are technically feasible, such strategies only intensify the inefficiencies inherent in the noncooperative behavior that caused the transfers in the first place. In addition, self-protection may be imperfect and incomplete, implying that another layer of risk is added to the problem (Shogren 1991).

Some environmental programs thought to be desirable within the framework of exogenous risk can be shown to accentuate the inefficiencies inherent in circumstances in which transferable risks and noncooperation prevail. For example, several eminent scholars have suggested that publicly sponsored hazard information programs often will be efficient alternatives to direct regulation of environmental hazards (National Research Council 1989; Viscusi and Magat 1987; Smith, Desvousges, and Johnson 1990). Although these authors readily grant that public information may cause individuals to self-protect, they do not address the possibility that this information may induce them to self-protect in inefficient ways. Shogren and Crocker (1991a) show that this information policy can reduce excessive levels of self-protection. The sufficient condition for this result, given unilateral self-protection, is an inelastic damage function, a function in which damages are highly unresponsive to changes in self-protection efforts.

Alternatively, policies that shift the time and space focuses of environmental hazards will prompt strenuous protection efforts on the part of recipients who have an elastic damage function. Limited empirical evidence supports the existence of an elastic damage function for environmental aesthetics when pollution levels are low and an inelastic damage function when pollution levels are high (Crocker 1985; Crocker and Shogren 1991; Smith and Desvousges 1987). Therefore, in terms of the aesthetic and health impacts of pollution, noncooperative environmental improvements could be self-defeating when pollution levels and risk are already low. Aggregate expenditures on protection may then outweigh the environmental benefits. In contrast, some pollutants, such as ambient carbon monoxide, exhibit inelastic damages at low levels and elastic damages at high levels. It follows that accurate assessments of the benefits of policies to reduce environmental hazards require precise and accurate knowledge of the responsiveness of damages to noncooperative forms of self-protection. Damage function elasticities are likely to be hazard-specific and activity-specific, as well as concentration-specific or level-specific.

Power, in the sense that some agents have more immediate, better informed, or less costly access to self-protection, can further accentuate the inefficiencies inherent in transferable risks. Power might consist of lawsuits, political lobbying, boycotts, and even the intentional generation of pollution. Better access implies a more powerful agent's self-protection is more productive than that of a rival (Dixit 1987). A successful transfer by the more powerful agent reduces the incentive of weaker agents to protect themselves from the pollution risk since their total return to protection effort declines. The threat that they pose to the more powerful agent therefore falls.

In general, in the absence of transferable risks, the economic case for social dictation of waste control technology is weak at best. With transferable risks, the regulatory case for insisting upon waste reduction technologies (waste prevention and recycling) rather than transfer technologies (waste disposal and treatment) is strong. It becomes even stronger when bits of reality like uncertainties about transferability and differences in power among agents are added to the basic stylized case. The agent who engages in prevention and recycling rather than disposal and treatment filters rather than transfers wastes. For example, the person who uses or sells recycled trash moves material flows from the free access environment to the claims of the economy, thus broadening the domain of efficient price signaling. Just as self-protection is excessive when transferable risks rule, it is too small when filtering governs. The agent who filters provides a benefit to other agents for which he or she goes unrewarded. A strong case can thus be made for subsidies to those who practice waste prevention and recycling. Conversely, economic sense requires that programs which encourage transfer be discouraged.

There are two broad ways to overcome the difficulties associated with transferable risk: legally imposed separability or institutionally sanctioned risk sharing. First, society can amend or reform the legal system to artificially impose separability. The structure of liability and negligence rules could be developed further to address transferable risk explicitly. New monitoring and enforcement schemes could be devised to detect and punish the guilty. This approach will require careful consideration of the new incentives that will be set in place and the new trade-offs that will develop. Second, if we acknowledge that critical interdependencies exist and that artificial separability may be costly to enforce, the alternative is to cooperate. While no panacea exists, societies could save considerable resources by developing risk-sharing institutions, perhaps with side payments, that foster cooperation in their environmental protection efforts. Failure to do so results in the expenditure of valuable protection resources at no gain in environmental quality.

Conclusion

An acknowledgment of the endogeneity of some environmental risks holds important ramifications for environmental program evaluations. Programs founded on an exogenous risk perspective are likely to be in error whenever endogenous risk prevails. The mistaken perspective induces errors in risk assessments, causes the benefits of risk reduction to be underestimated, and persuades those exposed to risk to spend excessively on self-protection. Society will expend scarce resources at little or no gain in environmental quality. Although endogenous risk has been implicitly discussed in terms of materials balance and energetic approaches to environmental policy, the behavioral implications have been ignored too long. In this chapter we offer a systematic framework to reconsider how we view risk reduction and the consequent implications for reform of environmental program evaluations. The key to reform is the creation of new incentives that reward those who attempt to unify risk assessment and management into an integrated system, estimate individual preferences for alternative risk reduction strategies, and foster risk-sharing institutions that explicitly address the nature of endogenous risk.

REFERENCES

Adams, R., T. Crocker, and R. Katz. 1984. "Assessing the Adequacy of Natural Science Information: A Bayesian Approach." *Review of Economics and Statistics* 66 (4): 568–75.

Blackorby, C., D. Donaldson, and D. Maloney. 1984. "Consumer Surplus and Welfare Change in a Simple Dynamic Model." *Review of Economic Studies* 51 (1): 171–76.

Bouzaher, A., R. Cabe, S. Johnson, A. Manale, and J. Shogren. 1995. "CEEPES: An Evolving System for Agroenvironmental Policy." In *Integrating Ecological and Economic Indicators,* ed. J. W. Milon and J. Shogren. New York: Praeger. 67–89.

Bradley, M., and D. Lehman. 1986. "Comparative Equilibrium versus Comparative Statics." *Canadian Journal of Economics* 19 (3): 526.

Burton, I., R. Kates, and G. White. 1978. *Environment as Hazard.* New York: Oxford University Press.

Crocker, T. 1984. "Scientific Truths and Policy Truths in Acid Deposition Research." In *Economic Perspectives in Acid Deposition Control,* ed. T. Crocker. Boston: Butterworth. 65–80.

———. 1985. "On the Value of the Condition of a Forest Stock." *Land Economics* 61 (3): 244–54.

Crocker, T., and J. Shogren. 1991. "Ex Ante Valuation of Atmospheric Visibility." *Applied Economics* 23 (1): 143–51.

———. 1993. "Dynamic Inconsistencies in Valuing Environmental Goods." *Ecological Economics* 7 (2): 230–54.

Cummings, R., D. Brookshire, and W. Schulze. 1986. *Valuing Environmental Goods: An Assessment of the Contingent Valuation Method.* Totowa, N.J.: Rowman and Allenheld.

Dixit, A. 1987. "Strategic Behavior in Contests." *American Economic Review* 77 (5): 891–98.

Ehrlich, I., and G. Becker. 1972. "Market Insurance, Self-Insurance, and Self-Protection." *Journal of Political Economy* 80 (4): 623–48.

Hirshleifer, J. 1970. *Investment, Interest, and Capital.* Englewood Cliffs, N.J.: Prentice-Hall.

Jones-Lee, M., M. Hammerton, and P. Philips. 1985. "The Value of Safety: Results of a National Sample Survey." *Economic Journal* 95 (377): 49–72.

Kates, R. 1978. *Risk Assessment of Environmental Hazard.* New York: John Wiley and Sons.

Keen. M. 1990. "Welfare Analysis and Intertemporal Substitution." *Journal of Public Economics* 42 (1): 47–56.

Kneese, A., R. Ayres, and R. d'Arge. 1970. *Economics and the Environment: A Materials Balance Approach.* Washington, D.C.: Resources for the Future.

Laffont, J.-J. 1980. *Essays in the Economics of Uncertainty.* Cambridge, Mass.: Harvard University Press.

Lichtenberg, E., and D. Zilberman. 1988. "Efficient Regulation of Environmental Health Risks." *Quarterly Journal of Economics* 103 (1): 167–78.

National Academy of Sciences. 1983. *Risk Assessment in the Federal Government: Managing the Process.* Washington, D.C.: National Academy Press.

National Research Council. 1989. *Improving Risk Communication.* Washington, D.C.: National Academy Press.

North, W., and T. Yosie. 1987. "Risk Assessment: What It Is; How It Works." *EPA Journal* 13 (9): 13–15.

Ruckelshaus, W. 1984. "Risk Assessment and Risk Management." *Risk Analysis* 4 (3): 157–62.

Shogren, J. 1990. "The Impact of Self-Protection and Self-Insurance on Individual Response to Risk." *Journal of Risk and Uncertainty* 3 (2): 191–204.

———. 1991. "Endogenous Risk and Protection Premiums." *Theory and Decision* 31 (2): 241–56.

Shogren, J., and T. Crocker. 1990. "Adaptation and the Option Value of Uncertain Environmental Resources." *Ecological Economics* 2 (3): 301–10.

———. 1991a. "Cooperative and Noncooperative Protection against Transferable and Filterable Externalities." *Environment and Resource Economics* 1 (2): 195–214.

———. 1991b. "Risk, Self-Protection, and Ex Ante Economic Value." *Journal of Environmental Economics and Management* 20 (1): 1–15.

Smith, V. 1985. "Supply Uncertainty, Option Price, and Indirect Benefit Estimation." *Land Economics* 61 (3): 303–7.

Smith V., and W. Desvousges. 1987. "An Empirical Analysis of the Economic Value of Risk Changes." *Journal of Political Economy* 95 (1): 89–111.

Smith V., W. Desvousges, and F. Johnson. 1990. "Can Public Information Programs Affect Risk Perceptions?" *Journal of Policy Analysis and Management* 9 (1): 41–59.

Stallen, P., and A. Tomas. 1984. "Psychological Aspects of Risk: The Assessment of Threat and Control." In *Technological Risk Assessment,* ed. P. Ricci, L. Sagan, and C. Whipple. The Hague: Nijhoff. 276–93.

Starr, C. 1972. "Risk Analysis." In *Perspectives on Benefit-Risk Decisionmaking.* Washington, D.C.: National Research Council. 123–50.

Swoboda, R. 1992. "Nitrate Problems." *Wallace Farmer* 63 (Spring): 11.

U.S. Environmental Protection Agency. 1984. *Federal Register, Parts II, III, V, and VI: Environmental Protection Agency.* Sept. 24.

Viscusi, W. K. 1979. "Insurance and Individual Incentives in Adaptive Contexts." *Econometrica* 47 (5): 1195–1207.
Viscusi, W., and W. Magat. 1987. *Learning about Risk.* Cambridge, Mass.: Harvard University Press.
von Neumann, J., and O. Morgenstern. 1944. *Theory of Games and Economic Behavior.* Princeton: Princeton University Press.
Zeckhauser, R., and W. K. Viscusi. 1990. "Risk with a Reason." *Science* 248 (4955): 559–64.

11

Pollution Prevention Frontiers: A Data Envelopment Simulation

KINGSLEY E. HAYNES, SAMUEL RATICK, AND

JAMES CUMMINGS-SAXTON

In a 1990 report to Congress, the U.S. Environmental Protection Agency (EPA) emphasized its strong support for pollution prevention by stating that it is the most effective way to reduce risks from pollution because the sources are reduced or entirely eliminated. Further, the EPA noted that pollution prevention offers the special advantage of synchronizing environmental protection with economic efficiency.

For pollution prevention to become a viable alternative to more traditional pollution control activities, measurement techniques must be developed that will allow the EPA to accurately assess the potential for pollution prevention, distinguish efficient plants from inefficient ones, and measure the progress of industry toward specific prevention goals. Because of a myriad of problems in setting and enforcing standards for a wide variety of hazardous pollutants, the EPA and Congress have committed to the development and implementation of policies and programs that emphasize pollution prevention strategies through the Pollution Prevention Act of 1990. A number of serious methodological issues are associated with the implementation of pollution prevention policies, not the least of which is a clear definition of "pollution prevention" and how it is to be measured (see Solomon in this volume). A number of legislative proposals have been implemented, or are under consideration, that imply both a definition of "pollution prevention" and some associated measures of success. House Bill 1457 and Senate Bill 585 represent such legislation (U.S. House Committee on Energy and Commerce 1990:13; U.S. Senate Committee on Environment and Public Works, 15). If enacted, these laws would require additional reporting of accidental release data, quantities of pollution recycled and treated, and a ratio of production in the reporting year to production in the previous year.

Many states have not waited for federal leadership and are already monitoring, encouraging, and collecting measurements of pollution prevention. Oregon's Toxic Use Reduction and Hazardous Waste Reduction Act ("Oregon" 1989) is representative of a modest approach, requiring each toxic user to submit an annual progress report noting the quantity of toxic chemicals used or generated as waste. The act's purpose is to force users to conform to future performance goals on pollution reduction, but efforts to assess pollution prevention progress using these data are limited to comparing the volumes of waste generated across facilities and over time. The Massachusetts Toxic Use Reduction Act also embraces the concept of anchoring pollutant release data to measures of economic activity and expands the concept to include materials accounting data ("Massachusetts"). Alternative institutional issues are discussed in more detail elsewhere in this volume (see Rosenbaum).

These initiatives underline the difficulties in achieving a commonly agreed-upon strategy for successful pollution prevention (Lave 1991). Although the initiatives are successful in forcing reportage of emissions, the data have shortcomings as aggregate pollution assessment measures. Such aggregate measures do not give credit for pollution prevention measures implemented before the accounting process began. Large emitters of toxic chemicals that have made and are continuing to make significant progress in pollution prevention may be penalized for their failure to normalize their pollution levels by some measure of output. Some reductions in the emission of toxic chemicals over time may be due to economic downturns and consequential reductions in output rather than to actual pollution prevention activity. Plant-specific information about technology, such as that available from mass and material balances as suggested in the Massachusetts legislation, will be needed to overcome many of these shortcomings by relating the amount of toxic chemicals emitted to the amount of material being used, produced, and combined into products (National Academy of Sciences 1990).

A throughput measure, or the ratio of pollution-related inputs to pollution-related outputs in a production process, is the basis of the pollution prevention evaluation methodology proposed by the Natural Resource Defense Council (NRDC). It is an intermediate step between general emissions data and plant specific technology assessments (Smith 1988). This measure using Toxic Release Inventory (TRI) data allows facilities to be compared to one another and to their own past performance. A number of measurement difficulties still remain, however. The proposed measure does not allow for meaningful comparisons across plants for a number of different chemicals with different hazardous attributes (e.g., toxicity level or environmental persistence). Therefore, the reduction in risk to society that would result when a plant substitutes one high-volume hazardous chemical for another at a lower volume cannot be fully captured. This measure does not yield a reasonable baseline for the potential amount of chemical pollution that can be prevented, or risks that may be reduced, with current technology and better management practices. Nor does it recognize what attainable best or better management practices are or what they can reasonably be expected to yield in pollution abatement.

Some comparison is necessary because although a few programs, such as targeted technology assistance, may be established on a facility-specific basis, attempts to allocate resources to areas of greatest importance require that facility performances be evaluated within some sort of context. Likewise, attempts to predict the effect of market incentives require special insights into the likely response patterns of various facilities.

Approach

To manage some of the problems embedded in alternative evaluation methodology (Jorgenson and Wilcoxin 1990), we propose to measure success in pollution prevention by using a new data envelopment analysis (DEA) approach (Haynes, Ratick, Bowen, and Cummings-Saxton 1993). The DEA technique, a nonparametric approach to operationalize concepts of production efficiency developed in economics management science (Shephard 1953; Farrell 1957; Charnes, Cooper, and Rhodes 1978), calculates the relative efficiency of a plant, defined as the amount of input the plant uses to produce its output. Data on both inputs and outputs can be measured in their natural units. For example, measures of weight, volume, toxicity, and dollars may be combined in the analysis. A plant's relative efficiency is obtained by calculating the ratio of the weighted sum of its outputs to the weighted sum of its inputs (Haynes, Stough, and Shroff 1990).

The DEA "efficiency" of a plant reflects the amount of pollution generated in its use of chemical inputs. The form of the DEA efficiency is similar to the ratio-based measure proposed by the NRDC. The DEA measure, however, permits simultaneous comparisons across a number of plants having a mix of pollutants and plant "output" measures. It also provides a performance measure for all plants based on the success achieved by efficient plants (leaders) and groups other plants (laggers) into sectors that are defined by one or more of the efficient plants. The results of the DEA procedure can be used to evaluate the total amount of pollution that can be prevented if all lagger plants were to become as efficient as the leaders.

Theoretical and methodological problems still must be addressed in using the DEA technique to measure success in pollution prevention activities (Stolp 1990). One of these is to characterize or combine pollutants and products in the efficiency measure. Another is to develop a means to test the validity of many of these measures with respect to their effectiveness in providing accurate and usable information about pollution prevention to inform decisions on the nature and structure of pollution prevention policies. Later studies will require in-depth analyses of plants selected as representatives of leaders and laggers on the basis of their DEA efficiency scores. That phase will be used to generate information that will help further refine the DEA measure, identify the characteristics that distinguish the industry leaders from the laggers, and delineate a framework in which information and technology may be transferred from the leaders to the laggers. Following an outline of the structural basis of the DEA application to pollution prevention frontiers, we will present a limited analysis of simulated data to demonstrate how pollution prevention performance may be compared across facilities.

Other efforts to assess pollution prevention progress at manufacturing facilities has been based on data collected via annual TRI surveys. The TRI results identify the quantities of inventoried chemicals released into the air and water and disposed of on-site or sent off-site by each surveyed firm. The collection of these data is useful for identifying the total quantities of chemicals of concern generated at manufacturing facilities and released to the environment and identifying the largest generators. These data cannot by themselves, however, provide insights into which firms are leaders or laggers in the arena of pollution prevention. This is the case even when the toxic release data are augmented by year-to-year indexes of production to take production variations into account (Ratick and Cummings-Saxton 1991). Hence our use of simulated data for demonstration purposes.

The methodology presented in this chapter appears capable of providing valid interfacility comparisons of pollution prevention performance. The methodology draws upon the economic concepts of production functions, efficiency frontiers, and DEA and shows how to provide credible results for interfacility comparisons and how these comparisons can be evaluated over a range of temporal considerations. The objective in developing this methodology and the basis for extending DEA analysis to this application is to establish a mechanism for assessing pollution-prevention status across facilities using data of the type being collected by or reasonably likely to be collected by the U.S. EPA and state agencies. Hence, the focus of our illustration is the chemical industry sector since appropriate data are most likely to be generated from there first (Brickman, Jasanoff, and Ilgen 1985; Lave and Omenn 1989).

Production Functions and Efficiency Frontiers

From an economics perspective, attributes of manufacturing activities are summarized in the form of production functions. Production functions are analytic expressions that describe how manufacturing activities convert the value inherent in resource inputs into the value represented by product outputs. Resource inputs usually are consolidated into three aggregate categories—capital (K), labor (L), and materials (M). The manufacturing activity is sometimes termed a decision-making unit (DMU) because management decisions operating at this level determine the precise mix of inputs employed and the nature and mix of product outputs (P). The production function is schematically depicted in figure 11.1.

The rationale for portraying manufacturing operations in the form of production functions is to provide an analytic construct for use in determining optimal operational configurations. The goal usually is either to maximize the value of product output for a given set of inputs or to minimize the cost of the resource inputs required to achieve a given level of product output. These are termed technical and economic efficiency, respectively.

The goal of pollution prevention is somewhat different from the traditional goals used in the production function. Nevertheless, the production function framework provides a useful perspective. In this case, preventing the generation and release of some subset of material residuals, such as toxic chemicals, is of particular interest. A revised schematic expression of the production function for breaking out the subset of chemical inputs (C) and their associated chemical residuals (CR) is given in figure 11.2.

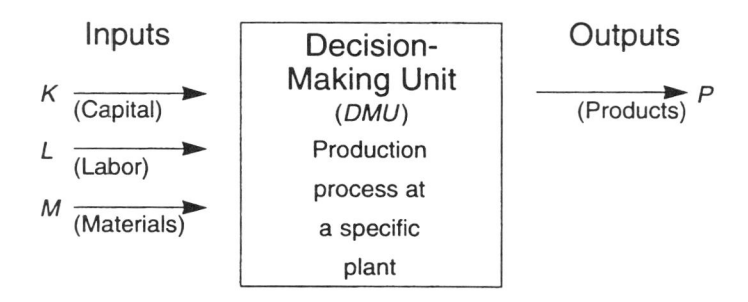

Figure 11.1. Simplified Production Function

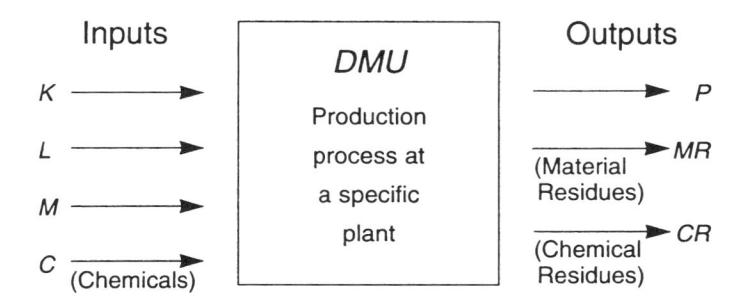

```
         Inputs          ┌──────────┐      Outputs
                         │   DMU    │
     K ─────────▶        │          │  ─────────▶ P
                         │Production│
     L ─────────▶        │process at│  ─────────▶ MR
                         │ a specific│  (Material
     M ─────────▶        │  plant   │   Residues)
                         │          │
     C ─────────▶        │          │  ─────────▶ CR
      (Chemicals)        └──────────┘   (Chemical
                                         Residues)
```

Figure 11.2. Production Function with Residual Outputs

In this schematic, material resources include all materials employed except the chemicals of concern. That is, material resources in figure 11.2 equal the material resources in figure 11.1 minus the chemical inputs (C). The material residuals (MR) are generated during the use of material resources (M). Material residuals could have been depicted in figure 11.1, but these residuals typically are overlooked in most production function analyses because they are considered of lesser importance. Also, residuals—material or chemical—may not be separately depicted, but they should be.

The goal of pollution prevention in the production function framework can be stated as minimizing the generation and release of environmental residuals for the materials or chemicals of concern, either in terms of per unit of output or in absolute terms (loading). However, what one wants to optimize is unclear in this case since undesired environmental residuals appear to accompany the desired products as joint outputs of the process. That is, resource inputs include the chemicals of concern and the outputs are joint products that include the undesired environmental residuals and the desired saleable product.

The problem can be conceptually addressed, however, by considering the nature of the production function. The amount of product generated by a given set of inputs is a function of the technology employed by that DMU. That is, as input variables K, L, M, and C are varied in the search for optimum combinations, the technology of the DMU determines the way in which the product output (P) varies. In the same way, the chosen technology of the DMU determines the nature of the environmental residuals generated from the slate of input chemicals. Thus, chemical residuals (and material residuals) are "unavoidable" subsets of the chemical (and materials) inputs when nearly all types of technology are employed. From this perspective, chemical and material residuals can be viewed as contained in the inputs to the DMU.

A convenient way to express the linkage between chemical and material residuals and chemical and material usage is to specify overall efficiencies in chemical use (*CE*) and in material use (*ME*). Chemical residuals or wastes (*CR*) are given by:

$$CR = (1 - CE)*C \qquad (11.1)$$

A similar equation holds for materials residuals (*MR*). Having established a production function for a given manufacturing operation, it is possible to identify optimal combinations of inputs *K, L, M,* and *C* for generating product outputs at minimal cost. These optimal values for *K, L, M,* and *C* form a multidimensional surface in product output–resource input space. This surface is termed the efficiency frontier for that production function. Facilities utilizing the technology described by the production function are termed efficient if they lie on the efficiency frontier and inefficient if they lie behind the frontier. The efficiency frontier as an economic concept was introduced by Michael J. Farrell (1957). An excellent review of Farrell's exposition is provided by Raymond J. Kopp (1981).

This economic approach is consistent with the engineering perspective that pollution prevention can be attained in many ways. These alternative methods can be divided into two primary types. One is to use chemicals of concern more efficiently (in an engineering sense) by instituting practices such as reducing leakage from processes, increasing recovery and recycling activities, increasing conversion yield, and changing the usage operation to minimize chemical requirements. The other is to eliminate the usage of chemicals of concern through replacing or through reconfiguring the overall process so that the chemical usage function is no longer required (Portney 1990).

The goal of pollution prevention is to minimize the chemical wastes generated per unit of chemical-using output from a given or an adjustable set of inputs. This often requires more detailed plant-specific information about technology, such as that available from mass and material balances, to calculate pollution prevention effectiveness at a given facility. Given the notation used in equation 11.1, the ratio of chemical residual wastes (*CR*) to output product (*P*) at given facility (*i*) can be expressed as the chemical usage (*C*) per unit output (*P*) times chemical residual waste (*CR*) per unit of chemicals used:

$$CR_i/P = (C_i/P_i)*(CR_i/C) \qquad (11.2)$$

The ratios of chemical usage per unit product and of product units per chemical usage have been introduced because in an engineering context the pollution prevention alternatives have differing effects on these two ratios and it is important that both be considered. Pollution prevention options that increase the efficiency of chemical usage affect the chemical waste to chemical usage ratio. They also affect the chemical input to product output ratio, but to a lesser extent, and a number of other factors may affect this ratio as well. Actions to eliminate chemical usage, on the other hand, directly affect the chemical input to product output ratio and have little effect on the chemical waste to chemical input ratio unless the efficiency of the improved usage function eliminated differs greatly from the average efficiency of chemical usage at the facility (Cummings-Saxton 1994).

Data Envelopment Analysis

DEA was developed to fill the gap between theoretical concepts of production functions, their empirical estimation, and their practical application. Classical economic and econometric studies of production functions rely on the estimation of parameters for a specified functional form (e.g., linear in the Cobb-Douglas case with constant elasticity of substitution or as a translog formulation in other cases). DEA, on the other hand, does not require an a priori specification of functional forms. In its practical usage it provides an objective methodology for obtaining efficiency frontiers from a set of data on the inputs and outputs of firms, processes, programs, or plants. It has been applied in a number of situations, e.g., site selection for the superconducting supercollider, evaluation of efficiencies in military units, and the estimation of the cost-effectiveness of breaking up AT&T, to name a few. DEA methods have also been used to assess the efficiency costs of environmental legislation (Fare and Grosskopf 1983; Fare, Grosskopf, and Pasurka 1986; Fare, Grosskopf, Lovell, and Pasurka 1989) by reformulating the DEA to account for unwanted outputs (environmental residuals) or regulated pollutants that cannot be disposed of without incurring an economic cost. Their technique allows for nonparametric estimation of the shadow prices of environmental residuals resulting from the production process. Kingsley E. Haynes, Roger R. Stough, and Homee L. Shroff (1990) have provided a set of reviews and applications in their special issue of *Computers, Environment, and Urban Systems* on DEA methodology.

Although DEA has had widespread acceptance and use, a number of criticisms have been raised on methodological and substantive issues, many of which may be ascribed to the fundamentally different perspectives of researchers and to the different purposes for which DEA is applied (Stolp 1990). It is appropriate to use a DEA structure for the development of a valid relative measure of success in pollution prevention in light of the variety and vintages of technologies employed, the level of detail and specificity of the data available, and the ways in which this measure can be applied to inform the decision-making process.

The relative efficiency (E_0) of a plant with respect to competing plants is defined to be the weighted sum of outputs divided by the weighted sum of inputs. The weights are necessary to allow for inputs and outputs to be measured in different units. Weights are not assigned a priori, but are obtained in the solution to a constrained optimization problem set up as a fractional linear program (see appendix 1 for program specification).

The optimization process itself determines the set of weights for input (V_{h0}) to output (W_{q0}) ratios that yield the maximum relative efficiency (from a minimum of 0 to a maximum of 1) for that plant. For any nonzero set of inputs and outputs a positive set of weights may be found to make $E_0 = 1$. The weights for the plant being analyzed (called virtual weights since they do not have any physical meaning) are used in the efficiency function for all other plants in the analysis. These efficiency fractions (ratios of weighted sums) form the constraints to the optimization as shown in appendix 1 and are restricted to be less than or equal to 1 for all the plants. If the optimization yields $E_0 = 1$, the plant is said to be relatively efficient. Further analysis of the full DEA output is required to determine why that plant is efficient. If $E_0 = 1$ then one or more of the efficiency fractions for the plant in the constraints were binding, i.e., equal to 1, with the weights V_{h0} and W_{q0}. To complete the computation of DEA efficiencies a separate optimization model is run for each plant in the analysis (see appendix 2 for model mechanics).

The result of the analysis is a ranking of relative efficiencies for each plant and for those plants with $E_0 < 1$ a specification of the degree to which the plant is inefficient. An indication of how inputs and/or outputs can be modified to achieve efficiency (can move to the efficiency frontier) is also provided in the output of the analysis. The objective and constraint equations given above (11.1 and 11.2) are nonlinear and require a reformulation into a linear programming model to obtain the efficiency scores and other analytical results. The method by which this reformulation is accomplished, and the interpretation of the output of the reformulated problem, are the operations research contribution made by Abraham Charnes, William W. Cooper, and Eduardo Rhodes (1978) to allow practical implementation of Farrell's theory of technical efficiency. Further, through various extensions, particularly those of Rajiv D. Banker, Abraham Charnes, and William W. Cooper (1984) and Rajiv D. Banker and Robert M. Thrall (1992), issues of returns to scale can be effectively managed.

Application

The DEA model can be adapted to assess pollution prevention. As in equation (11.2), two interacting terms are defined in the efficiency ratio for the amount of chemical residual generated per unit product output to imply different kinds of pollution prevention activities—the amount of chemical input used to produce the products (C_i/P_i) and the amount of chemical residual per unit of chemical input to the process (CR_i/C_i).

The first ratio (C_i/P_i) represents a measure of toxic use reduction through a decrease in the quantity of chemicals of concern. This can be analyzed in the DEA framework by characterizing P_{hi} as the amount of output chemical of type h used by plant i and I_{qi} as the amount of chemical of concern q used at plant i to produce its output. Both ratios may be used together in another possible formulation of the DEA model by characterizing outputs in the same way and inputs as both the chemicals of concern used as facility inputs and, as another vector of inputs, the chemical residuals of concern. Still another would be to characterize chemical residual wastes (CR_i) as inputs (I_{qi}) and the associated products (P_i) as outputs (P_{hi}).

The second ratio (CR_i/C_i) represents the efficiency measure suggested by the NRDC, i.e., the efficiency of chemical use. To expand this into the DEA framework and to allow comparisons across plants with multiple chemical wastes and multiple chemical inputs, P_{hi} can be defined as the amount of chemical of concern q used at plant i and I_{qi} as the types of related chemical residuals produced in its use. An initial assessment of this approach has been explored in a pilot study involving New Jersey chemical plants (Ratick and Cummings-Saxton 1991). In spite of severe data quality problems, the general approach described here appears feasible (Haynes, Ratick, and Cummings-Saxton 1994).

Although at this level of conceptualization these changes may appear straightforward, a number of significant methodological issues must be resolved. One is how chemical inputs or chemical residuals can be characterized from the available materials accounting data and which of these characterizations will be most effective and informative to the decision-making process. For example, the quantity brought on site, quantity shipped, quantity produced, quantity consumed, and change in inventory are available from some state data for a number of listed chemicals. For example, New Jersey requires this information in an addendum report to the TRI database for section 313 chemicals. More serious problems arise in the definition of "product," in which physical and nominal units may be used, and in the joint allocation of "inputs" to produce outputs. These issues are the subject of a number of ongoing studies at the

University of Massachusetts at Lowell (1991) and those sponsored by the EPA's Cincinnati Engineering Laboratory (U.S. EPA 1991b). The ability to consider multiple definitions of pollutants and products in the DEA may provide stability and robustness to any resulting measures and can help overcome some of these problems. This is a principal research question yet to be addressed.

Feasibility Assessment

The DEA-based pollution prevention frontiers model described above needs to be illustrated and assessed from a feasibility perspective. This section provides a simplified application with simulated data and compares the results of a DEA approach to that of the TRI ratio alternative.

Using two pollutants, data from thirty production facilities (plants) have been simulated for two time periods. For comparison purposes, t_0, the initial time period, is designated the baseline. The next time period is examined from static (t_1) and dynamic (t_1') perspectives. During the t_0 and t_1 periods the pollution prevention frontier is held static while during the t_1' period technological change is incorporated, which allows for dynamic adjustment of the pollution prevention frontier. During both the static and dynamic analysis average growth rates across the facilities is simulated at 10 percent. However, significant variation in growth has been incorporated, with one plant experiencing an output expansion rate of 47 percent and another plant experiencing an output decline of 41 percent. Distinctions between the static and dynamic frontiers are introduced for illustrative purposes only. In reality static and dynamic adjustments would occur simultaneously during any time period.

The pollution levels of the thirty plants vary across pollutants but have been divided into three classes (high, medium, and low) of pollution levels, with one-third of the plants falling into each category (see figure 11.3). In figure 11.4 the frontier facets A, B, C, D, and E, are designated by plants 5, 2, 1, 11, 16, and 29, which represent the facilities with the lowest combined pollution outputs in the initial, or baseline, time period (t_0). Holding this original low pollution output frontier static allows us to examine changes in pollution levels due to changes in total output production levels. In figure 11.5 it is clear that plants 18, 23, and 30 improve a little while plants 9, 12, 15, and 21 remain unchanged in pollution reduction. However in reality during this period the pollution frontier itself becomes dynamic due to competitive adjustments (see figure 11.6). In figure 11.6 the pollution prevention frontier is now defined by some of the old pollution prevention leaders (16, 29), some new ones (6, 11), as well as older leaders that have improved (1, 5). These plants (5, 6, 1, 11, 16,

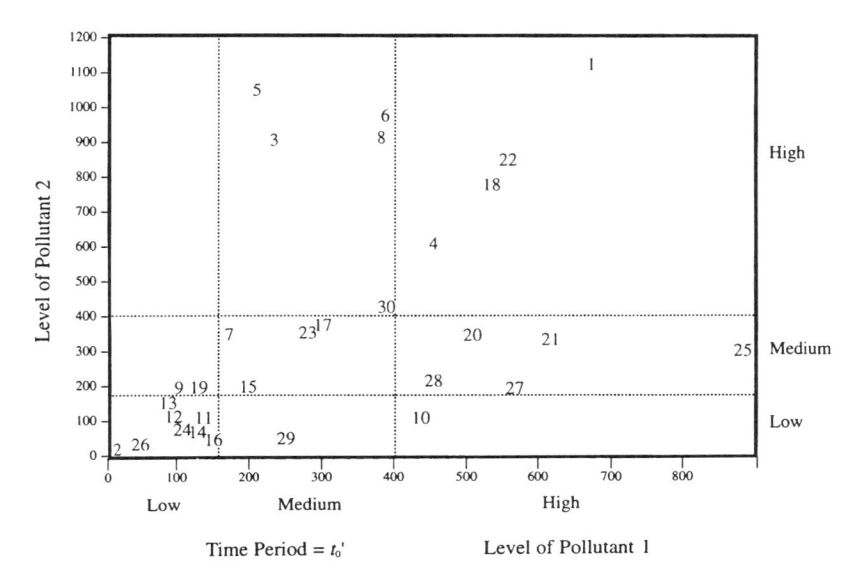

Figure 11.3. Pollutant Groupings (high-low matrix)

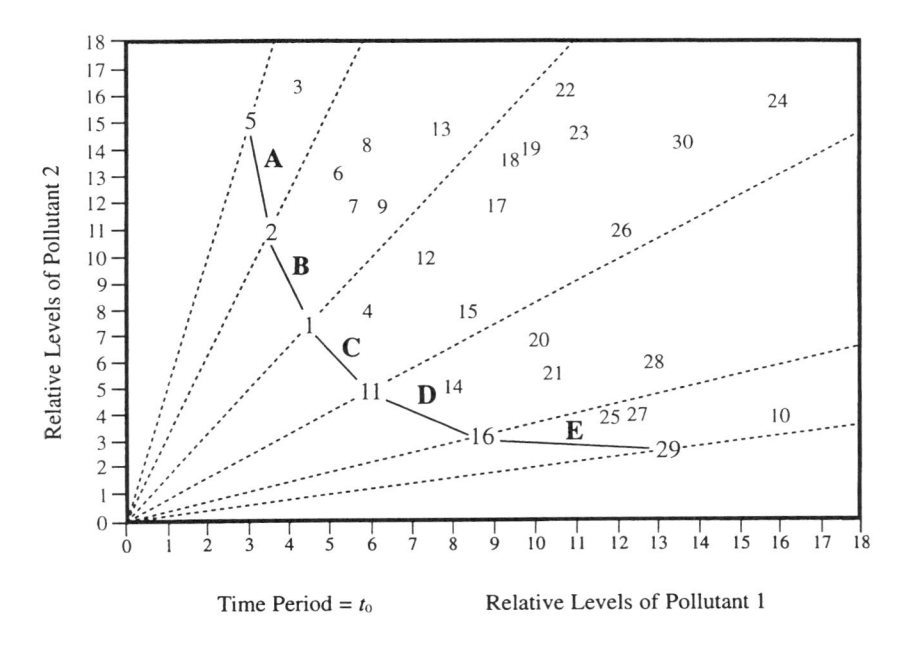

Figure 11.4. Pollution Prevention Frontiers

282   *Pollution Prevention Frontiers*

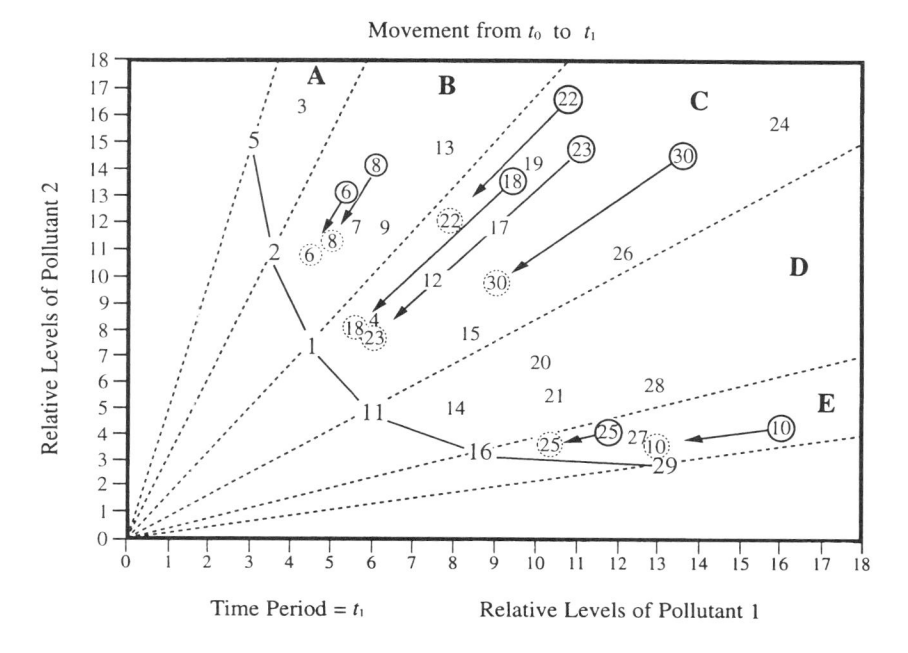

Figure 11.5. Movement toward the Pollution Prevention Frontier

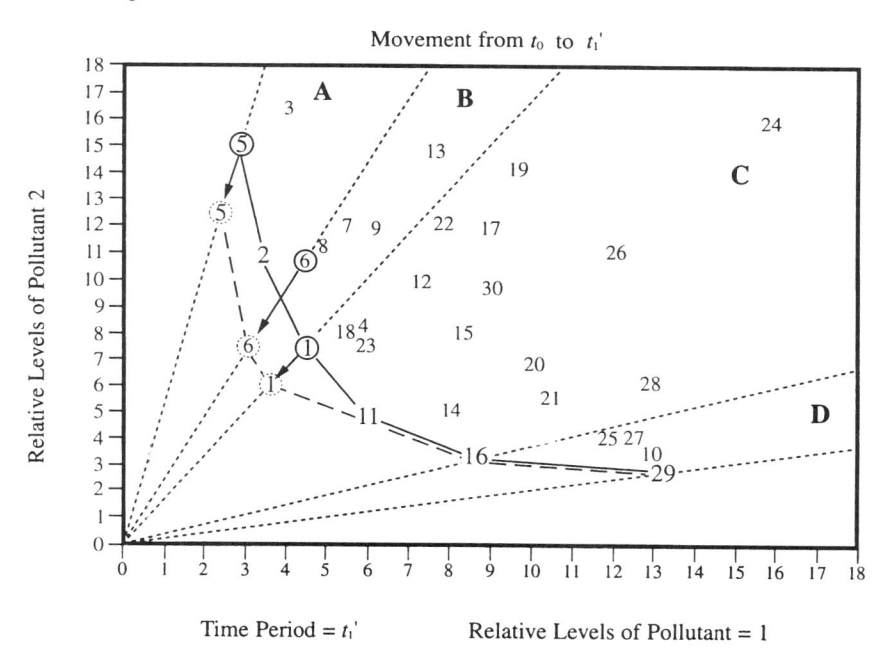

Figure 11.6. Inward Movement of the Pollution Prevention Frontier

29) now define the new frontier that determines pollution prevention leadership. Integrating the full dynamic of the adjustment process provides a picture of the adjustment of pollution prevention levels throughout our system of thirty plants. These adjustments are summarized in table 11.1.

To illustrate the advantage of the DEA pollution prevention frontier over the TRI system of dividing the second period's pollution levels into the initial period's pollution levels, the two approaches are compared for this simulated data set. The output for four plants are given in table 11.2. In this four-plant comparison total output growth varies from a 23 percent growth to a 21 percent decline and pollution prevention frontier efficiency varies from a leadership position of 100 percent relative efficiency to a 71.4 percent relative efficiency.

Table 11.1. Summary of Plant Rankings

Plant	Overall Efficiency (t_0)	Plant Rank (t_0)	Overall Efficiency (t_1)	Plant Rank (t_1)	Overall Efficiency (t_1')	Plant Rank (t_1')
1	100.0%	1	100.0	1	100.0%	1
2	100.0	1	100.0	1	81.4	10
3	77.9	10	77.9	16	64.7	19
4	83.3	8	83.3	14	74.5	14
5	100.0	1	100.0	1	100.0	1
6	73.6	12	91.2	9	100.0	1
7	73.6	13	73.6	17	58.8	22
8	66.3	19	84.8	13	65.8	17
9	68.4	16	68.4	19	56.4	24
10	73.0	14	93.8	7	93.8	7
11	100.0	1	100.0	1	99.8	6
12	67.4	18	67.4	21	60.0	21
13	55.6	22	55.6	26	45.5	28
14	86.0	7	86.0	12	85.9	9
15	68.2	17	68.2	20	66.0	16
16	100.0	1	100.0	1	100.0	1
17	55.2	23	55.2	27	49.5	26
18	50.8	24	90.4	10	79.2	12
19	49.1	25	49.1	28	43.1	29
20	65.7	20	65.7	22	65.6	18
21	71.4	15	71.4	18	71.4	15
22	43.4	28	60.8	25	52.8	25
23	44.8	27	90.2	11	80.8	11
24	35.2	30	35.2	30	33.6	30
25	79.0	9	92.5	8	92.5	8
26	47.6	26	47.6	29	46.5	27
27	76.8	11	76.8	16	76.8	13
28	61.1	21	61.1	24	61.1	20
29	100.0	1	100.0	1	100.0	1
30	40.3	29	62.0	23	58.5	23

Table 11.2. TRI Ratios versus PPF Efficiency Scores

	Plant 1 (+19%)		Plant 2 (–9%)		Plant 21 (–21%)		Plant 25 (+23%)	
Time period	t_0	t_1'	t_0	t_1'	t_0	t_1'	t_0	t_1'
Output	150.0	178.1	2.1	1.9	51.6	46.2	73.9	91.0
Pollutant 1	675.0	681.4	7.3	6.7	615.5	485.6	879.7	924.0
Pollutant 2	1125.0	1135.7	23.1	21.1	331.2	261.3	295.7	310.6
TRI ratio (poll. 1)	—	1.01	—	0.92	—	0.79	—	1.05
TRI ratio (poll. 2)	—	1.01	—	0.92	—	0.79	—	1.05
PPF efficiency (%)	100.0	100.0	100.0	81.4	71.4	71.4	79.0	92.5
TRI ratio interpretation relative to t_0	—	Worse	—	Better	—	Better	—	Worse
PPF interpretation relative to t_0	—	Better	—	Worse	—	Same	—	Better

Note: TRI ratio = (amount of pollutant *i* this period) / (amount of pollutant *i* last period); TRI ratio < 1 indicates improvement from last period; TRI ratio > 1 indicates decline from last period; TRI ratio = 1 indicates no change. PPF efficiency scores = changes in production output and technology; frontier is dynamic.

In the case of plant 1 the TRI ratio method indicates that the plant did worse in pollution reduction while the DEA pollution prevention frontier (PPF) efficiency measure indicates an improvement. Both are correct but the ratio is driven by slight increases in total pollution that does not take into account a 9 percent increase in total output. The DEA/PPF method adjusts from the total output increase in production and indicates that pollution levels did not increase proportionately (i.e., output increased and pollution increased but at a lower rate than the output increase). In contrast, plant 2 reduced overall pollution as noted by the TRI ratio method but it did this only by reducing output (–9 percent) and it did not reduce pollution levels in technological proportion to its decline in total output. Hence the DEA/PPF method indicates it has declined in its relative pollution reduction efficiency. Similar assessments can be made by examining the results of plants 21 and 25.

Overall the DEA/PPF allows for the original levels of pollution abatement of plants to be taken in account, i.e., it does not punish early pollution abatement leaders unlike the TRI ratio. Further the DEA/PPF adjusts for both growth and technology dynamics, but the TRI method does not. This feasibility assessment based on simulated data illustrates the applicability of the DEA/PPF approach and demonstrates its advantage over the NRDC-suggested TRI ratio procedures.

Conclusion

The information provided from the DEA-based measures can be used to identify leaders—firms that define the efficiency frontier—and laggers—firms that are not on the frontier in terms of their pollution prevention performance. This segregation of firms provides a basis for tailoring incentive measures to motivate laggers to move toward the efficient frontier and to encourage further innovations by the leaders to advance the frontier over time. Further, as has been demonstrated in our analysis of simulated data, this procedure produces better—valid, more accurate, sensitive to change, and equitable—results than the present EPA TRI ratio for assessing pollution abatement and efficiency among firms.

The foregoing discussion refers to chemicals of concern. In theory, pollution prevention is neither media nor chemical specific—i.e., the goals of source reduction and recycling are to minimize the quantity of pollutants either generated by or moving from the point of generation to the environment. Underlying motivations for the lack of media or chemical specificity are that released pollutants often transit between media, and assessors historically have failed to anticipate the difficulties some chemicals ultimately pose (U.S. EPA 1991a). For example, efforts to control chlorofluorocarbons (due to ozone depletion) and to reduce carbon dioxide (due to global warming) are aimed at chemical compounds that were once thought to be essentially harmless.

Realistically, resource allocation considerations provide substantial impetus for identifying some subsets of chemicals of greatest concern. Many chemicals are present in great abundance yet pose few human or environmental health hazards. If equal attention is paid to all such chemicals, problem definition becomes intractable. Thus, it is essential to focus some attention on a manageable set of chemicals for most pollution prevention activities. The hazards posed to society through the use and release of these chemicals are usually related to their acute effects on human health.

Other subsets of chemicals can be identified, depending upon the type of environmental concern attendant to their usage. For example, chemicals inducing low dose chronic effects, chemicals contributing substantially to global climate change, chemicals exhibiting environmental persistence, or chemicals released in significant quantities within a given geographic region can all be of potential interest. The DEA approach to measuring success in pollution prevention could be applied in this more general sense by ascribing measures of hazard attributes to those chemicals that pose a risk to society and substituting these risk-based measures for weight and volume measures. We believe that the DEA will provide a contribution to pollution abatement monitoring and guidance for pollution prevention progress at the level of the individual plant (DMU). Further, this approach has the potential for guiding diagnosis and technology leadership development in this field.

APPENDIX 1: PROGRAM SPECIFICATION

$$\text{Maximize } E_0 = \frac{\sum_{h} V_{h0} P_{h0}}{\sum_{q} W_{q0} I_{qi}} \qquad (11.3)$$

$$\text{Subject to: } \frac{\sum_{h} V_{h0} P_{h0}}{\sum_{q} W_{q0} I_{qi}} \leq 1 \quad \forall \quad i\ 1, 2, \ldots n$$

where P_{hi} is the amount of output type h produced by firm i; P_{h0} represents the amount of output type h produced by the firm whose DEA efficiency score we are trying to obtain. I_{qi} is the amount of input type q used by firm i; likewise I_{q0} represents the amount of input type q used by the firm whose DEA efficiency score we are trying to obtain. V_{h0} is the weight assigned by the optimization to output type h at the firm we are analyzing; $V_{h0} \geq 0$. W_{q0} is the weight assigned by the optimization to input type q at the firm we are analyzing; $W_{q0} \geq 0$.

APPENDIX 2: MODEL MECHANICS

The following numerical example with two inputs (pollutants) and one output is provided to illustrate this technique.

There are 10 DMUs, each using one chemical of concern measured in millions of tons per year (see table 11.3). These DMUs also generate two types of pollution: input 1 is a toxic air emission measured in tons per year; input 2 is a toxic water effluent measured in millions of gallons per year. The output and pollutants are all measured in their natural units, and the unit of measure in each column (pollutants and output) is the same for each DMU. The relative efficiency of DMU_j is measured by the weighted amount of output DMU_j produces divided by the weighted sum of inputs (pollutants) generated at DMU_j.

Table 11.3. Data for the Example

DMU_j	Input 1 (tons/year) (I_{1j})	Input 2 (10^6 gallons/year) (I_{2j})	Output (10^6 tons used/year) (O_{1j})	Input 1/ Output	Input 2/ Output
DMU_1	42.02	210.09	24.01	1.75	8.75
DMU_2	205.5	469.70	58.71	3.50	8.00
DMU_3	38.03	135.83	21.73	1.75	6.25
DMU_4	163.53	207.14	43.60	3.75	4.75
DMU_5	144.93	144.93	18.70	7.75	7.75
DMU_6	216.98	295.88	78.90	2.75	3.75
DMU_7	790.91	158.18	90.39	8.75	1.75
DMU_8	513.16	708.65	97.74	5.25	7.25
DMU_9	410.14	267.48	71.32	5.75	3.75
DMU_{10}	591.38	212.90	94.62	6.25	2.25

The optimization finds the set of weights (U_{1j}, W_{1j}, and W_{2j}) that makes the efficiency of DMU_j as large as possible given that none of the efficiency ratios for other DMUs exceed 1 when these weights are tested with their input and output values. Using the values in table 11.1 and implementing this for DMU_2 becomes:

Maximize $e_2 = (U_{12} \times 58.71) / (W_{12} \times 205.50 + W_{22} \times 469.70)$

Subject to the following constraints:

$(U_{12} \times 24.01) / (W_{12} \times 42.02 + W_{22} \times 210.09) \leq 1$: Testing DMU_2's weights for DMU_1

$(U_{12} \times 58.71) / (W_{12} \times 205.50 + W_{22} \times 469.70) \leq 1$: Assuring DMU_2's efficiency is ≤ 1

$(U_{12} \times 21.73) / (W_{12} \times 38.03 + W_{22} \times 135.83) \leq 1$: Testing DMU_2's weights for DMU_3

$(U_{12} \times 43.60) / (W_{12} \times 163.53 + W_{22} \times 207.14) \leq 1$: Testing DMU_2's weights for DMU_4

$(U_{12} \times 18.70) / (W_{12} \times 144.93 + W_{22} \times 144.93) \leq 1$: Testing DMU_2's weights for DMU_5

$(U_{12} \times 78.90) / (W_{12} \times 216.98 + W_{22} \times 295.88) \leq 1$: Testing DMU_2's weights for DMU_6

$(U_{12} \times 90.39) / (W_{12} \times 790.91 + W_{22} \times 158.18) \leq 1$: Testing DMU_2's weights for DMU_7

$(U_{12} \times 97.74) / (W_{12} \times 513.16 + W_{22} \times 708.65) \leq 1$: Testing DMU_2's weights for DMU_8

$(U_{12} \times 71.32) / (W_{12} \times 410.14 + W_{22} \times 267.48) \leq 1$: Testing DMU_2's weights for DMU_9

$(U_{12} \times 94.62) / (W_{12} \times 591.38 + W_{22} \times 212.90) \leq 1$: Testing DMU_2's weights for DMU_{10}

A set of weights that will optimize (solve) this problem (using DEA software) is given by: $U_{12} = 0.752$, $W_{12} = 0.177$, and $W_{22} = 0.071$. Substituting these weights into the objective function above yields:

$e_2 = (0.752 \times 58.71) / (0.177 \times 205.50 + 0.071 \times 467.70) = 44.15 / (36.37 + 33.21)$
$= 0.635$.

If these same weights are substituted into the constraints we obtain:

$(0.752 \times 24.01) / (0.177 \times 42.02 + 0.071 \times 210.09) = 0.808$: Testing DMU_2's weights for DMU_1

$(0.752 \times 58.71) / (0.177 \times 205.5 + 0.071 \times 469.70) = 0.633$: DMU_2's efficiency $\leq = 1$

$(0.752 \times 21.73) / (0.177 \times 38.03 + 0.071 \times 135.83) = 1.000$: Testing DMU_2's weights for DMU_3

$(0.752 \times 43.60) / (0.177 \times 163.53 + 0.071 \times 207.14) = 0.751$: Testing DMU_2's weights for DMU_4

$(0.752 \times 18.70) / (0.177 \times 144.93 + 0.071 \times 144.93) = 0.391$: Testing DMU_2's weights for DMU_5

$(0.752 \times 78.90) / (0.177 \times 216.98 + 0.071 \times 295.88) = 1.000$: Testing DMU_2's weights for DMU_6

$(0.752 \times 90.39) / (0.177 \times 790.91 + 0.071 \times 158.18) = 0.450$: Testing DMU_2's weights for DMU_7

$(0.752 \times 97.74) / (0.177 \times 513.16 + 0.071 \times 708.65) = 0.521$: Testing DMU_2's weights for DMU_8

$(0.752 \times 71.32) / (0.177 \times 410.14 + 0.071 \times 267.48) = 0.586$: Testing DMU_2's weights for DMU_9

$(0.752 \times 94.62) / (0.177 \times 591.38 + 0.071 \times 212.90) = 0.594$: Testing DMU_2's weights for DMU_{10}

Notice that when these weights are tested in the efficiency constraints for DMU_3 and DMU_6, they both equal 1.000. We cannot improve the efficiency of DMU_2 beyond 0.633—that is, change the weights (increase U_{12} and decrease W_{12} and W_{22}, making the efficiency of DMU_2 larger)—without causing the efficiencies of DMU_3 and DMU_6 to exceed 1.000, which is not permissible. Therefore, we have found the weights that maximize the relative efficiency of DMU_2. We have also found the two DMUs (DMU_3 and DMU_6) to which DMU_2 is relatively inefficient. The process is repeated for each of the other DMUs to obtain a complete set of pollution prevention efficiencies. The weights themselves have no physical or economic meaning: they are mathematical artifacts just used to find e_2 and are called virtual weights. The relative efficiencies for all DMUs found using this technique are given in table 11.4.

Table 11.4. Solution

DMU	Efficiency	Inefficient in Reference to	
1	1.00		
2	0.63	DMU_3	DMU_6
3	1.00		
4	0.78	DMU_6	DMU_{10}
5	0.45	DMU_6	DMU_{10}
6	1.00		
7	1.00		
8	0.52	DMU_3	DMU_6
9	0.79	DMU_6	DMU_{10}
10	1.00		

NOTE

The authors wish to express their gratitude to the EPA Office of Exploratory Research for the contract "Identifying Leaders and Laggers in Pollution Prevention" R820009-01-0. All analysis and interpretations are the responsibility of the authors and do not reflect agency policy or concurrence.

REFERENCES

Banker, R. D., A. Charnes, and W. W. Cooper. 1984. "Models from the Estimation of Technical and Scale Inefficiencies in Data Envelopment Analysis." *Management Science* 30 (8): 1078–92.

Banker, R. D., and R. M. Thrall. 1992. "Estimation of Returns to Scale Using Data Envelopment Analysis." *European Journal of Operational Research* 21 (1): 74–82.

Brickman, R., S. Jasanoff, and T. Ilgen. 1985. *Controlling Chemicals: The Politics of Regulation in Europe and the United States.* Ithaca: Cornell University Press.

Charnes, A., W. W. Cooper, and E. Rhodes. 1978. "Measuring Efficiency of Decision-making Units." *European Journal of Operational Research* 2 (6): 429–44.

Cummings-Saxton, J. 1994. "Pollution Prevention." In *Toxic Air Pollution Handbook,* ed. D. Patrick. New York: Van Norstrand-Reinholt. 440–58.

Fare, R., and S. Grosskopf. 1983. "Measuring Output Efficiency." *European Journal of Operations Research* 13 (2): 173–79.

Fare, R., S. Grosskopf, C. A. K. Lovell. 1989. "Multilateral Productivity Comparisons when Some Inputs Are Undesirable: A Nonparametric Approach." *Review of Economics and Statistics* 71 (1): 90–98.

Fare, R., S. Grosskopf, and C. Pasurka. 1986. "Effects on Relative Efficiency in Electric Power due to Environmental Controls." *Resources and Energy* 8 (2): 176–84.

Farrell, M. J. 1957. "The Measurement of Productive Efficiency." *Journal of the Royal Statistical Society,* ser. A, 120 (3): 253–90.

Haynes, K. E., S. Ratick, W. Bowen, and J. Cummings-Saxton. 1993. "Environmental Decision Models: U.S. Experience and New Approaches to Pollution Management." *Environment International* 19 (3): 200–220.

Haynes, K. E., S. Ratick, and J. Cummings-Saxton. 1994. "Toward a Pollution Abatement Monitoring Policy: Measurements, Model Mechanics, and Data Requirements." *Environmental Professional* 16 (4): 292–303.

Haynes, K. E., R. R. Stough, and H. L. Shroff, eds. 1990. "New Methodology in Context: Data Envelopment Analysis." Special issue of *Computers, Environment, and Urban Systems* 14 (2): 85–87.

Jorgenson, D., and P. Wilcoxin. 1990. "Environmental Regulation and U.S. Economic Growth." *Rand Journal of Economics* 21 (Summer): 314.

Kopp, Raymond J. 1981. "The Measurement of Productive Efficiency: A Reconsideration." *Quarterly Journal of Economics* 96 (3): 477–503.

Lave, L. B. 1991. "Benefit-Cost Analysis of Environmental Decisions: Does This Framework Optimize Social Decisions?" Carnegie-Mellon University Working Paper.

Lave, L., and G. Omenn. 1989. "Managing Toxic Chemicals: How Accurate Must Tests Be?" *Journal of the American College of Toxicology* 8 (6): 1081–89.

"Massachusetts Toxic Use Reduction Regulations." *Environmental Reporter* 1 (Sept. 6).

National Academy of Sciences. Board on Environmental Studies and Toxicology and Commission on Geosciences, Environment, and Resources. 1990. *Tracking Toxic Substances at Industrial Facilities: Engineering Mass Balance Information for Facilities Handling Toxic Substances.* Washington, D.C.: National Academy Press.

"Oregon Hazardous Waste Reduction Regulations." 1991. *Environmental Reporter* 1 (Nov. 15).

Portney, P. R., ed. 1992. *Public Policies for Environmental Protection.* Washington, D.C.: Resources for the Future.

Ratick, S. J., and J. Cummings-Saxton. 1991. "Analysis of Pollution Prevention Data and Measurement Techniques." Draft report to the U.S. Environmental Protection Agency.

Sheppard, R. W. 1953. *Cost and Production Functions.* Princeton: Princeton University Press.

Smith, N. C. 1988. *The Use of Mass Balance Data in the Natural Resources Defense Council Washington's Proposed Model Waste Reduction Program.* Washington, D.C.: National Resource Defense Council.

Stolp, C. 1990. "Strengths and Weaknesses of Data Envelopment Analysis: An Urban and Regional Perspective." *Computers, Environment, and Urban Systems* 14 (2): 103–16.

University of Massachusetts at Lowell. 1991. "Toxic Use Reduction Institute." Research report. Lowell, Mass.

U.S. Environmental Protection Agency. 1990. *Report to Congress, Pollution Prevention Strategy.* Washington, D.C.: GPO. ES-1.

———. 1991a. *Economic Incentives: Options for Environmental Protection.* Washington, D.C.: GPO. EPA-211-201.

———. 1991b. *Risk Reduction Engineering Laboratory.* Cincinnati: Cincinnati Engineering Laboratory. EP 1.89/2:600/S 2-901054.

U.S. House Committee on Energy and Commerce. Subcommittee on Fossil and Synthetic Fuels. 1990. *Waste Reduction Act.* 101st Cong., 2d sess. H. Rept. 1457.

U.S. Senate Committee on Environment and Public Works. *Waste Reduction Act.* 101st Cong., 2d sess. S. Doc. 585.

PART 5

ADMINISTRATION, OVERSIGHT, AND ASSESSMENT

12

Using Environmental Program Evaluation: Politics, Knowledge, and Policy Change

MICHAEL E. KRAFT

Since the early 1970s environmental policy has moved from the periphery of political life to a position of central and enduring concern both within the United States and globally. Federal, state, and local environmental policies now affect nearly every sector of the U.S. economy. Total national expenditures (public and private) for environmental protection in 1994 reached $140 billion, or about 2.1 percent of the U.S. Gross Domestic Product, and the U.S. Environmental Protection Agency (EPA) estimates it will rise to $171 billion annually (in 1990 dollars), or 2.6 percent of the Gross Domestic Product, by the year 2000 (U.S. EPA 1990). In the mid-1990s the federal government alone spent about $21 billion a year on all natural resource and environmental programs, or 1.4 percent of its budget. The General Accounting Office (GAO) has estimated that the nation's cumulative spending on the environment over the twenty-year period from 1972 to 1992 totaled in excess of $1 trillion (U.S. GAO 1992a). Yet questions about the effectiveness and efficiency of environmental policy are pervasive. In the eyes of its critics, environmental policy has produced few worthwhile gains, and at an excessively high cost. Such arguments were commonplace in congressional debates in 1995 over the regulatory reform proposals of the Republicans' Contract with America. In contrast, environmentalists generally applaud the formal policy commitments in U.S. environmental laws, but they express disappointment with the pace of implementation and the modest degree of improvement in environmental quality. Environmental program evaluation should appeal to critics as well as environmentalists.

Consistent with the spirit of "reinventing government" so prominent in the Clinton administration, environmental policies should benefit from systematic and sustained program evaluations and subsequent reappraisal of policy goals and means. Any consideration of future policy needs, from the massive job of cleaning up federal facilities and other hazardous waste sites to action on new global environmental threats such as climate change and loss of biological diversity, reinforces that conclusion. The reasons are simple. Governments at present are not able to fully fund the range of policy activities deemed essential to deal with these problems, and few additional resources can be expected in the near term. Under such conditions, a premium must be placed on better targeting of resources, a task that is dependent on the conduct and use of program evaluations. In addition, program failures attributable to causes other than insufficient resources may be correctable if properly diagnosed through such evaluation.

Enhanced concern for environmental program evaluation and reconsideration of policy directions are not entirely new to government, but have been evident since at least 1985. For example, the EPA has actively promoted a national debate over environmental risk priorities (U.S. EPA 1990; Davies 1996). Present statutory mandates often result in spending large sums of money without comparable gains in ecological and public health, in part because the U.S. Congress tends to reflect public concerns and fears about relatively low-level risks. Similarly, economists have long emphasized the gains in economic efficiency that could be achieved through more widespread use of market-based incentives for environmental protection, where applicable, as a supplement to conventional regulation (e.g., Portney 1990). During the early 1990s these arguments won widespread support (Freeman 1994; *Project 88* 1991). In much the same vein, political scientists and other policy scholars have suggested the potential to bolster environmental program effectiveness through improved implementation, administrative reforms, and new policy mechanisms (Mazmanian and Morell 1992; Tobin 1990; Bartlett 1994).

As these arguments have made clear, environmental program evaluation is a promising activity that potentially can help federal agencies (and other governmental and nongovernmental organizations) achieve much needed policy redirection in the rest of the 1990s and in the twenty-first century. Unfortunately, program evaluation in environmental policy has been quite limited to date, and in other areas of public policy it has long been afflicted with significant conceptual and empirical problems. Other chapters discuss these limitations. Of greater concern here is that the utilization of program evaluations has been severely disappointing (Weiss 1978; Rich and Oh 1994; Schneider 1986; Chelimsky 1991). Despite these problems, the imperative of improving environmental policy suggests the need to ask how meaningful environmental program evaluations might be designed and conducted to increase the likelihood that they can guide public decision making in the coming decades.

In this chapter I focus on factors affecting the utilization of program evaluation in federal agencies. First, I consider several issues in the design and conduct of environmental program evaluation. These bear on the political and institutional factors that are among the most important forces shaping the extent to which such evaluations are likely to be used by policymakers and administrators. Second, I review the varied purposes served by program evaluations to illustrate how they affect the utilization of such studies. Finally, I explore several key conceptual issues in the utilization of environmental program evaluations to identify the constraints on putting evaluations to work in a bureaucratic setting and how they might be dealt with. From this review it is clear that effective conduct and use of environmental program evaluations requires that evaluators give explicit and serious attention to the political character of environmental policy and to the organizational variables that affect the communication and use of knowledge in environmental policy making.

The Logic and Limits of Environmental Program Evaluation

The logic of environmental program evaluation is powerful. Limited resources and competing governmental programs demand that the nation determine which programs are working well and which are not and what might be done to improve policy performance. The conduct of program evaluation reflects a belief that the quality of public decision making can be improved through the systematic use of social science and other knowledge. Rigorous empirical study is not, of course, a sufficient condition for utilization of the findings and for effective policy change. However much scientists and policy analysts believe their dedication to scientific rationality accords them a superior claim to policy influence, their work and the policy recommendations that flow from it must be tested against other standards, such as technical feasibility, cost-effectiveness, and political and administrative feasibility.

To put the point slightly differently, program evaluation knowledge is not intended to replace experience, expertise, and managerial judgment in the agencies. Nor should it supplant the broader democratic discourse over policy goals we expect to see within the agencies and within the legislative bodies that have ultimate responsibility for legitimizing public policy. Rather, program evaluation aspires to inform political decision making by clarifying issues and producing, as Charles E. Lindblom and David K. Cohen (1979) put it, "usable knowledge." To succeed, such evaluation must be done well under what are by any standard difficult circumstances for an analytic enterprise.

Designing Environmental Program Evaluation: Conceptual Issues

For students of public policy, evaluation means judging the merit of governmental processes and programs, and especially determining whether and how programs affect the problems to which they are directed (Wholey 1991; Schneider 1986). Evaluators look for program impacts on individual behavior, institutional processes, and eventually environmental quality itself.

Systematic evaluations typically have four components: specification, measurement, analysis, and recommendations. To carry out such evaluations, one would expect to have specific criteria for success (such as clear program goals and objectives), evidence that is collected systematically from a representative sample of units of concern, and analysis of the data according to exacting scientific standards (Weiss 1972; Rossi and Freeman 1993). Systematic evaluations are far less common than other kinds of policy and program evaluations typically found for environmental policies, which vary significantly in their scope of inquiry and analytic rigor. The quality of an evaluation will affect its use by policymakers and administrators. Even more important, however, is that evaluations address major political and institutional issues of concern to government officials and other policy actors. Robert V. Bartlett (1994) usefully distinguishes three kinds of evaluation—outcomes, process, and institutional—with quite different implications for utilization.

Outcomes Evaluation Outcomes evaluation includes what is usually understood as program evaluation. Here analysts compare measures of environmental quality outcomes with policy objectives. The annual report of the Council on Environmental Quality compiles many such outcome measures reported each year by the EPA and other federal agencies. For instance, since 1973 the EPA's Office of Air Quality Planning and Standards has published an annual report that provides an extensive review of air quality and emissions trends, with an assessment of changes over time in key indicators of air quality (U.S. EPA 1993). The EPA's Office of Policy, Planning, and Evaluation assesses these and other program achievements and future needs as does the GAO, which produces a great many studies and reports each year on environmental programs managed by the EPA, the Department of the Interior, the Department of Energy, and other federal agencies and departments.

In 1995 the environmental research institute Resources for the Future embarked on a major study of the U.S. pollution control regulatory system in which it sought to measure progress toward stated goals (outcomes) for air quality, water quality, solid waste, and toxic chemicals. The project director, J. Clarence Davies, planned to identify the degree of change in environmental quality in each of the areas and the extent to which such change could be attributable to the government's regulatory programs. In addition, the validity of the goals themselves was to be studied in light of comparative risk analysis and the well-documented need for priority setting, demands for economic efficiency, prevailing American social values, and comparable activities in other nations.

Outcomes evaluation is not without significant limitations and potential abuses. For example, the quantity and quality of available data may be unacceptably low as a result of insufficient or uneven monitoring of environmental conditions. Thus it may not be known whether and in what ways those conditions are changing, and no meaningful conclusions about achievement of program goals and objectives are possible (Russell 1990). Sometimes the indicators used are not measures of environmental outcomes, but rather of bureaucratic tasks, such as enforcement actions, that are presumed to be causally related to changes in environmental quality. In addition, national data on policy or program results aggregate and mask what is often substantial variation among states and localities.

It is also important to note that environmental programs are a highly diverse lot. Even within the EPA, essentially a pollution control agency, there is considerable variation in program design from air and water quality regulation to cleanup of Superfund sites. Most of these programs also involve extensive interaction with state governments through a complicated system of intergovernmental relations and federal grants in aid. The states vary widely in their commitment to program goals as well as in their capacity to assist in their implementation (Lester 1994; Ringquist 1993). Thus generalization across policy areas and across the fifty states is problematic.

The diversity of federal programs increases sharply as the focus shifts from the EPA to environmental programs in the interior, agriculture, defense, and energy departments. Reliable data for the range of programs covered by these departments are often hard to come by, although there are good reasons to conduct evaluations. Critics often point to antiquated policies, for example on western land and water use, that result in inefficient use of natural resources and destructive environmental practices that may nevertheless be stoutly defended by politically powerful constituencies (Kraft 1996).

The programs run by the Department of Defense (DOD) and the Department of Energy (DOE) present another difficult evaluation challenge. These departments are increasingly important to environmental policy as they begin the cleanup of nuclear weapons production and other defense facilities, such as the highly contaminated Hanford Nuclear Reservation and the Savannah River and Rocky Flats plants. The best current estimates put the cost of such cleanup at between $200 and $300 billion over the next three decades (Russell, Colglazier, and Tonn 1992). Even in the short term the costs are high. In Fiscal Year 1993, the federal government spent close to $10 billion on cleanup ("environmental restoration") of DOD and DOE facilities, an amount that dwarfed the EPA operating budget of $2.5 billion. Yet there is no agreement to date on how clean such facilities need to be, and hence what indicators should serve to measure the progress of these high-cost programs.

Regardless of conceptual difficulties and limited data, outcomes evaluation has great political appeal. Such evaluations are used by both proponents and critics of environmental programs to demonstrate either impressive policy achievements or serious shortcomings, as suits the case, with implications for improved implementation, redirection of program priorities, or policy change of a more extensive form. Even the most systematic program evaluations are likely to be used in much the same way.

Unfortunately, outcomes evaluation leaves many important questions about program operations and long-term policy effects unaddressed. They may even be biased against showing policy success if the program's goals and objectives are not all that clear, consistent, or realistic, or if the program imposes large costs on government or society, all of which characterize environmental policy (Bartlett 1994). Economists often find environmental policy to fall short, for example, when measured against the standard of economic efficiency, irrespective of the degree of legislative and public support (Portney 1990; Freeman 1994). To some extent this deficiency may be corrected through the development of better indicators of program benefits, though there are no easy ways to do so, particularly for long-term and intangible benefits to both human and ecological health. A further limitation of such studies is that if program achievements are found to be nonexistent to modest, one may not learn *why* that is the case, and thus be left with no guide to appropriate policy redesign, reallocation of budgetary resources, changes in program leadership or direction, and other actions that might improve program operations.

An even more significant problem with outcomes evaluation, despite its popularity, is that it presupposes one's interest is chiefly in determining the degree to which program objectives are achieved. This outlook itself reflects an instrumentalist or problem-solving orientation. This is precisely the objective of most program evaluations, and the concerns that drive it are eminently defensible for all the reasons noted above.

Nonetheless, it is equally clear that public policy often involves more than rational problem solving. It may, for example, express environmental values and goals that are intensely held by the American public and especially by key interest groups, such as environmentalists. The inclusion of what are often symbolic values is crucial to gaining public support and legitimizing public policy in the first place as well as to maintaining the support of those publics over time. This is so even when such value statements have little direct relationship to legally specified policy goals and objectives. Examples include the National Environmental Policy Act (NEPA) of 1969 and the Endangered Species Act of 1973, both of which are heavily laden with symbolic value. Hence evaluations may well have broader purposes than measuring concrete policy outcomes, however important those outcomes may be. As Bartlett (1994) notes, these purposes are recognized explicitly in process and institutional evaluation.

Process Evaluation Process evaluations attempt to evaluate the merit of policy processes themselves—problem definition, agenda setting, formulation, legitimation, and implementation—using standards such as comprehensiveness, integration, promotion of ecological rationality, efficiency, responsiveness to public needs, or participation by significant interests. A good example is the kind of policy implementation analysis favored by political scientists (e.g., Tobin 1990; Rosenbaum in this volume). It is intended not only to determine the success and failure of public policies but even more importantly to examine the factors that help explain those outcomes. These might include insufficient scientific knowledge or technology, inconsistent statutory goals, inadequate resources, and weak support from the public or political leaders. Such knowledge is crucial if policy activists, legislators, and agency personnel are to improve program performance (Mazmanian and Sabatier 1983; Goggin, Bowman, Lester, and O'Toole 1990).

One example will suffice. The DOE's high-level nuclear waste repository program, authorized by the Nuclear Waste Policy Act of 1982, amended in 1987, is in many ways a case of policy failure (Clary and Kraft 1989). The DOE has admitted as much. Yet it is clearly insufficient to know only that the DOE is failing to site a repository at Yucca Mountain, Nevada, according to program objectives and time lines or that certain site characterization studies present methodological challenges or even that the department continues to face intense citizen and state opposition to its efforts. Without implementation analysis and a careful effort to identify the *reasons* for public and state opposition, there is no firm basis for redesigning nuclear waste policy or improving departmental performance. Many independent studies by social scientists (ironically, funded by DOE payments to Nevada for its own socioeconomic assessments) reveal that the DOE suffers from such a lack of credibility among important publics that it may be unable to salvage the nuclear waste program at all. At a minimum, doing so would require major changes in organizational culture and behavior within the department (Dunlap, Kraft, and Rosa 1993).

Institutional Evaluation Institutional evaluations cast an even broader net. They ask how institutional variables shape public policy processes and how institutions in turn are affected by public policy. For example, over a decade or more, environmental policies may shape bureaucratic culture, including prevailing beliefs, values, rules (e.g., decision criteria), procedures, and incentives. There is evidence that NEPA had such an effect and good reason to believe hazardous waste policies such as Superfund and the Resource Conservation and Recovery Act (RCRA) are now exhibiting similar effects on the way the nation handles toxic and hazardous chemicals.

The implication is that one must look beyond short-term measures of policy success and failure and ask about long-term changes in institutional behavior. Early reviews of NEPA were far more negative than later evaluations that documented the extent to which the act was altering organizational culture and routines (Caldwell 1982; Bartlett 1989). Much the same could be said of RCRA and Superfund. These policies are frequently criticized as failures (e.g., Mazmanian and Morell 1992). Yet over time they are likely to change in important ways both the public's attitudes toward toxic chemicals and the nation's use of hazardous and toxic materials, in part because of the high costs to industry for disposal of hazardous waste. The rise of pollution prevention as a major environmental goal is directly related to this policy experience.

Regrettably, program evaluations fail to give as much attention to institutional variables, particularly long-term institutional change, as they so clearly merit. As one example, William T. Gormley (1992) reviewed 120 GAO reports issued between 1980 and 1991 on federal and state environmental regulation and related activities. He found that the GAO ignored institutional and political factors affecting implementation, such as resources available to state agencies or the capabilities of state legislatures. These are variables that political science studies have shown to be highly relevant to successful environmental policy implementation (Lester 1994; Ringquist 1993). There is reason to believe that the GAO intentionally avoids study of political variables in an effort to restrict congressional criticism of its work.

The same pattern can be found within policy and program evaluation offices in federal departments. Some may try to constrain political controversy by putting such variables aside. Others may ignore them unintentionally as a result of organizational culture. For example, Judith Bradbury (1989) found that DOE officials consistently overlooked social science knowledge in assessing the department's nuclear waste programs. The DOE was far more likely to focus on scientific and engineering knowledge. As a result, it may have deprived itself of precisely the kind of information it needed to improve program effectiveness.

Political and Institutional Constraints on Using Environmental Program Evaluations

Whatever form of evaluation one examines, methodological problems abound. These include uncertainty over such basic questions as the criteria that should be used for measuring policy success or failure when the policies are highly complex, policy actors are numerous, and interpretations of policy purposes diverse. Equally important, how should one measure achievement of policy goals and objectives, and what time frame is most appropriate for the study? These limitations affect the utilization of policy analyses and program evaluations in bureaucratic agencies.

There is no shortage of examples to illustrate that support by career agency staff, and the existence of well-regarded studies conducted within and outside the agencies, is often insufficient to overcome opposition by elected and appointed political officials. These involve both policy analysis and program evaluation. Several cases in the early 1990s illustrate the policy conflicts and hint at the obstacles to utilization of program evaluations irrespective of their logic or quality.

Policy Analysis One notable conflict involved implementation of the 1990 Clean Air Act. In 1992, repeated disagreements erupted between the EPA and the White House Council on Competitiveness over the stringency of air quality regulations. The council's staff met regularly with EPA officials and urged the agency to make more than one hundred changes sought by the business community to the clean air regulations being developed, including some that Congress explicitly rejected in approving the act (Bryner 1995; Duffy 1994).

Another example concerns decisions by the Bush White House to override strong recommendations by the DOE and the DOD in formulating the 1991 National Energy Strategy. Following extensive public hearings and internal studies, reinforced by many external policy analyses, the DOE had recommended numerous measures to promote energy conservation and to reduce U.S. dependency on imported oil. The DOD strongly supported those recommendations out of concern for the national security implications of relying heavily on oil imported from the Middle East. The White House, however, significantly modified the National Energy Strategy before sending it to Capitol Hill. It chose to eliminate most of the energy conservation proposals that were viewed unfavorably by the president's conservative economic and political advisers. Neither the considered judgment of DOE's professional staff nor broad public backing for these programs was sufficient to persuade the president's staff to include the proposals. They preferred to emphasize energy production goals over conservation (Kraft 1996). Many other examples could be drawn from the early 1980s when Reagan administration political appointees clashed regularly with career staff of the EPA and other federal environmental and natural resource agencies (Vig and Kraft 1984; Portney 1984).

Another example of technical analysis that was either not used or underused involves the federally funded National Acid Precipitation Assessment Program (NAPAP). Begun in 1980, NAPAP involved several thousand scientists, dozens of institutions worldwide, and the expenditure of over $500 million in an effort to identify and systematically describe the causes and consequences of acid precipitation. Its published findings were widely viewed as first-rate scientific research and analysis that greatly improved understanding, and thus established a sound scientific basis for policy decisions. Yet in the eyes of many close observers, NAPAP was "less successful in influencing short-term policy initiatives . . . because it failed to address some important topics (or addressed them too late or with insufficient resources) to affect key decisions" (Russell 1993:56; cf. Moynihan 1993). The unaddressed questions included the costs and benefits of different levels of acid precipitation control and similar policy issues. By most accounts, NAPAP had much less effect on congressional passage of the Clean Air Act Amendments of 1990 than was expected by the scientific community.

Among the lessons drawn by critics is that scientific findings must be accompanied by assessments that evaluate the science and help policymakers and the public make difficult policy choices. Ideally, the assessments would be communicated to these lay audiences in a timely and effective manner, which was not done in the case of NAPAP. Publication of the assessment (which received less than 10 percent of NAPAP's budget) occurred far too late to aid policymakers, and even the major scientific findings were not released in periodic reports over the ten-year period, thus limiting their impact (Russell 1993).

Program Evaluation Similar problems affected the use of evaluations of EPA and DOE programs in the early 1990s. For example, the EPA's Superfund program has been criticized by the GAO for remaining highly vulnerable to waste, fraud, and abuse despite many previous warnings of problems with the EPA's reliance on cost-reimbursable contracts. The GAO and the Office of Technology Assessment (OTA) have criticized the EPA for inconsistent and ineffective enforcement of hazardous waste laws, reliance on an incompetent work force, and the choice of unproven but cheap remedies (U.S. GAO 1992b; U.S. Office of Technology Assessment 1988, 1989). Others have faulted the EPA for demanding high-cost, high-technology solutions that are difficult to justify in terms of benefits received. Despite these and other critical evaluations, the program continued largely unchanged, even if the message over time created significant pressures to revamp Superfund and related environmental restoration efforts, to which Congress responded as it considered reauthorization of the program in the mid-1990s.

Consider a second example. The GAO and the OTA have regularly issued reports critical of DOE's management of environmental restoration programs at its nuclear weapons production facilities. The management problems cover a wide range and include inadequate strategic planning and priority setting, insufficient recruitment of technically qualified personnel, and shortcomings in oversight and coordination of its own employees and the contractors upon whom the DOE is so heavily dependent for these programs (Rezendes 1991; U.S. Office of Technology Assessment 1991). To date, the reports do not appear to have had a major impact on the DOE's operation of these programs. Even the secretary of energy in the Bush administration, Admiral James D. Watkins, said as he was leaving office in January 1993 that the DOE's organizational culture remained a significant barrier to the department's implementation of these programs.

The Purposes of Environmental Program Evaluations

These cases of seemingly ineffective policy analysis and program evaluation indicate that evaluations serve many purposes, among them substantive, organizational, and political (Jones 1984; Schneider 1986; Goldenberg 1983). Academics and policy specialists focus on the substantive reasons for evaluating programs: to determine how well they are working and especially how they affect environmental problems. Government employees may perceive an additional benefit to evaluations: building support for the agency itself by demonstrating its effectiveness. Program critics may see evaluations as a way to control the behavior of agency personnel. Political reasons are omnipresent as evaluations are used to influence program supporters and significant constituencies. Program beneficiaries may need to be assured that their interests will continue to be served, and they in turn may use evaluations to further their political goals.

Such reflections suggest several conclusions. A program judged a dismal failure from one perspective may well be considered a ringing success from another. Programs that are politically and organizationally successful may be able to avoid evaluations. In contrast, those that are politically vulnerable may have evaluations thrust upon them by well-placed critics. Moreover, under such circumstances, even well-conducted, systematic evaluations that find considerable policy success may not persuade those who are unsupportive of policy goals.

Such an outcome was strikingly evident in the 104th Congress as the new Republican majority aggressively pushed its agenda to cut back sharply on environmental regulation (Kraft 1997). Balanced evaluations of environmental policy achievements, including several by the National Academy of Sciences in 1995, appeared to have a minimal impact on policymakers. Yet anecdotal "horror stories" about regulatory abuses by the EPA and others agencies (largely false) were repeatedly mentioned in speeches on the House floor (Cushman 1995). Similarly, legislation to advance the Contract with America called for agencies to follow an elaborate and time-consuming twenty-three-step regulatory review process involving extensive risk assessments and benefit-cost analyses that would provide little or no useful information. By one count, such proposals could increase the number of regulatory analyses conducted each year by thirty-fold. The result could be "paralysis by analysis" as agencies struggle to meet highly prescriptive congressional demands for cost assessments and other activities that offer but a dim hope of improving environmental policy (Portney 1995).

As these events in 1995 underscored, systematic research by itself is rarely likely to be a driving force in policy decisions, especially in a legislative setting. Such research is not likely to replace political judgment or ideology, particularly for macro-policy decisions, whatever it may do to help fine-tune those programs that are widely supported and not so vulnerable politically. This pattern is often found for controversial policies such as environmental protection, in which extensive conflict continues to exist over the magnitude of risks to public health and ecosystems and the policy goals and means used to deal with those risks.

The Endangered Species Act (ESA) illustrates many of these problems. The ESA has run into furious political opposition, most notably in Congress, because it symbolizes for many how government regulation can threaten property rights and restrict economic development. It is unlikely that even the best evaluations of the act's achievements will persuade the most vociferous opponents. All the commotion over the ESA and the unusual case of the northern spotted owl in the Pacific Northwest is particularly striking because careful assessments of the act indicate it has blocked few development projects since its adoption in 1973 (Tobin 1990). Indeed, a World Wildlife Fund study released in 1992 indicated that between 1987 and 1992, only 19 federal activities and projects out of 2,248 submitted for formal consultation under the act resulted in cancellation due to irreconcilable conflicts between development and species protection (World Wildlife Fund 1994). The restriction of timber harvesting on federal forests in the owl's habitat is an exception under the ESA. In the overwhelming majority of cases, federal agencies encountered no conflicts at all with the ESA, and in the several hundred cases in which a conflict did exist, agency officials found alternatives for moving the projects forward without ecological damage.

For a policy like the ESA, program evaluations can be useful to administrators in suggesting how to reconcile policy objectives with potentially conflicting goals of economic development and how to improve program effectiveness. This is particularly so when administrators are supportive of the act's goals, more evident in the Clinton administration than in the Reagan and Bush administrations. Such uses of program evaluations are also more likely to be found in administrations committed to improving government operations as opposed to those more interested in deregulating and privatizing environmental management. The Clinton administration's National Performance Review (1993a, 1993b) signals such a commitment, although it is too early to judge how many of the policy and management reforms supported by the review will be put into effect.

That assessments like those pertaining to the ESA take place in political and institutional contexts that limit their utilization, especially when programs are controversial, does not diminish the need for program evaluation. As Eleanor Chelimsky argues, the main value of such assessment "is not its capacity for political influence but its contribution to systematic, independent, critical thinking in the decision making process" (1987:11). However, if analysts want decision makers to pay attention to studies and use them, they need to understand the way politics influences decision making and adapt their research strategies in ways that promote utilization.

Utilization of Environmental Program Evaluation: Conceptual Issues

Asking about the use of program evaluation is a subset of broader questions concerning the use of science or policy knowledge in decision making. As Laurence Lynn (1978) has observed, there is an "uncertain connection" between knowledge and policy. That connection is a matter of considerable concern to policy analysts and policymakers, with two predominant views on the subject captured concisely by Lindblom and Cohen: "In public policy making, many suppliers and users of social research are dissatisfied, the former because they are not listened to, the latter because they do not hear much they want to listen to. Out of the discontent a river of mixed diagnosis and prescription now flows" (1979:1).

Much the same dissatisfaction is evident in environmental policy. After twenty-five years of policy expansion at federal and state levels, environmental policy is often found lacking by policy analysts and scientists who see political and bureaucratic forces overriding rational consideration of policy and program needs (National Academy of Public Administration 1995). At the same time, agency officials often see little value in environmental policy analysis, particularly in the work of social scientists.

Although extensive evaluations have been conducted for health and social services programs, environmental program evaluations are rare in comparison. Nonetheless, there are reasons to suppose the same kinds of obstacles to utilization exist in the environmental policy area. If environmental program evaluations have not been used, or might not be used in the future, one might well look for the explanations in two places: the studies and their authors and the policymakers who are the intended audience of these studies. A prior question helps to clarify the whole process. What do we mean by utilization of program evaluations?

Michael Quinn Patton (1978) notes that most of the literature on program evaluation does not explicitly define *utilization*. There is, instead, a presumption that "utilization occurs when there is an immediate, concrete, and observable effect on specific decisions and program activities resulting directly from evaluation research findings" (24). Such a narrow definition of utilization leads invariably to the observation that much social research has been underutilized and to the presumption that the same will characterize other kinds of program evaluation, for example in environmental policy. The skepticism is warranted, but the definition of utilization itself is partly to blame.

Scholars have long distinguished different concepts of knowledge use in the social sciences that are equally applicable to environmental policy (Rich 1981; Rich and Oh 1994). One of the most common distinctions made is between instrumental and conceptual utilization. In instrumental utilization, the approach taken by Patton, the underlying model of research use is social engineering. Knowledge is presumed to drive the policy process. The information is considered by policymakers to have short-term relevance or instrumental utility for the decisions they face, hence the expectation that utilization means short-term and specific effects on program decisions. Problems in program operations are discovered and agency decision makers take corrective action to improve program performance. If this model applies at all in the real world of bureaucratic politics, it fits best those policy areas characterized by a high level of political consensus and relatively technical and delimited policy decisions. Examples include the distribution of municipal wastewater treatment funds under the Clean Water Act for most of the 1970s and 1980s and, until recently, the issuing of leases for livestock grazing on western public lands.

In contrast to the social engineering view of knowledge use, some scholars have proposed an enlightenment model. Here knowledge may be important, but less for immediate decisions than for clarifying the causes and scope of policy problems and suggesting longer-term courses of action. This knowledge may be said to have conceptual rather than instrumental utility. Over time such research may well alter important perceptions, beliefs, and values, and eventually the attractiveness of policy alternatives and bureaucratic structures that would have been ignored earlier. This is especially likely to be the effect in policy areas characterized by high levels of conflict and significant scientific uncertainty.

A good example is the concept *sustainable development*. Made popular in the 1987 Brundtland Commission report, *Our Common Future*, sustainable use of natural resources was widely endorsed by the mid-1990s. It served as the cornerstone of the National Commission on the Environment's influential report *Choosing a Sustainable Future* (1993) as well as the United Nations' *Agenda 21*, the key strategy document emerging from the 1992 U.N. Conference on Environment and Development. By 1993, President Bill Clinton had established the Council on Sustainable Development in the White House to explore institutional and policy actions to connect long-term goals of sustainability to short-term governmental actions (President's Council on Sustainable Development 1996).

The very nature of such indirect impacts means that such uses are more difficult to detect or measure than obvious, immediate effects on program decisions. Yet they may well have a powerful effect on our understanding of policy problems, alternatives, and institutional capacities. This is particularly the case with process and institutional evaluations. Policy research of this kind must be understood in light of what John W. Kingdon (1984) calls the problem and policy streams in his model of the agenda-setting process. That is, utilization of program evaluations is a long-term and complex process, not a one-time and simple occurrence, and it must be assessed accordingly.

Further insight into the nature of utilization is provided in the distinctions made by Carol H. Weiss (1978), who identified eight meanings to *research utilization*. These included problem solving, conceptualization, political ammunition, manipulation, and advancement of self-interest as well as a language of discourse and interaction between researchers and policymakers. Most of these could be categorized as instrumental or conceptual, but Weiss highlights the subtle variations on these themes. It should be noted as well that *utilization* as defined in these ways is difficult to measure or document. It is as problematic as determining the influence of technical knowledge on decision making or on policymakers' perceptions and attitudes (Pollard 1987; Sabatier 1978). Most empirical studies rely on interviews to describe such influence, although difficult methodological problems present themselves in any such enterprise.

In one careful empirical effort to examine such uses by policymakers, David Whiteman (1985) studied the way reports of five OTA projects (on medical technology, nutrition, coal pipelines, residential energy conservation, and railroad safety) were handled by committees in Congress. He interviewed OTA and congressional committee staff with no preconception of the types of uses he would find, but with a focus on "any subsequent effect that the information had on the individual's cognition or behavior" as a measure of use of the reports (1985:297). In this way he identified three kinds of use: *substantive* (used to develop an issue position where no previous commitment had been made), *elaborative* (extension and refinement of the components of a position within boundaries previously established), and *strategic* (used to reinforce and advocate existing positions). Whiteman found "a wide variety of types and areas of use," with strategic being the most common (1985:308). Variables affecting the type of use included the levels of issue saliency and group conflict. David H. Greenberg and Marvin B. Mandell (1991) have suggested that the strategic category should be further divided into symbolic (purely a reaffirmation of existing positions) and persuasive (where references to policy research affect the outcome of policy deliberations). There are some hints in these studies of what one might expect for utilization of environmental program evaluations in bureaucratic and legislative settings.

Putting Environmental Program Evaluations to Work

These conceptual and empirical studies reflect a widely shared recognition, as Robert Haveman has observed, that "social research studies are part of an adversary process in which policymakers . . . will use whatever data are at hand to support their case, regardless of the methodological purity by which it has been developed" (1987:6). Others have been equally blunt in dismissing the relevancy of the quality of program evaluations and methodological rigor and emphasizing the political needs of key administrators. James Q. Wilson expressed the point well: "When [organizations] use social science at all, it will be on an ad hoc, improvised, quick-and-dirty basis. A key official, needing to take a position, respond to a crisis, or support a view that is under challenge, will ask an assistant to 'get me some facts.' . . . Social science is used as ammunition, not as a method, and the official's opponents will also use similar ammunition. There will be many shots fired, but few casualties except the truth" (1978:92).

Wilson concludes that the "resource in shortest supply in the development of good programs is not good research, but wise, farseeing, shrewd, and organizationally effective administrators" (92). Although Haveman and Wilson refer to experience with social programs rather than environmental policy, there is no reason to assume environmental programs will be any less political. There is just as much conflict over program goals and operations despite greater use of scientific and technical information in policy making. Thus the use or nonuse of environmental program evaluations or other studies would seem to depend on the same factors: the characteristics of the analysts and the policymakers, the interaction between them, and the political and organizational contexts in which both find themselves. The quality of program evaluations may make some difference, but much less than often supposed by social and environmental scientists.

The program evaluation literature suggests a number of conclusions about factors that tend to limit the use of evaluations and other policy analyses. Four in particular merit review here: methodological weaknesses, organizational resistance, inadequate dissemination of research, and a constrained definition of professional roles. So do some possible strategies environmental program evaluators may use to overcome these obstacles.

Methodological Weaknesses

As seen by potential users, a major constraint on utilization is the methodology of evaluation research. An inappropriate study design, insufficient or questionable data and analysis, or conclusions that do not address questions held to be important by policymakers understandably are barriers to utilization (Weiss 1972; Chelimsky 1991). This is particularly so when the policy in question is highly controversial and the evaluation is vulnerable to criticism on methodological grounds. As noted above, there are genuine methodological impediments to evaluating the impacts of most environmental programs given inadequate or nonexistent monitoring and data, delayed effects of environmental changes, and related problems.

There are a number of solutions one might consider for environmental evaluations. One is long-term and involves improvements in environmental monitoring and databases. The EPA and other agencies are making some progress here. The proposed department of the environment would likely include a center for environmental statistics that would be far more able than the Council on Environmental Quality to compile, integrate, and assess trend data. It should be able to draw not only from the agency's own activities (e.g., its Environmental Monitoring and Assessment Program) but also from the U.S. Geological Survey's National Ambient Water Quality Assessment and similar efforts by other federal agencies, the states, and private organizations. In 1992, the EPA identified eighty-three different environmental data programs in twenty-five separate federal agencies. Hence there is a need to integrate the various streams of data and to improve the nation's ability to assess interrelationships among different environmental stressors and overall environmental quality and to relate the findings to public policy actions (National Commission on the Environment 1993). Yet budgetary shortages continue to plague research and development operations within the EPA and other agencies. The EPA in particular is frequently attacked for lacking a sound scientific foundation for its environmental regulations (Carnegie Commission on Science, Technology, and Government 1992). So too are other agencies and departments, although perhaps to a lesser extent. The environmental restoration programs run by the DOE and the DOD suffer from widely disparate estimates of the scope of the problem, the probable costs of remediation, and the efficacy and acceptability of remediation technologies (Kraft 1994).

Aside from improved databases, environmental program evaluations depend upon building consensus on appropriate measures of program success. Even where they are available, outcome measures such as the number of hazardous waste sites remediated or the improvement in groundwater quality may be insufficient in the absence of agreed-upon standards. What constitutes an acceptable level of environmental quality for a particular program? To address such concerns, program evaluations need to go beyond outcomes evaluation and include elements of process and institutional evaluation, which give more attention to contested standards, disagreement or confusion over program goals and objectives, and institutional capacities for program implementation. Only in this way can some of the methodological obstacles be dealt with.

Even where it is possible, improving the rigor of program evaluations may not help if the result is a more costly and time-consuming study that fails to yield recommendations for practical courses of action that improve policy performance at a reasonable cost. There is little gained in emphasizing methodological rigor if the findings are likely to be ignored by key policymakers. It may be more important in some cases to answer the questions facing policymakers, even if imperfectly and tentatively, than to answer definitely questions that are not the central concern within an agency or the legislature. In short, evaluators need to tailor methodologies to fit the particular political and organizational context of the program. One recommendation from the general literature on program evaluations applies here. It is that greater reliance should be placed on relatively simple evaluation designs that can be carried out in a reasonable amount of time (Patton 1978; Weiss 1972) and that evaluators use multiple methods and approaches; the strength of one mitigates the weakness of another (Smith 1988). These suggestions seem equally applicable for environmental programs, particularly where outcome measures are in doubt or little agreement exists on their meaning and utility.

Organizational Resistance

Another set of factors involves resistance within agencies to new and possibly unwanted information that could pose a threat to program survival or expansion. It may be that program evaluations are separated too much from policy planning and policy-making functions within agencies, a common state of affairs that makes integrated decision making difficult. Agency resistance may reflect public apprehensions over a program's effects (e.g., use of incineration to dispose of toxic chemicals) or the concerns of a constituency group, such as contractors or industry. Evaluation research also may be seen as competing with other demands for organizational resources, whether the assessments are internal or external to the agency (Moran 1987). In addition, there may be conflicts among different agency personnel (e.g., environmental engineers, lawyers, and economists) attributable to distinctive professional backgrounds, perspectives, and functions. Such a state of affairs may lead to skepticism toward, or rejection of, studies that reflect other perspectives.

To cite one example, the DOE and the EPA rely heavily on contractors for hazardous waste programs. The GAO frequently criticized the two agencies for poor management and inadequate supervision of contractors that cost the federal government hundreds of millions of dollars a year. Yet the critiques did not lead to any evident change in agency-contractor relations. The apparent short-term lack of impact was somewhat deceptive. Over time a powerful case was being made against continuation of lax contractor supervision, and appropriate policy and organizational changes were more likely to follow when the political climate became ripe. Indeed, in July 1994 Secretary of Energy Hazel O'Leary announced a new process of competitive bidding on DOE contracts that she hoped would address such concerns. Emphasis was to be placed on smaller, better-defined contracts and improved cost controls (Cushman 1994). As Paul A. Sabatier (1987, 1988) has wisely noted, policy change is dependent upon more than mere critiques of agency performance. Change occurs through active competition of advocacy coalitions (e.g., environmental and industry groups), which in turn increases the value of program and policy evaluations that may be used in exchanges among those coalitions and their allies in policy-making bodies such as Congress.

The evaluation literature offers some modest hope for addressing pervasive organizational resistance, even in the short term. For example, Carol T. Mowbray (1988) suggests that analysts should try to market evaluation as a worthwhile service to change decision makers' perceptions of their utility. Program managers themselves might actively encourage evaluations that can be conducted at minimal cost and yet yield useful knowledge (Wholey 1991).

More useful insights come from an appreciation of the organizational and political contexts of program decisions. Evaluations may be perceived to be in conflict with policy priorities of top agency officials or White House preferences (Chelimsky 1991; Jones 1984). Evaluators need to be alert to such internal organizational conflicts and political factors and be flexible enough to promote appropriate evaluation results and recommendations that suit the opportunities that present themselves within agency policy making (Chelimsky 1987). They could also make use of political channels external to the agency (e.g., Congress and allied interest groups) to promote their assessments and recommendations.

Recognition that program evaluations, like other forms of policy analysis, invariably become part of a larger societal process of policy learning and change (Jenkins-Smith 1988; Sabatier 1988; Sabatier and Jenkins-Smith 1993) suggests other paths to building program support. State-level program evaluations and pilot programs may demonstrate the value of formal evaluations and hence increase support for similar efforts. The emergence of the states as "laboratories of democracy" that have proven to be adept at formulating and adopting innovative environmental and energy policies underscores the potential of this particular avenue (Kushler 1989; Rabe 1997). DeWitt John (1994) examined several innovative environmental policies at the state level and demonstrated that through the use of nonregulatory and cooperative approaches, state governments reduced the use of agricultural chemicals in Iowa, helped to devise a plan for restoring the Florida Everglades, and conserved electricity in Colorado. His endorsement of "civic environmentalism" is pertinent to efforts on a national basis to seek alternatives to cumbersome and inefficient environmental regulation. Similarly, Daniel Mazmanian (1992) explains how astute political and corporate leadership, a collaborative decision-making process, and new energy markets helped to create an "energy-environmental revolution" in California. Between 1973 and 1988, the state's population increased by 37 percent and its economy by 46 percent, yet energy consumption went up by only 8 percent.

Some kinds of organizational resistance can be overcome only with policy intervention at a higher level that alters existing agency motivations. One classic case is the long-standing practice of the U.S. Forest Service to sell timber on public land at a price lower than the cost of building and maintaining access roads needed to harvest the timber. Numerous evaluations of the program criticized it on grounds of economic inefficiency. "Below-cost" timber sales make no sense to evaluators, but for years they made a great deal of sense to Forest Service policymakers because they generated cash for the agency's programs. In early 1993, the Clinton administration identified such timber subsidies as one of several controversial federal natural resource programs it was eager to eliminate.

Inadequate Dissemination of Research

One of the most common explanations for limited use of program evaluations and other social science research is inadequate dissemination of the results—within and outside of agencies. Douglas S. Lipton has captured the general perspective: "The more complex the evaluation, the more jargon in the language, the more hidden or equivocal the conclusion, the more caveats in the preamble, the thicker the report, the more obscure the evaluator, the more sensitive the issue, the more apt the policymaker is to discard, ignore, or attack the evaluation. If he or she attacks it, the evaluator at least has a fighting chance to defend it, but, unfortunately, the policymaker usually shelves the report—the worst possible outcome" (1992:184).

This outcome may flow from the style of communication (e.g., lack of clarity, excessive use of technical or bureaucratic jargon, inappropriate format, and the sources of information used) as well as the timeliness and extent of distribution of the evaluation, particularly to important political constituencies (Weiss 1977; Sabatier 1978). Among the many sources of impaired communication between evaluators (whether social scientists or natural scientists and engineers) and decision makers are differences in worldviews or ideologies, education and training, professional language and concepts, problem definitions, frames of reference (substantive and temporal), and constituencies. The gulf can be so wide that the evaluator's message may be lost entirely.

The communication problem is so pervasive that proposed remedies abound. Even with the highly technical information associated with environmental policy, analysts could aim for clear, understandable presentations of the results and strive for timely and effective dissemination of reports. The GAO and the Office of Management and Budget, among others, have issued guidelines intended to enhance effective communication of evaluation studies. These focus on making evaluations "utilization focused" (where planning for utilization is an integral part of the evaluation planning), setting relative rather than absolute criteria for the relevance of program evaluations, and encouraging more interaction between evaluators and policymakers (Lipton 1992).

This kind of advice reflects a basic truth about policy making and the role of science. Evaluators need to remember that utility of findings is a matter to be judged by the *policymaker* and not the analyst. Analysts could profit by spending time with policymakers before initiating a study to be certain they understand their perceptions of policy problems and of program goals and needs (Patton 1978). The critical element here is *genuine dialogue* between evaluators and decision makers. Mowbray (1988), among many others, argues that utilization can be improved with explicit attention to policymaker needs throughout the evaluation process. This involves an effort by evaluators to understand organizational routines and procedures, policymaker values and attitudes, and political influences that will shape decision making, and thus the dissemination and use of program evaluation information.

Consistent with this view, Timothy W. Hegarty and Douglas L. Sporn (1988) propose specific techniques that can be used to forge such links between evaluators and decision makers. They recommend doing so at five points in the evaluation process: during identification of candidate programs for evaluation, during selection among them, while the evaluation is being conducted, during the reporting of evaluation results, and in the assessment of the impacts of the completed evaluations. Others offer similar advice to the would-be evaluator to cope with problems of misunderstood policymaker needs and the consequent lack of utilization (Smith 1988; Lipton 1992; Pollard 1987). The trick is for the evaluator to learn how to achieve such worthy objectives in the often complex and fast-moving agency setting. This is particularly so for environmental policy that is undergoing rapid change in areas such as cleanup of contaminated federal facilities.

Constrained Professional Roles

A fourth major category of factors shaping use of evaluation studies pertains to the analysts themselves, particularly perceptions of their professional role. Studies by Weiss (1977) and others suggest that a posture of rigid neutrality and aloofness detracts from utilization. Whatever may be said of the virtue of pursuing knowledge for its own sake or maintaining a firm attachment to professional neutrality, it does not seem to facilitate the use of evaluation studies. In contrast, active promotion of research findings and recommendations appears to improve utilization. This is not to say that evaluators should be partisan or advocates of particular agendas. The question is more one of professional style. Work by Kingdon (1984) on policy entrepreneurs and studies of the effectiveness of GAO evaluations suggest the value of considering such an activist orientation (Johnston 1988).

The GAO approach is instructive and may serve as a model for executive agencies. The GAO actively pushes for program changes in line with its recommendations, and it is aggressive in keeping its recommendations in front of congressional committees and agency officials. It has a formal tracking system for determining agency compliance, and its staff members are required to make frequent checks on recommendations considered still valid but not implemented. Perhaps this posture helps to explain the high rate of utilization of GAO evaluations, put at between 51 and 77 percent (Johnston 1988).

The Promise of Environmental Program Evaluation

The literature on program evaluation offers considerable documentation of a poor track record of utilization in the federal bureaucracy and other venues, such as Capitol Hill and the White House. Much of that literature focuses not on environmental program evaluation, but rather on social policies such as education, mental health, and welfare. Does environmental program evaluation offer more promise than the general record reveals? And are the various recommendations for improving utilization, derived in part from this literature, equally, less, or more pertinent for environmental policy?

Those questions are difficult to answer. In many ways, environmental policy is distinctive, suggesting that generalizations from the program evaluation literature may not be fully appropriate. For example, environmental policy involves a greater degree of scientific uncertainty (e.g., over risks to public health and ecosystems) than do many other policy areas. And policy actions are often more complex administratively, in part because environmental programs rely heavily on regulatory or command-and-control mechanisms for achieving goals and objectives. In several areas (e.g., DOD and DOE cleanup programs and Superfund) environmental policy is strongly dependent on outside contractors. Even if their reliability and competence were not in question, such a dependency raises questions of political accountability. All these characteristics suggest that environmental program evaluation may face some obstacles not found in other policy areas.

At the same time, at least some environmental policies (e.g., portions of the Clean Air Act, RCRA, and Superfund) have highly specific objectives and deadlines, made increasingly so over the past decade by a Congress suspicious of the EPA's dedication to achieving mandated policy goals. That should make certain kinds of program evaluations easier than in policy areas characterized by diffuse and inconsistent goals and objectives.

Depending on the particular program, then, experience with program evaluation and problems of utilization in other areas of public policy may be highly pertinent to environmental policy. If nothing else, this experience is a suitable point of departure for those seeking to engage in program evaluation or to develop strategies for its use in bureaucratic and other institutional settings.

In this chapter I have highlighted two overriding considerations for the conduct and use of environmental program evaluations. One is that outcomes evaluation is insufficient as a measure of program success or failure. A fuller and more appropriate assessment of environmental programs requires elements of process and institutional evaluations as well. The second is that the political character of environmental decision making must not be overlooked in pursuit of a seemingly objective evaluation of program operations. Applied to concern over utilization of the evaluation findings, the implications are clear. Evaluators should be alert to political and institutional variables affecting utilization. In particular, the design, conduct, and reporting of environmental program evaluations must reflect a firm understanding of policymaker and agency needs, and efforts to promote utilization must be an integral part of the evaluation process. To ignore these considerations is to risk losing much of the evaluation game.

REFERENCES

Bartlett, R. V., ed. 1989. *Policy through Impact Assessment: Institutional Analysis as a Policy Strategy.* New York: Greenwood Press.

———. 1994. "Evaluating Environmental Policy Success and Failure." In *Environmental Policy in the 1990s,* ed. N. J. Vig and M. E. Kraft. 2d ed. Washington, D.C.: Congressional Quarterly Press. 167–88.

Bradbury, J. 1989. "The Use of Social Science Knowledge in Implementing the Nuclear Waste Policy Act." Ph.D. diss., University of Pittsburgh.

Bryner, G. C. 1995. *Blue Skies, Green Politics: The Clean Air Act of 1990.* 2d ed. Washington, D.C.: Congressional Quarterly Press.

Caldwell, L. K. 1982. *Science and the National Environmental Policy Act: Redirecting Policy through Procedural Reform.* University: University of Alabama Press.

Carnegie Commission on Science, Technology, and Government. 1992. *Environmental Research and Development: Strengthening the Federal Infrastructure.* New York: Carnegie Commission on Science, Technology, and Government.

Chelimsky, E. 1987. "The Politics of Program Evaluation." *Evaluation Practice* 8 (1): 5–21.

———. 1991. "On the Social Science Contribution to Governmental Decision-Making." *Science* 254 (Oct. 11): 226–31.

Clary, B. B., and M. E. Kraft. 1989. "Environmental Assessment, Science, and Policy Failure: The Politics of Nuclear Waste Disposal." In *Policy through Impact Assessment: Institutionalizing Analysis as a Political Strategy* ed. R. V. Bartlett. Westport, Conn.: Greenwood Press. 37–61.

Cushman, J. H. 1994. "Department of Energy Pushes Competitive Bids." *New York Times,* July 7, C2.

———. 1995. "Tales from the 104th: Watch Out, or the Regulators Will Get You!" *New York Times,* Feb. 28, A10.

Davies, J. C. 1996. *Comparing Environmental Risks: Tools for Setting Government Priorities.* Washington, D.C.: Resources for the Future.

Duffy, R. J. 1994. "The Politics of Regulatory Distrust: The Quayle Council on Competitiveness and the Clean Air Act." Paper presented at the annual meeting of the Western Political Science Association, Albuquerque, Mar. 10–12.

Dunlap, R. E., M. E. Kraft, and E. A. Rosa, eds. 1993. *Public Reactions to Nuclear Waste: Citizens' Views of Repository Siting.* Durham, N.C.: Duke University Press.

Freeman, A. M., III. 1994. "Economics, Incentives, and Environmental Regulation." In *Environmental Policy in the 1990s,* ed. N. J. Vig and M. E. Kraft. 2d ed. Washington, D.C.: Congressional Quarterly Press. 189–208.

Goggin, M. L., A. O'M. Bowman, J. P. Lester, and L. J. O'Toole Jr. 1990. *Implementation Theory and Practice: Toward a Third Generation.* Glenview, Ill.: Scott, Foresman/Little Brown.

Goldenberg, E. N. 1983. "The Three Faces of Evaluation." *Journal of Policy Analysis and Management* 2 (4): 515–25.

Gormley, W. T. 1992. *Political Methodology at the GAO.* Washington, D.C.: Georgetown University Public Policy Program.

Greenberg, D. H., and M. B. Mandell. 1991. "Research Utilization in Policymaking: A Tale of Two Series (of Social Experiments)." *Journal of Policy Analysis and Management* 10:633–56.

Haveman, R. 1987. *Poverty Policy and Poverty Research: The Great Society and the Social Sciences.* Madison: University of Wisconsin Press.

Hegarty, T. W., and D. L. Sporn. 1988. "Effective Engagement of Decisionmakers in Program Evaluation." *Evaluation and Program Planning* 11:335–39.

Jenkins-Smith, H. C. 1988. "Analytical Debates and Policy Learning: Analysis and Change in the Federal Bureaucracy." *Policy Sciences* 21 (2–3): 169–211.

John, D. 1994. *Civic Environmentalism: Alternatives to Regulation in States and Communities.* Washington, D.C.: Congressional Quarterly Press.

Johnston, W. P., Jr. 1988. "Increasing Evaluation Use: Some Observations Based on the Results at the U.S. GAO." In *Research Utilization,* ed. J. A. McLaughlin, L. J. Weber, R. W. Covert, and R. B. Ingle. New Directions for Program Evaluation 39. San Francisco: Jossey-Bass. 79–84.

Jones, C. O. 1984. *An Introduction to the Study of Public Policy.* 3d ed. Monterey, Calif.: Brooks/Cole.

Kingdon, J. W. 1984. *Agendas, Alternatives, and Public Policies.* Boston: Little, Brown.

Kraft, M. E. 1994. "Searching for Policy Success: Reinventing the Politics of Site Remediation." *Environmental Professional* 16 (Sept.): 245–53.

———. 1996. *Environmental Policy and Politics: Toward the Twenty-First Century.* New York: HarperCollins.

———. 1997. "Environmental Policy in Congress: Revolution, Reform, or Gridlock?" In *Environmental Policy in the 1990s: Reform or Reaction?* ed. N. J. Vig and M. E. Kraft. 3d ed. Washington, D.C.: Congressional Quarterly Press. 119–42.

Kushler, M. G. 1989. "Use of Evaluation to Improve Energy Conservation Programs: A Review and Case Study." *Journal of Social Issues* 45 (Spring): 153–68.

Lester, J. P. 1994. "A New Federalism?: Environmental Policy in the States." In *Environmental Policy in the 1990s,* ed. N. J. Vig and M. E. Kraft. 2d ed. Washington, D.C.: Congressional Quarterly Press. 51–68.

Lindblom, C. E., and D. K. Cohen. 1979. *Usable Knowledge: Social Science and Social Problem Solving.* New Haven: Yale University Press.

Lipton, D. S. 1992. "How to Maximize Utilization of Evaluation Research by Policymakers." *Annals of the American Academy of Political and Social Science* 521 (May): 175–88.

Lynn, L., ed. 1978. *Knowledge and Policy: The Uncertain Connection.* Washington, D.C.: National Academy Press.

Mazmanian, D. 1992. "Toward a New Energy Paradigm." In *California Policy Choices,* ed. J. J. Kirlin. Vol. 8. Los Angeles: University of Southern California, School of Public Administration. 195–215.

Mazmanian, D., and D. Morell. 1992. *Beyond Superfailure: America's Toxics Policy for the 1990s.* Boulder: Westview Press.

Mazmanian, D. A., and P. A. Sabatier. 1983. *Implementation and Public Policy.* Glenview, Ill.: Scott, Foresman.

Moran, T. K. 1987. "Research and Managerial Strategies for Integrating Evaluation Research into Agency Decision Making." *Evaluation Review* 11 (5): 612–30.

Mowbray, C. T. 1988. "Getting the System to Respond to Evaluation Findings." In *Evaluation Utilization,* ed. J. A. McLaughlin, L. J. Weber, R. W. Covert, and R. B. Ingle. New Directions for Program Evaluation 39. San Francisco: Jossey-Bass. 47–58.

Moynihan, D. P. 1993. "Acid Precipitation and Scientific Fallout." *Forum for Applied Research and Public Policy* 8 (Summer): 61–65.

National Academy of Public Administration. 1995. *Setting Priorities, Getting Results: A New Direction for EPA.* Washington, D.C.: National Academy of Public Administration.

National Commission on the Environment. 1993. *Choosing a Sustainable Future: The Report of the National Commission on the Environment.* Washington, D.C.: Island Press.

National Performance Review. 1993a. *Accompanying Report of the National Performance Review.* Washington, D.C.: GPO.

———. 1993b. *From Red Tape to Results: Creating a Government That Works Better and Costs Less.* Washington, D.C.: GPO.

Patton, M. Q. 1978. *Utilization-Focused Evaluation.* Beverly Hills, Calif.: Sage Publications.

Pollard, W. E. 1987. "Decision Making and the Use of Evaluation Research." *American Behavioral Scientist* 30 (July–Aug.): 661–76.

Portney, P. R., ed. 1984. *Natural Resources and the Environment: The Reagan Approach.* Washington, D.C.: Urban Institute Press.

———, ed. 1990. *Public Policies for Environmental Protection.* Washington, D.C.: Resources for the Future.

———. 1995. "Beware of the Killer Clauses inside the GOP's Contract." *Washington Post*, national weekly ed., Jan. 21, 23–29.
President's Council on Sustainable Development. 1996. *Sustainable America: A New Consensus*. Washington, D.C.: GPO.
Project 88—Round II, Incentives for Action: Designing Market-Based Environmental Strategies. 1991. A public policy study sponsored by Senators T. E. Wirth and J. Heinz. Washington, D.C.
Rabe, B. G. 1997. "Power to the States: The Promise and Pitfalls of Decentralization." In *Environmental Policy in the 1990s: Reform or Reaction?* ed. N. J. Vig and M. E. Kraft. 3d ed. Washington, D.C.: Congressional Quarterly Press. 31–52.
Rezendes, V. S. 1991. "Long Road to Recovery Begins at DOE Plants." *Forum for Applied Research and Public Policy* 6 (1): 19–26.
Rich, R. F. 1981. *Social Science Information and Public Policy Making*. San Francisco: Jossey-Bass.
Rich, R. F., and C. H. Oh. 1994. "The Utilization of Policy Research." In *Encyclopedia of Policy Studies*, ed. Stuart S. Nagel. 2d ed. New York: Marcel Dekker. 69–92.
Ringquist, E. J. 1993. *Environmental Protection at the State Level: Politics and Progress in Controlling Pollution*. Armonk, N.Y.: M. E. Sharpe.
Rossi, P. H., and H. E. Freeman. 1993. *Evaluation: A Systematic Approach*. 5th ed. Newbury Park, Calif.: Sage Publications.
Russell, C. S. 1990. "Monitoring and Enforcement." In *Public Policies for Environmental Protection*, ed. P. R. Portney. Washington, D.C.: Resources for the Future. 243–74.
Russell, Milton. 1993. "NAPAP: A Lesson in Science, Policy." *Forum for Applied Research and Public Policy* 8 (Summer): 55–60.
Russell, M., E. W. Colglazier, and B. E. Tonn. 1992. "The U.S. Hazardous Waste Legacy." *Environment* 34 (6): 12–15, 34–39.
Sabatier, P. A. 1978. "The Acquisition and Utilization of Technical Information by Administrative Agencies." *Administrative Science Quarterly* 23 (Sept.): 396–417.
———. 1987. "Knowledge, Policy-Oriented Learning, and Policy Change." *Knowledge: Creation, Diffusion, Utilization* 8 (4): 649–92.
———. 1988. "An Advocacy Coalition Framework of Policy Change and the Role of Policy-Oriented Learning Therein." *Policy Sciences* 21 (2–3): 129–68.
Sabatier, P. A., and H. C. Jenkins-Smith, eds. 1993. *Policy Change and Learning: An Advocacy Coalition Approach*. Boulder: Westview Press.
Schneider, A. L. 1986. "The Evolution of a Policy Orientation for Evaluation Research: A Guide to Practice." *Public Administration Review* 46 (July–Aug.): 356–63.
Smith, M. F. 1988. "Evaluation Utilization Revisited." In *Evaluation Utilization*, ed. J. A. McLaughlin, L. J. Weber, R. W. Covert, and R. B. Ingle. New Directions for Program Evaluation 39. San Francisco: Jossey-Bass. 7–19.
Tobin, R. 1990. *The Expendable Future: U.S. Politics and the Protection of Biological Diversity*. Durham, N.C.: Duke University Press.
U.S. Environmental Protection Agency. 1990. *Environmental Investments: The Cost of a Clean Environment*. Washington, D.C.: Environmental Protection Agency. EPA 230-11-90-083.

———. 1993. *National Air Quality and Emissions Trends Report, 1992*. Research Triangle Park, N.C.: EPA Office of Air Quality Planning and Standards. EPA 454/R-93-031.

U.S. General Accounting Office. 1992a. *Environmental Protection Issues, Transition Series*. Washington, D.C.: GPO. GAO/OCG-93-16TR.

———. 1992b. *Superfund: Actions Needed to Correct Long-Standing Management Problems*. Washington, D.C.: GPO. GAO/T-RCED-92-78.

U.S. Office of Technology Assessment. 1988. *Are We Cleaning Up?: Ten Superfund Case Studies*. Washington, D.C.: GPO. OTA-ITE-362.

———. 1989. *Coming Clean: Superfund Problems Can Be Solved*. Washington, D.C.: GPO. OTA-ITE-433.

———. 1991. *Complex Cleanup: The Environmental Legacy of Nuclear Weapons Production, Summary*. Washington, D.C.: GPO. OTA-0-485.

Vig, N. J., and M. E. Kraft, eds. 1984. *Environmental Policy in the 1980s: Reagan's New Agenda*. Washington, D.C.: Congressional Quarterly Press.

Weiss, C. H., ed. 1972. *Evaluation Research: Methods of Assessing Program Effectiveness*. Englewood Cliffs, N.J.: Prentice-Hall.

———, ed. 1977. *Using Social Research in Public Policy Making*. Lexington, Mass.: Lexington Books.

———. 1978. "Improving the Linkage between Social Research and Public Policy." In *Knowledge and Policy: The Uncertain Connection*, ed. L. Lynn. Washington, D.C.: National Academy Press. 23–81.

Whiteman, D. 1985. "The Fate of Policy Analysis in Congressional Decision Making: Three Types of Use in Committees." *Western Political Quarterly* 38 (June): 294–311.

Wholey, J. S. 1991. "Using Evaluation to Improve Program Performance." *Bureaucrat* 20 (Summer): 55–59.

Wilson, J. Q. 1978. "Social Science and Public Policy: A Personal Note." In *Knowledge and Policy: The Uncertain Connection*, ed. L. Lynn. Washington, D.C.: National Academy Press. 82–92.

World Wildlife Fund. 1994. "Old-Growth Forests, Ecosystem Management, and Option 9." *Conservation Issues* 1 (May–June): 5–8.

13

Congressional Oversight and Program Evaluation: Substitutes or Synonyms?

GARY C. BRYNER

Program evaluation centers on the implementation and impact of public policies. It measures a number of characteristics of public policies, such as the process of implementation, the efficiency of policy efforts and resource use, the effectiveness of the programs, and the extent to which they accomplished their goals. Program evaluation focuses on assessments of the extent to which these policy interventions bring about change in the behavior of individuals and organizations and, in turn, to what extent these changes affect environmental quality (see the introduction).

Program evaluation is particularly challenging in environmental policy. The uncertainty over the causes and consequences of ecological problems makes it difficult to assess the impact of public policies. The long lag time between exposure and evidence of a problem makes it difficult to identify the factors responsible for environmental problems. Pollutants may be transferred to another environmental medium instead of actually being reduced in volume, making evaluation of programs incomplete.[1] The challenges in changing human behavior in ways that are more protective of the environment are inextricably intertwined with a host of other concerns, from economic growth to individual freedom. Some environmental hazards are especially risky, since their effects are largely irreversible, either in terms of loss of human life or ecological changes. As other authors in this collection have emphasized, there is little agreement over what indicators of environmental quality should be used in establishing goals and evaluating progress (see Solomon, Minear and Nanny, and Harris and Scheberle in this volume).

There is also little agreement over how much needs to be known about the health and environmental effects of pollutants before taking regulatory action.

What should be the balance between waiting for more research and intervening to prevent possible harm? A central issue is how risks should be calculated: some argue that all persons should be protected, including the most susceptible to the effects of pollution, while others insist that the risk posed to members of the community in general or on average be the basis of regulatory action (see Crocker and Shogren in this volume). Once risks are estimated, a second round of questions addresses how reducing environmental risks should be balanced with other values, such as individual freedom and corporate autonomy, and how political factors constrain the use of these evaluations (see Rich, Kraft, and Mangun in this volume for a discussion of the broader political context in which program evaluation takes place).

The distribution of the consequences of technological advances poses a particular challenge to democratic policy making. Many of the adverse environmental consequences of industrial activity will fall on future generations, while the benefits are largely confined to the current generation. It is not clear how the interests of future generations or subgroups of the population that have the fewest economic and political resources can be protected in a political system dominated by well-heeled interests (U.S. EPA 1992).

Much of the emphasis of environmental program evaluation is on improving the management of environmental programs. Just as important, however, are efforts to strengthen the role of program evaluation in the formulation of environmental policy by Congress and the executive branch (and, similarly, at the state level by legislators and governors). Program evaluation is a prerequisite for effective action in the reauthorization of legislation, the appropriation of funds required to achieve statutory goals, and the formulation of new laws. Program evaluation should not be seen as primarily an administrative function because it must also play a central role in the legislative process.

As many scholars and observers have recognized, Congress has created tremendous expectations for regulatory programs in the environmental statutes it has enacted, largely since 1970. More than 150 environmental laws and major amendments to them have been enacted by Congress. The laws that have created the most aggressive regulatory programs are quite new, and this area of law and policy, in comparison with some other areas, is rather young and still evolving. Evaluating programs is a critical congressional concern as these laws are continually being reviewed and reauthorized and as funds are appropriated each year. Congressional oversight is the forum for these evaluations. Oversight permeates the activities of Congress as members are constantly collecting information about how existing programs are working as they reauthorize laws, appropriate funds, and investigate problems (Aberbach 1990).

In this chapter I examine how program evaluation might be used to improve the formulation of environmental policy. I focus primarily on Congress and, in particular, on congressional oversight efforts to assess existing environmental laws and regulatory programs, to reauthorize these laws and programs, and to enact new laws to address new problems. Many of the same issues, however, apply to program evaluation and policy development in the executive branch. I will examine the goals of oversight, the mechanisms involved, criticisms of oversight, and strategies for Congress to make better use of program evaluation in formulating environmental policy.

Defining Oversight

Congress defined *oversight* in the Legislative Reorganization Act of 1946 (60 Stat. 832) as "continuous watchfulness of the execution by the administrative agencies ... of any laws." Rule X (b) (1), adopted by the House of Representatives in 1974, charges each committee with the responsibility to "review and study, on a continuing basis, the application, administration, execution, and effectiveness of laws" and to "determine whether such laws and the programs thereunder are being implemented and carried out in accordance with the intent of the Congress and whether such programs should be continued, curtailed, or eliminated."[2]

Central to the definition of oversight are the purposes of those who engage in it. For some in Congress, oversight seeks to assure that the intent of Congress is followed, that agencies are responsive to the will of the legislature, and that the spirit as well as the letter of the laws are being carried out. In theory, oversight involves a careful, objective examination of the statutes under which agencies are acting and a comparison of those statutory provisions with the actions actually taken by the agency. Although overseers may appeal to the original intent of the legislation, oversight may also be triggered by a concern with how well the agency is responding to current needs and problems.

Oversight activities may be used by members of Congress to protect and shore up political support for programs of interest to them. Much oversight takes place in an environment of policy advocacy. Committee members and staffers are generally drawn to assignments that mesh with their personal policy interests. Tension arises when a program or policy that members of Congress favor is not being pursued in a way that satisfies them. There may be significant differences of opinion among members of Congress and agency officials over how statutes should be implemented. Political appointees who are chosen because of their commitment to a president's promise to reduce regulatory burdens, for example, will likely clash with congressional champions of vigorous regulatory intervention. Oversight is also used to protect congressional prerogatives and powers against executive branch usurpation. Oversight is an essential means for members of Congress to ensure that agencies consult with them before embarking on major regulatory initiatives or changing established policies.

Another purpose of oversight is to permit members of Congress to respond to criticisms, complaints, and crises in which agencies have become embroiled. Oversight allows members of Congress to focus attention on administrative actions that are unpopular or misguided or illegal. Members of Congress take advantage of information uncovered by its staff or by outside groups to examine specific agency actions and to exert their influence in areas of interest to them. Oversight may be pursued primarily for publicity and political benefits. Or, as an executive branch official observed, "Where there is publicity to be gained, there is oversight to be had" (Ogul 1976:37).

Oversight is often aimed at ensuring that programs are implemented efficiently and that administrative waste, fraud, and abuse are identified and eliminated. Estimates of waste and inefficiency by the General Accounting Office (GAO), President Reagan's Grace Commission (a group of business executives headed by the industrialist J. Peter Grace that examined ways of reducing government spending), Vice President Gore's National Performance Review project, and a variety of other studies have fueled congressional oversight of federal agencies. Oversight of the U.S. Environmental Protection Agency (EPA) has focused on management issues, such as the way in which travel and other funds are expended and charges of excessive reliance on external contractors who are not well monitored by agency officials. A March 1993 report by the EPA inspector general and a hearing by the House Energy and Commerce Committee's Oversight and Investigations Subcommittee during that same month decried "massive" cost overruns in agency contracts, poor management of contracts, lack of competitive bidding, and an "intolerable waste of taxpayers' money" ("EPA" 1993:A18) (for an assessment of the EPA, see Rosenbaum in this volume).

Of most relevance here are oversight efforts that seek to use program evaluations and other information in reauthorizing or amending the statutes under which agencies operate. This may take place as part of a routine, annual reauthorization or may be aimed at particular problems and controversies. Oversight may also be designed to evaluate the effectiveness of an agency's pursuit of a particular policy and to consider ways of increasing the agency's ability to accomplish its tasks. Program evaluation has increasingly characterized the kind of oversight performed by the GAO; the GAO established in 1980 a unit specifically responsible for conducting program evaluations. These kinds of evaluations may also be conducted by committee and subcommittee staffs (see Solomon in this volume).

Since the jurisdiction, authority, and tasks delegated to regulatory agencies greatly exceed the resources provided, agency officials must make basic choices concerning the setting of priorities, the balancing of competing goals of economic growth and regulatory protection, and the regulating of complex and poorly understood phenomena. This places additional pressure on Congress to examine how well regulatory programs are working and how they might better respond to the problems to which they are directed. Oversight also gives members of Congress an opportunity to monitor how agencies proceed and to direct them in ways consistent with legislative intent. Since regulations affect important constituencies in virtually every district, members of Congress naturally seek to ensure that they can scrutinize carefully the actions agencies take. As environmental laws have become increasingly specific and detailed, informing congressional decision making with good program evaluations has become even more important. Congress is much less likely to delegate broad and vaguely defined powers to agencies than it has sometimes done in the past. Members write laws in excruciating detail, making it even more important for them to understand how well existing programs are working and how they can be improved through statutory changes.

Oversight Processes and Tools

Oversight permeates the activities of Congress. It is central to the budget process, to formulating and enacting legislation, to the Senate's confirmation of presidential appointees, and to constituent service—even sometimes to impeachment. In each activity, Congress possesses a powerful array of oversight tools and mechanisms to collect information from agencies and direct agency activity. Each of them offers an opportunity for members of Congress and their staffs to evaluate executive branch programs (U.S. Congressional Research Service 1984).

The Authorization Process

The authorization process provides the primary opportunity for members of Congress to review and redirect the implementation of the laws they pass. Congress has a variety of mechanisms for collecting information to be used in reauthorization deliberations. Hearings are the most visible means of obtaining information from agencies. Congress has required by law more than two thousand reports from the executive branch, many of which can be used to assist committees in preparing reauthorization legislation (National Academy of Public Administration 1988). Authorization subcommittees also receive information through informal meetings with agency staffs, letters, other oversight arenas, GAO reports, media investigations, constituent complaints and casework funneled through members' personal staffs, and other sources. The level of scrutiny given to agencies and programs in reauthorization proceedings varies. Some agencies must undergo annual reauthorizations and the accompanying oversight proceedings, while others enjoy multiyear or permanent authorization. The EPA, for example, operates under a permanent authorization with the exception of its research and development programs, which require annual reauthorization (U.S. Senate Committee on Governmental Affairs 1977).[3]

The Appropriations Process

The appropriations process provides one of the most direct opportunities for oversight. Appropriations hearings, agency budget submissions, reports, and meetings between agency and appropriations subcommittee staffs provide the fodder for congressional scrutiny of agency activity. As the budget process has become the most time-consuming element of Congress's agenda, however, the appropriations process has been weakened. Budget resolutions are often passed so late that authorization committees have little time to act. Other trends have similarly weakened the power of the appropriations process. Legislation proposed by appropriations subcommittees that in the past were largely ratified by Congress as a whole have been increasingly challenged through floor amendments to appropriations bills. The appropriations process has increasingly been used, especially when authorizing legislation has not been enacted, as a way for Congress to respond to concerns raised in the oversight process and instruct agencies to take or refrain from specific actions. The growing use of riders—provisions attached to appropriations bills that include substantive mandates and affect policy—reduces the influence of authorizing committees (Haas 1988; Schick 1976:89–90).

Auditing

The final step in the budget process, auditing, is also a part of oversight. The GAO audits a sample of agencies and programs to ensure that agency expenditures are consistent with congressional mandates. These reviews usually take up to a year and provide Congress with detailed, though somewhat delayed, information on the administration of the programs and agencies it has authorized.

General Oversight and Investigations

Legislative committees also engage in oversight actions that are not directly part of a reauthorization process but serve to give the committees general information about what is going on in the agencies under their jurisdiction and an opportunity to redirect agency activity. Committees respond to crises, scandals, and other problems affecting the agencies by calling hearings, asking for briefings by agency officials, writing letters, making phone calls, and commissioning special studies by the GAO and outside contractors. Much of this review takes place after the agency has taken action that arouses some controversy. Because congressional committees lack the staff to monitor agency activity comprehensively, they rely largely on criticisms and concerns raised by affected parties and others to identify areas for oversight investigation (McCubbins and Schwartz 1984).

In the House, most committees have established oversight subcommittees to oversee the agencies and programs within their jurisdiction. In theory, these subcommittees are tied less closely to the agencies and programs and are less likely to be advocates of the activity they oversee than are their legislative subcommittee counterparts. In the Senate, only a few committees have established investigations subcommittees. The House Government Reform and Oversight Committee and the Senate Committee on Governmental Affairs have a governmentwide responsibility to oversee executive branch activities as well as to examine broad issues of executive management, personnel practices, and organization. These committees have legal power to extract information, including taking depositions and issuing subpoenas. This power is restricted, however, in cases of national security and other areas in which presidents claim executive privilege in withholding information, but court decisions have placed significant constraints on presidential prerogatives and claims.[4] The House Republican leadership, in creating the Government Reform and Oversight Committee, emphasized the importance of a broad-ranging assessment of current laws, programs, and agencies.

The GAO is a major arm of Congress in overseeing the executive branch. Its auditing and investigative responsibilities were originally defined in the Budget and Accounting Act of 1921. The GAO was directed to review agencies and programs on the initiative of the comptroller general, at the request of any committee of Congress with jurisdiction over a program or activity, or by either house of Congress. In recent years, the GAO issued annually some seven hundred reports to members of Congress and committees and ninety-five reports to agencies. From 80 to 90 percent of its audits, evaluations, and investigations are traditionally requested by Congress. Much of the agency's attention is directed to program evaluation: of the recommendations it made in 1987, for example, 496 suggested changes to improve program effectiveness; 79 dealt with financial management; 78 were aimed at reducing costs or increasing revenues; 72 addressed congressional oversight and legislation concerns; and 191 covered other matters (U.S. GAO 1986, 1987).[5]

Oversight of Environmental Regulation and Programs

Environmental regulation takes place in an extremely complex political arena. Responsibilities for environmental regulation in the executive branch are widespread and include a White House office for environmental policy; the Office of Management and Budget; the Nuclear Regulatory Commission; and several departments, including Interior (public lands, energy, minerals, national parks), Agriculture (forestry, soil, conservation), Commerce (oceanic and atmospheric monitoring and research), State (international environmental agreements and concerns), Justice (environmental litigation), Defense (civil works, dredge and fill permits, pollution from Defense facilities), Energy (energy development and allocation), Transportation (airplane noise, oil pollution, transportation of hazardous substances), Housing and Urban Development (urban parks, planning), Health and Human Services (health), and Labor (occupational health).

A great number of congressional committees have at least some jurisdiction over environmental regulation, reflecting the division of responsibility in the executive branch. Legislative responsibility in the Senate is focused primarily in the Environment and Public Works and the Agriculture Committees. In the House, the Commerce, Resources, Transportation and Infrastructure, Science, and Small Business Committees share legislative jurisdiction. The Appropriations, House Government Reform and Oversight, and Senate Governmental Affairs Committees have major oversight and budget interests in the EPA and related agencies. The EPA, for example, is confronted by an extremely diverse and fragmented set of overseers: thirty-four Senate and fifty-six House committees and subcommittees exercise jurisdiction over the agency (U.S. EPA 1987). In the 1980s and early 1990s, agency officials have made 50–80 appearances before congressional committees, responded to 4,000–5,000 written inquiries, held countless informal meetings and telephone conversations with congressional staff members, and briefed members of Congress on major rule makings and enforcement activities each year.

Congressional oversight has been increasingly fueled by concern over the influence of officials in the Office of Management and Budget (OMB) and elsewhere in the Executive Office of the President and the White House in regulatory agencies. The Reagan administration's Task Force on Regulatory Relief and the OMB's regulatory review process were particularly aimed at regulations issued by the EPA and contributed to a growing mistrust between Congress and the executive branch (U.S. OMB 1987, 1990). Under Executive Order 12291, issued in 1981, the OMB was required to approve regulatory analyses accompanying all major regulations. These OMB reviews generated enormous controversy, as OMB overseers sought to push the agency in directions different from those suggested by congressional committee staffs, reports, and, in some cases, statutory language. Throughout the mid-1980s, more than two-thirds of proposed regulations submitted to the OMB by the EPA were changed after the review; by the end of the decade, as EPA officials apparently became more able to anticipate OMB concerns, the percentage of regulations changed fell to about one-third. The number of regulations reviewed fell from more than seven hundred at the beginning of the 1980s to about two hundred by the end, reflecting a significant reduction in rule-making activity.[6]

The Bush administration continued the review process and established in 1991 the President's Council on Competitiveness to reduce the burdens of regulation that threatened the competitiveness of U.S. industry. The council was widely attacked by Democratic members of Congress and public interest groups for encouraging agencies to pursue regulatory options that were inconsistent with congressional intent, displacing agency authority to determine the content of regulations, and relying on secret meetings with industry officials in which arguments and complaints were raised that contravened the procedural protections provided for in administrative law and failed to give others the opportunity to rebut the contentions (Clarke 1992; Rauch 1991; Woodward and Broder 1992).

President Clinton, in one of his first official acts, abolished the council and modified the regulatory review process. Under Executive Order 12866 (1993), each regulatory agency is required to submit to the OMB's Office of Information and Regulatory Affairs a regulatory plan outlining the most important regulatory actions it intends to take during the coming year. If elements of the plan are found by OMB officials to be inconsistent with the president's twelve principles of regulation, negotiations are triggered until the problems are resolved. When specific regulations having a major impact (an economic effect of $100 million or more or other major economic repercussions) are proposed, they must be screened by OMB desk officers to ensure that agencies design regulations that are the most cost-effective means of achieving the regulatory objective; provide a "reasoned determination that the benefits of the intended regulation justify its costs"; rely on the "best reasonably obtainable scientific, technical, economic, and other information concerning the need for, and consequences of, the intended regulation"; identify and assess alternative forms of regulation and specify, whenever possible, performance objectives rather than specific behaviors required; tailor regulations in ways that impose the least burden on society; and, consistent with achieving the goals of the regulation, take into account the cumulative cost of regulation.

These oversight efforts from OMB and the White House created great tension between Congress and the executive branch. While they have been softened considerably in the Clinton administration, they occasionally flare up and trigger congressional inquiries.

Criticisms of Congressional Oversight

EPA officials and others in the executive branch have been quite critical of Congress's oversight activities. Oversight is often castigated for being excessively burdensome. Former EPA associate administrator Milton Russell has argued that "without the umbrella kind of oversight—someone dealing with the organic EPA, there is so much oversight, that it absorbs a large amount of time, a very large amount of resources" (National Academy of Public Administration 1988:26). Others emphasize that oversight is obsessed with embarrassing agency officials and exposing problems that make "good press" instead of exploring more basic problems and concerns with the implementation of environmental statutes (National Academy of Public Administration 1988:26). A second set of criticisms of oversight, and of actions in the broader context of congressional-EPA relations, is that Congress fails to provide flexibility and a sufficient degree of administrative discretion to permit the agency to accomplish its statutory tasks efficiently and effectively. Russell argues that "oversight and litigation drive 90 percent of the agency's priorities, and there is very little opportunity to do anything else. . . . The agency's work is driven by outside pressure—litigation, hammers, some 600 statutory deadlines, and public pressure" (National Academy of Public Administration 1988:27). For former EPA administrator Lee Thomas, oversight often becomes

> what I term a kind of a witch hunt oversight, which is kind of a "I gotcha oversight," that is, "They didn't do what I wanted on this so I'm going to have a hearing and blast them and get my statement to the press ahead of his statement and make sure I get a lot of good press on my blast of the agency." . . . That is motivated, sometimes to a large extent, by an overzealous staff, sometimes by a member who is particularly interested in publicity, and sometimes by members and staff who feel that the way to get their direction on something is through intimidation. I don't think that's very constructive. As a matter of fact, I think that it has a chilling effect on decision-making in the agency, and in an agency like ours, you need to be able to make decisions; you need to be able to make reasonable decisions with the understanding that you're accountable for them, you may well hear about them, but that it's not going to end up as some kind of posturing, personal attack. (National Academy of Public Administration 1988:27)

Oversight efforts must strike a delicate balance. An agency doing its job efficiently and effectively is not the stuff news is made of. Problems need to be uncovered, and members of the Clinton administration have identified some major problems with the operation of the EPA. But too much criticism undermines support for a regulatory program. Effective oversight will root out problems and maintain a skeptical perspective but will do so in ways that ultimately contribute to the goals of the statutes the agency is seeking to implement.

From the perspective of congressional staff members, there is never enough oversight. One purpose of oversight is to protect the independence of the EPA from other elements of the executive branch. One House committee staff member argued that oversight "is often done to support the EPA against the executive establishment—against intrusion by OMB, to ensure the ability of the EPA to enforce the law against other agencies, especially large agencies like the Department of Defense and Department of Energy. [Congress's] role has largely been to support the EPA's authority, to reinforce the EPA's weak position in the executive branch" (National Academy of Public Administration 1988:29).

The tension and aggressiveness of congressional oversight, members argue, has been a result of the Reagan and Bush administrations' "tendency to ignore the will of Congress" (National Academy of Public Administration 1988:29). The frustration expressed by many congressional staff members is that oversight is very limited in directing agencies and in getting them to follow the will of members of Congress rather than the president. Given the limited congressional staff, oversight "is really most suited to kind of ad hoc fine tuning. When you have an administration that really just doesn't want to follow the law, oversight is just not going to correct that across the board" (National Academy of Public Administration 1988:29). Oversight can bring about some changes—the EPA has reversed itself in some rule-making and enforcement decisions, and oversight led to the resignation of virtually all of the top EPA officials in 1983—but Congress is still unable to monitor all the agency's activities (National Academy of Public Administration 1988).

Oversight is also driven by congressional lack of confidence in regulatory agencies. As one House staff member put it:

> Lots of people lost faith with agencies in the 1980s. That led to an increased amount of vigilance. There were lots of concerns—were the agency's concerns the same as those of Congress? Why are all the programs not being carried out? Congress believes that the agency should be doing all the tasks given it. There are more mandates than the agency has resources, and the Hill believes that the administration ought to put the resources that are needed there. The Hill wants the agency to do everything. The administration should put the money in the

agency that it needs. The EPA says it doesn't have enough. Congress is frustrated because the agency hasn't used the authority we gave them. They are not implementing the law. . . . We can't trust EPA with discretion. (National Academy of Public Administration 1988:30)

In areas where congressional subcommittees, interest groups, and agencies have developed close, mutually advantageous relationships, oversight has been criticized for being insufficiently rigorous. Members of Congress usually operate from a perspective of advocacy for the programs they oversee and have little interest in investigating and criticizing agencies in a way that might jeopardize funding and other support for them. These subsystems of government are unable themselves to provide careful, critical analyses of the appropriateness of policies and the effectiveness of their implementation (Dodd and Schott 1979).

Oversight, other critics argue, fails to have a real impact on the executive branch, since there is usually little follow-up on initial efforts and agencies know that they will likely not be required to respond to congressional criticisms. The oversight of authorizing committees is rarely regularized and comprehensive, but is usually ad hoc and spotty. The review of agencies in the appropriations process frequently occurs, but is not sufficiently thorough and in-depth to provide effective review (U.S. Senate Committee on Governmental Affairs 1977). Oversight done by the government operations committees is not usually integrated with that done elsewhere, and the lessons learned by these committees are often lost, for the committees have no power to alter enabling statutes under the jurisdiction of other committees. Congress, it is argued, lacks the information it needs from the executive branch to evaluate adequately the implementation of the laws it passes. Agencies resist providing information to Congress because it takes time and resources to compile and might be embarrassing or used against them. And even when information is forthcoming, Congress may not have the institutional capability to process it into practical, usable form. High staff turnover and lack of staff experience on Capitol Hill in many of the complex and technical areas exacerbate the difficulties of oversight (Pianan 1987).

These kinds of political pressures dominate the congressional oversight agenda and impinge on the production and use of program evaluation by Congress and the executive branch that could help improve environmental policy making. While some program evaluation takes place in oversight, as some recent legislative enactments such as the Clean Air Act Amendments of 1990 demonstrate (Bryner 1993), much more can be done to orient oversight toward the kind of program evaluation that can provide guidance in reauthorizing legislation (National Academy of Public Administration 1988).

Improving Congressional Oversight of Regulatory Agencies

There are two primary approaches to oversight. The first is narrow in focus, ad hoc, and episodic in response to an immediate, discrete problem or event in administrative activity. It often results from congressional concern that agencies are failing to take appropriate actions or have been unduly influenced by external sources. Such oversight is designed to identify (and remedy) a specific problem. It may also, at times, be driven simply by the possibility of political gain, such as a "quick-hit" hearing that embarrasses agency officials and gives members of Congress some media coverage.

Nevertheless, ad hoc oversight can be of considerable value. It can highlight problems and get information out to Congress and the public. It can call attention to particular problems and, while it may embarrass agency officials, may also be the only way to get their attention. Administrators are naturally critical of this kind of oversight, as are most members of Congress, who complain that they cannot get the attention of agency officials or counter OMB's influence unless television cameras are present.

Much of the oversight of the EPA throughout the 1980s and into the 1990s followed this ad hoc approach. Given the problems in the agency, this kind of oversight has been quite important. The challenge has been in how to ensure that it actually produces change. However, ad hoc oversight, while valuable and necessary, falls short of addressing many of the most important and pressing problems facing the EPA. It serves to limit administration discretion, but congressional vetoes of specific decisions, whether formal or informal, indicate only what the agency should not do but does not address broader problems, such as allocating scarce resources and balancing conflicting goals and purposes. The EPA's regulatory agenda is defined by court orders and statutory deadlines rather than explicit congressional or executive attempts to set priorities and make judgments about what problems can and should be given the greatest attention. Ad hoc oversight sometimes leads to, but in all cases should be complemented by, more systematic, comprehensive congressional inquiries.

While ad hoc oversight proceedings are necessary and politically inevitable, they should be integrated with the second kind of oversight—systematic and long-term analysis of laws and programs that goes beyond specific problems to evaluate the statutory framework from which agencies operate, reviews the environmental problems that led to government intervention in the first place, and considers anew whether laws are appropriately matched with the problems they are designed to remedy. It asks whether the specific mechanisms of intervention are accomplishing their purposes. Most importantly, it examines implementation by the EPA of laws with an eye toward refining these statutory mandates to address more effectively the environmental problems we face.

As Congress gives more attention in oversight to the setting of priorities, the allocation of resources, and the basic policy choices that must be made rather than specific EPA decisions, it can increase the likelihood that its legislative purposes are achieved and that basic policy choices are made by elected officials. As Congress focuses more attention on the EPA's planning activities, it can play a larger role in the formulation of EPA's regulatory agenda instead of waiting until policies are finally applied to the parties and purposes for which they were intended before intervening. Clear expressions by Congress of agency priorities can be combined with some administrative flexibility and discretion.

Resources for Oversight

Improving oversight of the EPA and other agencies requires creating incentives in Congress to engage in oversight activities and sufficient resources to ensure that they are effective and productive. While there are powerful incentives to engage in some kinds of oversight, systematic, comprehensive, policy-oriented, and long-term evaluation of environmental and other statutes and their implementation that offers few immediate political benefits often does not take place. Oversight resources are not always carefully allocated, resulting in overlapping and competing investigations. House and Senate leaders, as well as the chairs of specific committees, can do much to provide incentives for more systematic, comprehensive oversight that is aimed at evaluating statutes and programs. They can urge the allocation of resources and other benefits that encourage (or discourage) oversight. They can seek to define more clearly the relationship between the governmental affairs, government operations, appropriations, legislative, and budget committees to improve coordination and integration and reduce overlap and duplication. They can more effectively integrate the findings of their own investigations with those by other committees and congressional support agencies, follow oversight hearings with inquiries and studies, and assure that reauthorization and other legislation is informed by oversight findings and conclusions. Congress could fund more systematic evaluation of specific statutes, agencies, and programs. Funds could be made available for an entire Congress, or longer when appropriate, to encourage long-term studies.

Funding some oversight activity on a project-by-project basis would be a significant departure from current practice. Committees would still receive a general appropriation to conduct their business. But the House and Senate Rules Committees would, under this recommendation, set aside funds for special oversight projects proposed by the other committees. Ideally, committees would establish their oversight agenda in the December after each election so that the Senate and House committees could entertain requests for these funds in January. House and Senate leadership could meet with the administrative committees to establish guidelines for the awarding of these funds and to ensure strong leadership support for the idea. While these changes could improve oversight in general, Congress could begin with oversight of the EPA.

Coordinating Oversight in Congress

While oversight is an essential element in ensuring accountability of the administration of government to elected officials and to the laws they pass, it may sometimes further diffuse instead of pinpoint accountability. As indicated above, the EPA is subject to review by a great number of authorization and appropriation subcommittees and committees from both chambers of Congress; by the White House and OMB; by the public through participation in hearings, pressure on government agencies and officials, and lawsuits; by the media; and by federal courts. As a result it has often been unclear who ultimately has responsibility for agency decisions—OMB officials, federal judges, subcommittee members or their staff, or agency officials (Carnegie Commission on Science, Technology, and Government 1994).

Congressional leaders, committees, and support agencies can do much to increase the likelihood that oversight efforts of the EPA utilize scarce resources efficiently and effectively. Overlaps among authorizing, appropriations, governmental affairs/government operations, and budget committees should be creatively exploited for they can serve as a check on insufficiently vigorous oversight. But their plans and activities should be coordinated to ensure that all agencies and programs are reviewed regularly and that each committee benefits from the work of the others. Improved coordination of oversight faces major barriers, as few committees are willing to sacrifice jurisdictional turf. However, the major restructuring of the House of Representatives in 1995 demonstrates that such changes are possible when there is a major shift in power. In a few cases, the overlaps can pose difficulties for agencies that must respond to a myriad of overseers. Committees that share jurisdiction over agencies could meet at the beginning of each session with the leadership of each house of Congress to discuss oversight activities and examine issues that cut across agency statutes and committee responsibilities.

When each new Congress begins its first session, committee and subcommittee chairs from both chambers could meet with the EPA administrator to discuss areas of agency activity that should be the focus of specific oversight efforts. The members of Congress could express their concerns and give the administrator an opportunity to indicate areas where congressional direction is needed. Committees could encourage coordination of oversight activities and follow-up by requiring that committees or subcommittees prepare a report for each oversight project; agencies inform the committee of actions taken or planned on recommendations; those reports and agency responses be circulated to all relevant House and Senate committees; and the oversight findings and recommendations be included in the committee reports accompanying all authorization and appropriations bills.

Much of the systematic oversight that takes place is done by the GAO, although the other congressional support agencies play major roles. The GAO has conducted oversight of areas neglected by congressional committees, although it primarily assists Congress in the oversight it undertakes. Under the 1974 Congressional Budget and Impoundment Control Act, the GAO is required to report to Congress on the extent of program evaluation activities. Congress needs to ensure that the GAO has the technical and financial resources to keep pace with the rising demand for information and analysis (Carnegie Commission on Science, Technology, and Government 1991).

The amount of systematic review and evaluation of programs that Congress and its support agencies can do is limited, and Congress must also encourage executive branch efforts to evaluate the management and effectiveness of environmental programs in the EPA and other agencies. But such evaluations should be of great interest to members of Congress as they evaluate the adequacy of environmental statutes as well as the way in which they are implemented. Program evaluations can be a primary source of information for more systematic oversight activities by Congress. Members of Congress do not suffer from a shortage of information, but they often lack the specific kind of information they need. Funding for and review in Congress of program evaluations conducted by the EPA and the executive branch ought to be given high priority. Setting aside funds for program evaluation may appear to some to be a luxury, and pressure to reduce discretionary spending is likely to be aimed at program evaluation funds, but the benefits of program evaluation far exceed their costs. Precisely because budget constraints are so severe, Congress should provide for more program evaluation to assist it in making the difficult decisions about which programs should be expanded and contracted.

Congress could require the EPA to engage in a thorough, systematic, and comprehensive evaluation of the programs it administers and share the results of that analysis with Congress by earmarking funds for program evaluation, providing criteria by which to measure program effectiveness, and setting deadlines for submitting evaluation reports to committees.

Congress and the Regulatory Review Process

Much of the congressional frustration with the EPA lies in members' concerns about the executive branch's regulatory review process. Despite the respite in divided government with the election of Bill Clinton in 1992, the separation of powers still resulted in tension between the two branches over the direction of environmental policy. Congress ought to set the parameters of the regulatory review process to ensure that regulations are coordinated, to avoid duplication and conflicts, and to improve regulatory analysis by facilitating interagency communication and sharing of information.

Part of the solution to the tension between the executive and legislative branches over EPA rule making may also lie in initiatives in Congress to develop its own regulatory review process that will permit it to confront more directly the conflict with the executive branch over control of environmental policy and increase its ability to oversee and give direction to the implementation of the laws it enacts. This would require Congress to expand its capacity to deal with policy evaluation. It already has considerable resources in the congressional support agencies (Office of Technology Assessment, Congressional Research Service, Congressional Budget Office, and the GAO), but needs to develop new ways of adapting this analytic infrastructure to evaluating environmental laws and the regulations written by agencies in implementing them.

Instead of getting involved at the end of particular rule makings, as it has tried to do with the legislative veto in its various forms, Congress should become more involved in the shaping of regulatory policy, in setting priorities, and in making choices about what regulations are most important. The regulatory agenda that the EPA (and other agencies) are already required by executive order to prepare can be used by Congress and the White House as the vehicle for more proactive oversight. Congress could require that the agenda be adopted by the authorizing committees. Such consultation and review would avoid surprises, make more explicit the setting of agency priorities, and help Congress see the importance of allocating adequate resources to accomplish the tasks it delegates to the EPA.

The primary problem with the regulatory process is not that there is insufficient scrutiny by Congress, the White House, and the courts but that this oversight is narrow, ad hoc, episodic, and belated because it usually comes at the end of policy formation when agencies have already committed considerable resources. Agencies issue draft regulations that generate opposition from those at whom the regulations are aimed and Congress obligingly responds. Much more valuable is oversight efforts that are redirected toward the formation of the regulatory agenda of the EPA and other agencies. Perhaps the most frequently voiced criticism of environmental policy is that the goals established by environmental statutes and the way they are implemented by the EPA fail to address the most serious ecological problems. The environmental goals around which laws are constructed fail to require reduction of the most serious risks, but are fragmented and poorly integrated (Adams 1985; Borelli 1988). In 1987 the EPA completed a study that compared the risks of thirty-one major categories of environmental and health problems—ranging from global warming to oil spills—challenged some of the agency's earlier thinking about risks, and found major differences between the seriousness of risks and the agency's agenda (U.S. EPA 1987). The EPA commissioned in 1989 a review of its comparative risk assessment by the agency's Science Advisory Board that aimed a number of criticisms at the EPA and, indirectly, at Congress and environmental laws. The Science Advisory Board found that the EPA has largely been a "reactive" agency, insufficiently oriented toward reducing the greatest amount of risk. Not all risks can be reduced, but not all problems are equally serious, and the agency has failed to set priorities for reducing the most important problems (U.S. EPA 1990).

The 104th Congress renewed the need for good program evaluation in congressional oversight as the House of Representatives, in its Contract with America, took aim at environmental regulations. On February and March of 1995, the House passed a risk assessment bill and a benefit-cost measure that were combined with other environmental bills and then passed as HR 9. It would require the EPA and eleven other agencies, when proposing regulations with an expected cost of $25 million or more, to assess the nature of the risk being regulated using scientific data, including the range of risk; compare the risk at issue with similar risks, including everyday risks such as auto accidents; assess how effectively the rules would reduce risks to the public, based on detailed scientific data; assess risks that might occur if alternatives are substituted for the substance or practice to be regulated; and demonstrate that the costs of compliance are justified by the expected benefits and that the proposal is the most cost-effective alternative. Judicial review of agency actions would be expanded so that any party could sue the agency for failing to follow any of the risk or benefit-cost assessments. The benefit-cost requirement would override any existing statutory provision: if a less costly alternative than the proposed rule was identified, it would have to be substituted by the agency. The agencies must convene external committees, including representatives of regulated industries, to review the risk and cost assessments accompanying regulations with an impact of $100 million or more. All federal agencies are required to perform a series of analyses for any proposed rule with an expected impact of $50 million or more to demonstrate that the benefits outweigh the costs. Only emergency situations, military readiness–related activities, and federal approval of state programs and plans would be exempt from these requirements. The Senate failed to pass its risk assessment/regulatory reform bill in 1995 and the effort died well before Congress adjourned in 1996 (U.S. Senate Committee on the Judiciary 1996).

Given the prescriptiveness of environmental statutes, however, the EPA cannot make the kinds of changes in its regulatory activities anticipated by these reports and bills without the involvement of Congress. Congressional oversight that focuses on the formulation of the EPA's regulatory agenda can facilitate setting of priorities, while still giving agency officials sufficient discretion to manage their efforts. These oversight efforts can be structured by Congress to ensure that the kind of information required for effective program evaluation be included in the information agencies provide.

The Role of the Executive Branch

Many of the recommendations outlined above require the involvement and support of the president and the head of the EPA. Oversight efforts by Congress cannot as effectively evaluate existing statutes and their implementation without the commitment of executive branch officials to view oversight as a two-way street and maintain an interest in improved executive-legislative communication and cooperation. The executive branch must share responsibility for the state of executive-legislative relations. While the primary focus of the recommendations above is on Congress, presidents and their staffs and the EPA can do much to contribute to good oversight as they accept Congress as a partner in the policy-making process.

Both Congress and the executive branch can do much more to reorient oversight toward a consideration of the adequacy and appropriateness of the statutory framework under which the EPA operates and to explore options for rewriting laws to provide a more integrated regulatory program. The EPA can also use this opportunity to indicate to Congress what priorities it proposes and how it plans to allocate scarce resources in pursuing their statutory mandates. As a result, oversight is more likely to engage Congress and the EPA in an examination of basic policy directions than simply a review of specific agency decisions. Congress can do that by formulating an oversight agency that provides for a regular, rotating review of EPA programs as well as the overall agency. Congress could require that the EPA submit analyses of the appropriateness of relevant statutes, its capacity to carry out their mandates, and any recommendations for amendments. The EPA could also establish a mechanism for developing and providing suggestions to committees for areas of agency activity and statutes that would benefit from oversight proceedings.

Both the executive and legislative branches must allocate adequate resources for oversight. The EPA must provide resources to respond to congressional requests in a timely fashion. It must be willing to keep Congress apprised of its decisions and engage members of Congress in discussions of policy choices instead of simply announcing its decisions. Agencies that develop means of keeping congressional agencies abreast of regulatory initiatives and avoiding surprises are much less likely to find themselves being attacked in oversight hearings and investigations.

Most importantly, members of Congress and executive branch officials must move away from the idea of oversight as an afterthought, as a peripheral activity, and as a second-guessing of agencies in an adversarial atmosphere to oversight as the foundation of good legislation and central to everything Congress does. As Congress evaluates the structure of environmental statutes, oversight will be at the heart of the process by which it will make the important choices it confronts. Too little emphasis has been given in both of these branches to broad program evaluation that considers whether the basic structure of existing laws is appropriate, whether their purposes are still relevant, whether their provisions are well suited to the problems they are expected to remedy, and whether they should be rewritten. In a complex, rapidly changing world, the laws that Congress enacts require frequent review, and oversight is a primary, fundamental responsibility of Congress in constantly reexamining the framework of environmental regulation it has put in place.

The separation of powers is the essential constitutional underpinning for oversight, but it can be interpreted in many ways (Goldwin and Kaufman 1986). From one perspective, the separation of powers is inherently adversarial in nature. The legislative and executive branches are pitted against each other, and the emphasis is on the checking of one branch by another, particularly when different political parties control the two branches. Thus, oversight is expected to take place in an atmosphere of tension and mistrust.

From another view, the separation of powers is more balancing than checking, more aimed at bringing different perspectives to bear on common concerns and reflecting the interests of different constituencies than on blocking proposed actions. While there is a general division of legislative and executive labor, both the legislature and the executive branch are involved in making and implementing policy. Because a basic equality between the two branches is acknowledged, there is a willingness to share power rather than divide it and a recognition that policy making and policy implementation cannot easily be separated. The complexity of the environmental policy tasks confronting government realistically requires a sharing of powers by separate institutions.

The more that congressional-executive relations approximate this second model, the more likely that oversight can take place in a constructive atmosphere. Some minimum level of trust between the two branches is essential for effective government, even when they are controlled by different political parties. Congress needs to delegate some discretion and flexibility to agencies and to the president. The White House and agencies need to be clearly committed to faithfully executing the laws as written by Congress. In general, the more mature statutes are, and the more agreement there is concerning what laws mean and how they are to be implemented, the more likely it is that the two branches can work reasonably well together.

The Prospects for Increased Program Evaluation in Congress

How can Congress increase its use of program evaluation in formulating legislation, reauthorizing existing laws, and appropriating funds? The most commonly identified problems are that these evaluations fail to address the questions that members of Congress are really interested in and that they are not available in a timely fashion. For program evaluation to be more useful in Congress, it needs to be more closely attuned to congressional timetables. Interest groups, trade associations, and other policy advocates have learned to operate within these constraints and provide timely program evaluations in defense of their policy proposals. The successful advocates are sensitive to the kinds of questions members of Congress are interested in and tailor their evaluations accordingly.

At one level of analysis, this kind of program evaluation is quite defensible. It contributes to the expectation that ultimate policy-making authority rests with political officials who set the policy agenda. It recognizes the importance of political constraints in program evaluation and ensures that evaluations are not wasted because of their incompatibility with political concerns. Program evaluation can strengthen the efforts of policy entrepreneurs in Congress who seek to go beyond narrow functions of representing their districts and states to fashioning national policies to address important problems.

At another level, however, program evaluation needs to challenge the way members of Congress are thinking about environmental policy and consider whether the questions Congress is asking are the right ones. Program evaluation needs to be able to ask whether existing regulatory programs are addressing the right problems or are treating root causes rather than symptoms of ecological damage and environmental threats to public health. Program evaluation is a challenge to interest group politics. It threatens congressional acquiescence to the demands of powerful interests. It promises to help members of Congress consider a public interest that is not simply the aggregation of private interests. Program evaluation can provide a means of escaping the preoccupation with short-run results that permeates politics and encourage long-run thinking.

Members of Congress are not always ready for this kind of evaluation. Timing is important. As policy evaluation is done, it can be delivered to members of Congress in ways that explain clearly what the problems are, how they can be remedied, and why Congress ought to act. Program evaluation can contribute to demands for change, but evaluations themselves are rarely sufficient to launch Congress in a new direction. Until members of Congress are ready to change, or until they are overwhelmed by strong political forces that demand change, program evaluation will have to plod along, patiently identifying problems, offering solutions, and building an analytic record. The thirteen years of oversight of the Clean Air Act that culminated in the 1990 amendments is an example of how such a record can be built. But it, like other efforts, fell short of providing a sufficient basis for a major overhaul of the law. Just as program evaluation can focus attention on the long-run impact of policies, it must itself be part of a long-term enterprise to strengthen the capacity of government to assess its environmental policy efforts, learn from experience, and make policies more efficient and effective.

NOTES

1. The Pollution Prevention Act of 1990 (PL 101-58) recognizes the importance of pollution reduction rather than transferring it across media (from land to water to air, for example) and requires program effectiveness studies to be conducted by states that receive pollution prevention grants from the EPA. But the act does not appear to provide the kinds of incentives necessary to bring about a real shift toward pollution prevention (see Rosenbaum in this volume).

2. This discussion of oversight relies heavily on *Congressional Oversight of Regulatory Agencies,* a report produced from a project I directed at the National Academy of Public Administration in 1988, and research I conducted for *Blue Skies, Green Politics,* a book on environmental policy (1993).

3. Some agencies, such as the Nuclear Regulatory Commission and the International Trade Commission, must be reauthorized annually; other agencies, such as the Federal Communications Commission, the Securities and Exchange Commission, and

the Federal Trade Commission, are reauthorized every two to four years. Some agencies, such as the FCC, enjoyed permanent authorizations until the early 1980s, when Congress moved to a biennial authorization to increase its ability to monitor commission activities and to check the independence of agency heads unsympathetic with or opposed to the intent of Congress (Starobin 1988). Other members of Congress have pushed for longer reauthorization periods to help reduce their workload (Haas 1988).

4. Some review takes place as part of administrative decision making. Under some statutes, agencies must submit proposed actions to congressional committees for review. Report and wait provisions require that proposed rules be submitted to committees and not go into effect until thirty or sixty days after the congressional notification; the regulation goes into effect unless legislation is enacted rejecting the initiative. Other provisions require that Congress pass a joint resolution of approval, to be signed by the president, before the proposed rule goes into effect; still others call for the proposal to go into effect unless Congress passes a joint resolution of disapproval, also requiring presidential concurrence. These are the only forms of the legislative veto formally permitted under the Supreme Court's *Chadha* ruling. The Supreme Court's ruling against the legislative veto in 1983 limited Congress from employing one of its favorite oversight tools and stimulated congressional interest in considering alternative ways to facilitate executive-legislative relations and combining grants of discretionary authority with means of holding executive officials accountable for their actions (*INS v. Chadha,* 462 U.S. 919). Succeeding decisions of the Court in separation of powers cases have called into question other compromises Congress and the executive branch have developed for the sharing of power. In 1986 the Court struck down the Gramm-Rudman-Hollings deficit reduction act's trigger mechanism because it delegated executive authority to an official of the legislative branch and argued that the Constitution "does not contemplate an active role for Congress in the supervision of officers charged with the execution of the laws it enacts" and that "once Congress makes its choice in enacting legislation, its participation ends" (*Bowsher v. Synar,* 478 U.S. 714, 1986). Congressional participation in the appointing of officials to the Federal Election Commission was rejected by the Court in 1976 (*Buckley v. Valeo,* 424 U.S. 1) and cases challenging the constitutionality of independent regulatory commissions, special prosecutors, and guidelines issued by the U.S. Sentencing Commission have been raised. As these cases have sought to impose more of a separation between the executive and legislative branches, they have pushed Congress toward a more formal and adversarial stance vis-à-vis administrative agencies. In March 1995, the Senate resurrected the legislative veto and passed a bill that would require federal agencies to send to Congress proposed major regulations; Congress could then pass a joint resolution of disapproval that would be subject to presidential veto (Cushman 1995a, 1995b).

5. While some 80 percent of the GAO's work is in response to direct congressional requests, the GAO has some opportunity to establish and pursue its own independent oversight and evaluation of government programs. Its general management reviews, for example, are designed to study the organization and operation of entire departments and agencies and produce recommendations for improving management (U.S. GAO 1987).

6. These figures do not include a number of EPA regulations, such as pesticide tolerances (except those in which an existing tolerance is made more strict), other actions concerning pesticide regulation, and rules in which the EPA gives unconditional approval to state actions, such as state implementation plans under the Clean Air Act and state water standards. The figures include both proposed and final rules (U.S. OMB 1990:appendix).

REFERENCES

Aberbach, J. D. 1990. *Keeping a Watchful Eye: The Politics of Congressional Oversight.* Washington, D.C.: Brookings Institution.

Adams, J. H. 1985. *An Environmental Agenda for the Future.* Washington, D.C.: Island Press.

Borelli, P. 1988. *Crossroads: Environmental Priorities for the Future.* Washington, D.C.: Island Press.

Bryner, G. C. 1993. *Blue Skies, Green Politics: The Clean Air Act of 1990.* Washington, D.C.: Congressional Quarterly Press.

Carnegie Commission on Science, Technology, and Government. 1991. *Science, Technology, and Congress: Analysis and Advice from the Congressional Support Agencies.* New York: Carnegie Commission on Science, Technology, and Government.

———. 1994. *Science, Technology, and Congress: Organizational and Procedural Reforms.* New York: Carnegie Commission on Science, Technology, and Government.

Clarke, D. 1992. "Point of Darkness: The White House Council on Competitiveness." *Environmental Forum* (Jan.–Feb.): 31–34.

Cushman, J. H. 1995a. "Senate in Accord on Plan to Alter Rule-Making Role." *New York Times,* Mar. 29, A-1.

———. 1995b. "Senate Offers Plan to Curb Agency Rules." *New York Times,* Mar. 28, A-1.

Dodd, L., and R. Schott. 1979. *Congress and the Administrative State.* New York: Wiley.

"EPA in Sad Shape, New Boss Testifies." 1993. *Washington Post,* Mar. 11, A18.

Goldwin, R. A., and A. Kaufman. 1986. *The Separation of Powers: Does It Still Work?* Washington, D.C.: American Enterprise Institute.

Haas, L. J. 1988. "Unauthorized Actions." *National Journal* 20 (Jan. 2): 17–21.

McCubbins, M. D., and T. Schwartz. 1984. "Congressional Oversight Overloaded: Police Patrols versus Fire Alarms." *American Journal of Political Science* 28 (Fall): 165–79.

National Academy of Public Administration. 1988. *Congressional Oversight of Regulatory Agencies.* Washington, D.C.: National Academy of Public Administration.

Ogul, M. S. 1976. *Congress Oversees the Bureaucracy.* Pittsburgh: University of Pittsburgh Press.

Pianan, E. 1987. "Bleak House." *Washington Post,* national weekly ed., Dec. 7, 32.

Rauch, J. 1991. "The Regulatory President." *National Journal* 23 (Nov. 30): 2904–6.

Schick, A. 1976. "Congress and the 'Details' of Administration." *Public Administration Review* 36 (Sept.–Oct.).

Starobin, P. 1988. "FCC and Congress Clash over Proper Roles." *Congressional Quarterly Weekly Report* 46 (27 Feb.): 479–84.

U.S. Congressional Research Service. 1984. "Congressional Oversight Manual."

U.S. Environmental Protection Agency. Office of Legislative Analysis. 1987. "Congressional Hearings Held, 1985, 1986."

———. 1990. *Reducing Risk: Setting Priorities and Strategies for Environmental Protection.* Washington, D.C.: U.S. Environmental Protection Agency, Science Advisory Board. SAB-EC-90-021.

———. 1992. *Environmental Equity: Reducing Risk for All Communities.* Washington, D.C.: Environmental Protection Agency.

U.S. General Accounting Office. 1986. *Comptroller General's Annual Report, 1986.* Washington, D.C.: General Accounting Office.

———. 1987. *Federal Evaluation.* Washington, D.C.: General Accounting Office.

U.S. Office of Management and Budget. 1987. *Regulatory Program of the United States.* Washington, D.C.: GPO.

———. 1990. *Regulatory Program of the United States.* Washington, D.C.: GPO.

U.S. Senate Committee on Governmental Affairs. 1977. *Principal Recommendations and Findings of the Study on Federal Regulation.* Washington, D.C.: GPO.

———. 1986. *Office of Management and Budget: Evolving Roles and Future Issues.* Washington, D.C.: GPO.

U.S. Senate Committee on the Judiciary. 1995. *The Comprehensive Regulatory Reforms Act of 1995.* Washington, D.C.: GPO.

———. 1996. *Hearing on the Regulatory Reform Act of 1996.* 104th Cong., 2d sess. S. Doc. 343.

Woodward, B., and D. S. Broder. 1992. "Quayle's Quest: Curb Rules, Leave 'No Fingerprints.'" *Washington Post,* Jan. 1.

CONCLUSION

Environmental Program Evaluation: Promise and Prospects

GERRIT J. KNAAP AND TSCHANGHO JOHN KIM

If history offers glimpses into the future then in the final years of the twentieth century more volumes of environmental program evaluations of one kind or another will appear. Just as the Great Society of the 1960s stimulated advances in and applications of program evaluations in education, poverty, and unemployment, the environmental movement of the 1970s and 1980s should stimulate similar advances and applications in the evaluation of programs that address the environment. Driven by pocketbook environmentalism, however, most evaluations of environmental programs will focus on program efficiency; far fewer will address program implementation or environmental outcomes. Ironically, focusing evaluation narrowly on economic efficiency is inefficient. Economic efficiency in the evaluation enterprise requires that resources be allocated such that the marginal benefit of every type of evaluation is equal to its marginal cost. Channeling all evaluation efforts toward estimating the costs and benefits of proposed programs, for example, places at risk the opportunity to improve existing programs through changes in program implementation or through a better understanding of environmental processes.[1] Unfortunately, in the current political climate, there is no assurance that every form of evaluation will be pushed to the appropriate margin.

Despite wavering and uneven political support, the authors in this volume suggest, program evaluation has matured as a field and as a profession. They further suggest that significant progress has been made toward understanding how program implementation is influenced by institutional context; how biological, chemical, and ecological processes shape the natural environment; how to incorporate environmental services, environmental risks, and pollution prevention in economic analysis; and how to ensure that program evaluations, once completed, are effectively used. In short, the authors suggest that the subfield of environmental program evaluation has far greater potential than is currently being used. According to a national panel formed to evaluate the performance of the U.S. Environmental Protection Agency (EPA), for example:

> EPA should establish a program evaluation function for the agency as a whole.... Further, on-going evaluation should be a routine responsibility of managers in every EPA program.... The findings of such evaluations could be used to support program planning and decision making, budget requests or decision memoranda submitted to the chief operating officer.... EPA should structure external peer-reviews of its major operating programs on a recurring basis. Congressional and OMB staff, as well as staff from other federal agencies, states, and local governments, and the regulated community could participate in such reviews. (National Academy of Public Administration 1995:170)

Such effective use of program evaluation by the EPA, or other environmental agency, requires insights into the art and practice of program evaluation, the applicability of program evaluation to environmental issues, and the keys to putting evaluations to work. It is those issues that we address here.

The State of the Art

As Robert F. Rich describes in some detail, fashions in the art of program evaluation have changed over time. Early program evaluations were prepared by social scientists, who were perplexed to find their recommendations rejected or ignored by program administrators. Disillusioned with the work of academics, program administrators subsequently conducted their own evaluations with highly participatory methods. These evaluations were naturally better received, but often provided few objective or useful insights. Rich characterizes the present as a time of movement to the middle, in which scientific integrity is not compromised for practicality and the interests of the program administrator are not forsaken for the interest of scientific objectivity. Environmental policy may indeed be the domain where this middle ground can be found. Then again, it may not.

Environmental policy issues are characterized by scientific uncertainty; benefits that are diffuse and difficult to measure; costs that are concentrated, often large, and easy to measure; and core values and beliefs that vary widely among the population. While there is hope that program evaluations can be both rigorous and well received, it is not clear that environmental policy is the natural domain for this to occur.

The history of program evaluation suggests that the enterprise itself is strongly influenced by socioeconomic factors and the resulting political pressures. Program evaluation flourished in the 1960s and 1970s in response to demands that social programs be made more effective. Benefit-cost analysis flourished in the 1980s and 1990s in response to demands that environmental regulations be made less onerous. As the participants in the political process know quite well, the ramifications of a program evaluation depend significantly on the evaluation methodology—estimates of program costs and benefits do not yield the same insights as assessments of program implementation or outcomes. If environmental programs are to become both less onerous and more effective, they must be subject to evaluations of varying types—despite political pressures that favor certain methodologies over others.

The Place of Practice

Whereas the state of the art in program evaluation is in flux, the art of environmental program evaluation has no state at all. It has only artists. The home of many such artists is the General Accounting Office (GAO). As the agency created by Congress to evaluate the performance of the executive branch, the GAO offers perspectives on how environmental programs are currently evaluated by federal agencies and how its evaluations are influenced by institutional context.

According to Lawrence S. Solomon, the GAO faces severe obstacles when evaluating environmental programs. In addition to obstacles peculiar to environmental programs, the GAO frequently faces the "fear of evaluation." Like an audit by the Internal Revenue Service, an audit by the GAO is typically unwelcome, which can make information difficult to obtain. Thus the objectivity of an external audit may come at the cost of inaccurate or incomplete transfer of information. As a result, evaluations of environmental programs by the GAO frequently focus as much on the quality of information as on the quality of the program under evaluation.

Solomon's examples of the GAO's work provide clear illustrations of the difficulties that arise when environmental programs by one federal agency are evaluated by another. The GAO focuses largely on data—their adequacy, appropriateness, and use in analytical exercises. The GAO's findings suggest that data used by the EPA in programs for managing hazardous waste, reproductive and developmental toxicants, and groundwater quality are woefully inadequate. They tend to be incomplete, unreliable, and generally unsatisfactory. The programs themselves may also fail to satisfy, but without better data the

GAO is often unable to provide such an assessment. These examples suggest that the practice of environmental program evaluation within the federal government needs improvement and that for many environmental programs the data necessary to begin a program evaluation have yet to be developed.

Despite the urgent need for better data, Walter A. Rosenbaum argues that federal agencies, and the EPA in particular, will not soon develop data useful for evaluative purposes. The bases for Rosenbaum's skepticism are clear and compelling: the data necessary for assessing the efficacy of environmental programs are difficult and costly to obtain; data development and collection are among the least valued and supported administrative activities; and agencies have far more to gain from advertising initiatives than from gathering data for evaluation. Some of these factors may be specific to the EPA's pollution prevention program and not endemic to other programs of the EPA or other agencies. But institutional obstacles to data development by federal agencies are not easily overcome—despite repeated admonitions by the GAO and others.[2]

Data development, however, is not the only problem. Environmental programs involve both process and product. Although data on outcomes provide perhaps the ultimate measures of program success, they offer little of use for making programs successful. Constructive criticism requires not only an assessment of what is achieved but also an understanding of why achievement suffers. Such an understanding requires knowledge of the institutional context. According to Rosenbaum, for example, the EPA's pollution prevention program has struggled because of the EPA's historical orientation toward end-of-pipe regulation, the Office of Pollution Prevention's organizational weaknesses, and the difficulty of articulating program goals. One could take issue with Rosenbaum's assessment of the vigor with which the EPA pursued pollution prevention, question whether the structure of the EPA explains its lack of vigor, and doubt the generalizability of his findings. None of these issues, however, could be resolved with better data on outcomes.

Many environmental programs in the United States require federal mandates, federal funding, and state implementation. This intergovernmental structure is intended to allow states to tailor their programs to local environments while meeting national environmental goals. To pursue these dual goals, the federal government typically sets national environmental quality standards and grants responsibility for implementation to states that have adopted programs designed to meet or exceed those standards. The federal government then monitors compliance.

This intergovernmental structure influences the practice of environmental program evaluation. Since federal program administrators are primarily interested in whether states meet federal mandates and since federal mandates typically contain administrative rather than outcome requirements, evaluations of state environmental programs tend to focus on procedure instead of outcomes (a practice William R. Mangun describes as "bean counting"). Thus, for example, federal reviewers will examine whether a state has a drinking water program and how many permits have been issued under the program, rather than focus on the quality of the drinking water. To satisfy federal reviewers, state agencies tend to focus on the same procedural measures. As a result, the criteria for primacy and federal funding become the criteria for most program evaluations. This illustrates once again how the institutional context in both state and federal agencies influences the practice of environmental program evaluation.

In the introduction we suggested that program evaluations tend to focus on one of three issues: process, impacts, or efficiency. Process evaluations, we suggested, focus on the behavior of institutions and individuals; they assess how environmental programs cause the behavior of institutions or individuals to change. Solomon, Rosenbaum, and Mangun all examine institutional responses to environmental programs and all three find that, to varying degrees, institutions indeed respond to environmental programs. The dominant theme in all three of these chapters, however, is how institutions affect program evaluations and that the results of a program evaluation can be understood only by understanding the institutional context of the evaluation process. That is, it makes a difference who is conducting the evaluation, what is being evaluated, and how the evaluation is conducted. Subjecting all environmental programs to a benefit-cost test administered by a single government agency fails to yield the insights possible from multiple perspectives and multiple approaches. It may in fact do more harm than good.

The How Come of Outcomes

The impact of environmental programs on environmental quality is perhaps the most critical but least addressed issue by program evaluators. Somewhat ironically, environmental impact statements are required for all federal programs that affect the environment—except programs specifically designed to do so. The analysis of such impacts requires the development of appropriate environmental indicators; as Roger A. Minear and Mark A. Nanny, James Karr, and Hallett J. Harris and Denise Scheberle illustrate, rapid progress is being made on such indicators.

According to Minear and Nanny, many environmental concerns can be traced to developments in analytical chemistry. The ability to detect PCBs and their subsequent ban in manufacturing use provides but one vivid example.

Increased abilities to detect chemical substances need not, of course, lead to ever-stringent environmental regulations. Improved methods of detection can also be used to assess progress toward existing standards. These methods can be used to improve risk assessments and calibrate chemical transport models. Chemical transportation (or mass balance) models hold considerable promise for future environmental program evaluations. Such models will enable chemists to understand not only the movements and fates of toxic substances but also how environmental programs can and should intervene in these processes. As a result, the work of chemists and other natural scientists will become much more closely linked to program evaluation and policy development.

Although fishable and swimmable rivers, lakes, and streams were the original goals of the Clean Water Act, chemical indicators have been used much more than biological indicators as measures of anthropogenic influence and the success of environmental programs. Karr's index of biological integrity represents a tool for assessing water quality based on biological principles. The index has several advantages over other indicators: it provides a quantitative measure of water quality, its cardinal value indicates divergence from pristine quality, and it incorporates professional knowledge of biological systems. As a result, it can be used to compare streams to each other and to assess the influence of environmental programs. These advantages have led several states to use similar indexes to evaluate water quality programs.

No matter how refined, however, no single indicator will suffice as a measure of environmental quality. Directly or indirectly, the choice of indicators represents an expression of environmental management objectives. Indicators must provide an assessment of ecosystem integrity; but in diverse and democratic societies they must address social factors as well. For the Green Bay remedial action plan, for example, indicators include measures of sport fishing, recreation, and water-borne transportation. The inclusion of social variables as part of an environmental indicator program has significant managerial advantages. Including these makes it possible to identify potential trade-offs between alternative management goals, such as between fish abundance and hydroelectric power. This information could then be used by policymakers for considering such trade-offs directly or as input to benefit-cost analyses.

Analyses of environmental programs that focus on environmental impacts are rare and sorely needed. Most explanations for why this is so point to the lack of environmental indicators. But data collection is only part of the evaluation process. Once collected, the data must be analyzed to identify program impacts that must then be used to formulate programmatic change. With limited understanding of ecosystem processes and potentially large variations in indicator values that occur with these processes, identifying appropriate programmatic change remains a difficult statistical task. Data development, once again, is not the only problem.

Diminishing Marginal Productivity

The infusion of economic analysis into program design and evaluation has been one of the most significant trends in environmental management since the 1980s. This trend has at least two explanations. One is that increasing fiscal constraints and economic competition forced environmental programs to compete against other costly yet highly valued programs. The other is that economics has been used as an instrument to weaken environmental programs by conservative administrations. Both arguments have merit.

Regardless of cause, however, the clamor for economic analysis has led to rapid developments in analytical techniques. Benefit-cost analysis as an analytical framework has existed for decades, but Reagan's Executive Order 12291 imbedded the framework firmly into the policy-making process. With their work suddenly viewed as relevant, economists refined their techniques for estimating environmental benefits. These techniques involve examinations of how much people will pay for access to environmental amenities, how much people will pay for properties with environmental amenities, and how much people say they will pay for hypothetical amenities. Each of these techniques still lacks precision and widespread applicability; but that economists are now willing to consider hypothetical bids revealed through surveys is itself a remarkable development and has helped remove disciplinary barriers between economists and other social scientists.

Despite advances in benefit estimation, benefit-cost analysis continues to have severe limitations. Although not his intention, many limitations of benefit-cost analysis are illustrated by David H. Moreau. As the director of North Carolina's Water Resources Research Institute, Moreau sought a benefit-cost evaluation of North Carolina's soil-erosion control program and Water Protection Act. He got something well short. For the soil erosion control program, researchers were able to evaluate neither costs nor benefits; for the drinking water program, researchers were able to estimate only costs. Faced with these results, Moreau concludes that benefit-cost analysis is often reduced to cost-effectiveness analysis, in which programs are compared on the basis of costs but not benefits. But since even costs could not be estimated for the soil-erosion program, Moreau's examples raise doubts about even this limited form of economic evaluation.

Risk analysis is another technique for evaluating environmental programs to which economists have made significant contributions. The economic logic of risk analysis is again compelling: resources should be devoted toward those programs most successful at reducing risks to humans and their environment. The Toxic Substance Control Act and the Federal Insecticide, Fungicide, and Rodenticide Act, in particular, are founded on this logic. But the use of risk analysis for program evaluation remains controversial. As prescribed by the EPA, risk analysis involves two complex tasks: evaluating the source, nature, and extent of environmental risks and evaluating the significance or value of such risks. The first task involves the difficult assessment of how humans and other organisms are exposed and respond to environmental risks. The second task raises all the complexities involved in benefit-cost analysis—including the valuation of risks to human and environmental health. According to a 1995 National Academy of Public Administration report, "The unfortunate reality that the EPA and congress must confront is that neither risk assessment nor economic analysis can answer most of the other crucial questions about environmental programs" (47).

Thomas D. Crocker and Jason F. Shogren provide an even more penetrating critique of risk analysis. They point out that environmental risks are both endogenous and exogenous, or that some risks can be reduced by individual evasive action. The risk of toxic poisoning, for example, can be reduced by refusing to catch and eat fish from contaminated waters. This implies that when risk evaluation and management are separated, as suggested by the National Academy of Science and the EPA, risk assessments will be misspecified, the value of risk reduction will be underestimated, and environmental programs may promote excessive expenditures on self-protection. By integrating risk assessment and risk management, these pitfalls can be avoided. But until such integrated forms of risk analysis are better developed, Crocker and Shogren's critique raises further doubts about the legitimacy of economic analysis—used in isolation—for evaluating environmental programs.

A third issue introduced by John Braden and Chang-Gil Kim involves the use of economic incentives. As the authors state, analyses of economic incentives address the strategy of implementation rather than program outcomes. For nearly as long as economists have advocated benefit-cost analysis, they have also advocated replacing environmental regulations with economic incentives. In the 1980s, however, policymakers began to heed these economic arguments and created environmental programs that replaced command and control regulations with fees for wastewater permits, markets for air emissions, and liability charges for toxic waste contamination. In theory, these policy instruments create incentives for firms to adopt the most efficient pollution abatement technologies and for firms most efficient at abating pollution to abate the most pollution. The efficacy of such policies, however, remain largely a matter of faith.[3]

The lack of evidence supporting the efficacy of economic incentives reflects two limitations: they have rarely been tried and they are difficult to evaluate. The problem in evaluation arises for two reasons: pollution prevention is difficult to measure, since it requires a baseline for comparison, and pollution rises with production, thus a reduction in productivity can be mistakenly viewed as progress toward pollution prevention. The work by Kingsley E. Haynes, Samuel Ratick, and James Cummings-Saxton offers a promising strategy toward resolving both of these problems.

Haynes, Ratick, and Cummings-Saxton have developed a method for assessing pollution prevention by constructing pollution prevention frontiers. These frontiers are capable of providing valid interfacility and intertemporal assessment of progress toward preventing pollution. In essence, pollution prevention frontiers provide a ranking of firms (or any other production unit) based on a weighted sum of outputs divided by a weighted sum of chemical discharges. These rankings can then be used to compare firms to each other at particular points in time and to compare a firm to itself at different points in time.

Although problems remain in the application of pollution prevention frontiers—such as choosing the appropriate set of pollutants to consider and obtaining accurate data on those chosen pollutants—these frontiers have several advantages over traditional economic analyses of pollution prevention. Inputs and outputs can be measured in their natural units; the values of inputs and outputs need not be impacted; and no restrictive assumptions are necessary on the residuals of regression equations. Further, pollution prevention frontiers can distinguish leaders from laggers in pollution prevention, suggest a framework for transferring technology from leaders to laggers, and provide guidance for structuring economic incentives for continued progress in pollution prevention. In short, pollution prevention frontiers—despite their rather modest economic foundations—offer opportunities to resolve problems in evaluating pollution prevention programs identified by Solomon and Rosenbaum.

Economic analysis is concerned with allocating scarce resources to meet competing human desires. With the rise in pocketbook environmentalism, economists have had an increasingly influential voice in both the design of environmental programs and their evaluation. Economists have devoted their considerable intellectual resources toward improving economic approaches to evaluation. Yet pure economic analysis of environmental programs may have reached the point of diminishing returns. Despite years of methodological refinement, benefit-cost and risk analyses are still highly imprecise and sensitive to underlying analytical assumptions, and input-output and computable general equilibrium models are not demonstrably superior to carefully constructed ratios for many types of environmental evaluations.[4]

The Meaning of Use

No matter how high the quality of an environmental program evaluation, evaluation serves little purpose if its results are not used. This is not a new insight nor one limited to evaluations of environmental programs. As described by Rich, the lack of utility, or at least such a perception by program administrators, led to the development of utilization-focused and participation-oriented evaluation methodologies. These methodologies are not yet common for evaluating environmental programs. In part for this reason, the issue of use remains relevant in this domain.

According to Michael E. Kraft, the issue of use cannot be meaningfully addressed without defining the purpose of evaluation. A narrow definition—e.g., one that defines *use* in terms of clear and direct evidence of induced programmatic change—will clearly lead to a pessimistic view of the extent to which program evaluations are used. Few program evaluations will pass this litmus test. Alternatively, if the definition of *use* includes problem solving, conceptualization, political ammunition, manipulation, and advancement of self-interest, few program evaluations would fail to satisfy one of these criteria. The point is not that *use* should be more broadly defined, but rather that program evaluators need to consider the intended use of the evaluation in decision making about evaluation design.

Despite the admitted subjectivity of use, Kraft identifies four obstacles to use that program evaluators must strive to overcome; these include methodological weaknesses, organizational resistance, inadequate dissemination, and constrained professional roles. Each of these have particular implications for evaluations of environmental programs. Methodological weaknesses in evaluations of environmental programs often stem from limitations of data. As discussed by many previous authors, environmental program evaluations—and their use—will continue to suffer until better data are developed and made widely available. Organizational resistance to environmental program evaluations reflect the organization of environmental agencies, many of which are structured by environmental media, staffed by natural scientists most experienced in pollution regulation, and heavily dependent on outside contractors. To enhance utilization, environmental program evaluators need to consider how these organizational factors condition organizational resistance. Excluding the most technical evaluations, dissemination of findings should not be a particularly acute problem for evaluations of environmental programs. Environmental programs have high visibility, and unlike other regulatory agencies, environmental agencies tend not to be "captured" by the regulated industry. Given a public with growing interests in environmental issues, dissemination should be largely a matter of effort. Constrained professional roles, on the other hand, may pose a larger problem for evaluations of environmental programs than those in other policy domains. This will be true if a significant portion of environmental program evaluations are conducted by natural scientists and if natural scientists are more constrained than others in their professional roles.

Many of the issues raised by Kraft are revisited by Gary C. Bryner, who focuses on congressional oversight as a form of program evaluation. Bryner defines *congressional oversight* as the continuous watchfulness by Congress of the execution of law by administrative agencies. For environmental programs, oversight involves several executive agencies, over ten departments, the Nuclear Regulatory Commission, and several congressional committees. Like other forms of evaluation, congressional oversight has several purposes, which include increasing political support for programs, responding to criticisms of programs, and assuring legislative intent.

Bryner criticizes the practice of most oversight as narrow in focus, ad hoc, lacking in substance, and serving political motivations; he also offers suggestions for oversight's improvement. Bryner's major contribution, however, lies in his characterization of oversight as a continuous, interorganizational form of program evaluation. Whereas others have viewed program evaluation as a single or perhaps repeated form of study or administrative exercise, Bryner takes a much broader view, one in which environmental program evaluation is part of an integral cycle of program design, evaluation, and redesign.

Kraft and Bryner offer new insights for increasing utilization in the executive and legislative branches. Such insights are indeed welcome, especially when the executive and legislative branches are ideologically divided. At present, environmental program evaluations are used primarily as roadblocks to the adoption of new environmental programs. If used well, such roadblocks may limit the adoption of poorly designed programs, but they do little to help programs become better designed or to improve the programs currently in place. Capturing these potential benefits of environmental program evaluation will require their utilization at many more levels of the policy-making process than either federal or state statues currently require.

A Research Agenda

No single collection of essays can assimilate and synthesize all the material relevant to an emerging subfield of knowledge. No sweeping generalizations will likely withstand the test of time. Yet it is reasonably clear that the demand for environmental program evaluation in the late 1990s is driven by pocketbook environmentalism and thus the practice is dominated by a focus on economic efficiency. Resources are indeed scarce and must be used wisely. Yet as Giandomenico Majone asserts: "This fact justifies neither the identification of public decisions with the economist's allocation decision, nor the elevation of allocative efficiency to a position of unique importance" (1986:63). If environmental programs are to become more effective, program evaluators must examine their implementation and their environmental impacts as well as their economic efficiency.

The chapters in this volume that examine implementation suggest that understanding institutional structures is necessary for understanding why programs succeed or fail. They also suggest that institutions affect how programs are evaluated, that environmental program evaluation suffers from a lack of appropriate data, and that the paucity of data reflects the influence of institutional structure. These are useful insights, but they fail to address the most salient questions in environmental program implementation. These issues involve the relative efficacy of alternative agencies at implementing environmental programs, the relative efficacy of alternative policy instruments, and the appropriate delegation of responsibility for implementation to state and local governments. Specific questions, for example, include these: Would the Department of Defense serve as a better manager of hazardous waste than the Department of Energy? If so, why? Are pollution taxes and fees better policy instruments than direct regulations, fines, and prison sentences? For which types of pollution is this true? Should responsibility for environmental protection be delegated to the states? These are issues on which further research on program implementation could offer a great deal.

The chapters in this volume that address environmental impacts focus heavily on environmental indicators. They suggest that indicators can and have influenced the direction of environmental programs, though not always in desirable ways. The use of environmental indicators for evaluating environmental programs, however, requires an understanding of environmental systems, facility with imperfect data and research designs, and knowledge of the public policy process. No single discipline can claim preeminence in all these areas. Progress in environmental evaluation will thus require a multidisciplinary effort and sustained communication among the multiple disciplines. Environmental program evaluation is perhaps a rubric under which such communication can take place.

Environmental economists have more work to do as well. Benefit-cost and risk analysis techniques can always use more refinement; but at present, their theoretical development remains considerably ahead of their useful application. More work toward application would not only serve to increase the precision of estimates but also help to identify types of environmental programs for which economic evaluation provides the greatest insights and those for which policymakers should focus on alternative criteria. The treatment of equity could also use further development. Environmental equity seems to concern everyone these days except economists. Further work is also needed on the integration of risk assessment and risk management. For advancement in these and other areas, economists might contribute most effectively by collaborating with scholars trained in other disciplines.

Finally, additional work is needed on how to stimulate the use of environmental program evaluations. Use at present is largely confined to the debate over proposed programs. The use of program evaluation for improving existing programs must proceed by making evaluations part of the process of program design and development. Can monitoring and evaluation procedures be institutionalized before political and institutional impediments are firmly entrenched? Can program administrators find a way to use evaluations conducted by researchers who use scientific methods? These questions can be answered only with a better understanding of the cycle of program design, implementation, evaluation, and development.

In closing, we acknowledge that our focus on outcomes, process, and efficiency might be viewed by some as narrow. Robert Bartlett (1994) suggests that a blindered focus on action and outcomes—while ignoring societywide changes in rules, routines, beliefs, cultures, and knowledge—can lead program evaluators to find policy failure everywhere. Such is the risk of the evaluation enterprise. But if environmental program evaluations fail to find evidence of environmental impacts, procedural change, or potential gains in efficiency, then something is wrong with the environmental programs, the environmental program evaluations, or both. Further, it is incumbent upon those who care about either environmental quality or the evaluation enterprise to find out which is the case. We hope this collection inspires concerted efforts to do so.

NOTES

1. For a promising example of an environmental program evaluation that combines assessments of environmental outcomes, economic efficiency, and administrative processes see Center for Risk Management (1995).

2. For more on EPA-sponsored monitoring and assessment programs see National Research Council (1995).

3. According to a 1994 report by the National Academy of Public Administration: "The long-term effectiveness of economic instruments and their widespread adaptation hinge on integrating evaluation into the design and operation of economic incentive programs. Currently, the majority of economic instruments are administered without systematic measurement of their performance, either through self-examination or external oversight. . . . A major conclusion of this study is the importance of making systematic evaluation an integral component of the design of any economic instrument" (178).

4. See Freeman (1994) for an alternative perspective on the contribution and role of economics in environmental policy and policy analysis.

REFERENCES

Bartlett, R. 1994. "Evaluating Environmental Policy Success and Failure." In *Environmental Policy in the 1990s,* ed. N. J. Vig and M. E. Kraft. 2d ed. Washington, D.C.: Congressional Quarterly Press. 167–88.

Center for Risk Management. 1995. "The Pollution Control Regulatory System: An Evaluation." Draft outline. Washington, D.C.: Resources for the Future.

Freeman, A. M., III. 1994. "Economics, Incentives, and Environmental Regulation." In *Environmental Policy in the 1990s,* ed. N. J. Vig and M. E. Kraft. 2d ed. Washington, D.C.: Congressional Quarterly Press. 189–208.

Majone, G. 1986. "Analyzing the Public Sector: Shortcomings of Current Approaches—Part A. Policy Science." In *Guidance, Control, and Evaluation in the Public Sector,* ed. F.-X. Kaufmann, G. Majone, and V. Ostrom. New York: Walter de Gruyter. 61–70.

National Academy of Public Administration. 1994. *The Environment Goes to Market: The Implementation of Economic Incentives for Pollution Control.* Washington, D.C.: National Academy of Public Administration.

———. 1995. *Setting Priorities, Getting Results: A New Direction for EPA.* Washington, D.C.: National Academy of Public Administration.

National Research Council. 1995. *Review of EPA's Monitoring and Assessment Program.* Washington, D.C.: National Research Council.

Contributors

JOHN B. BRADEN is a professor of environmental economics in the Department of Agricultural and Consumer Economics and the director of the Illinois Water Resources Center at the University of Illinois at Urbana-Champaign. Braden is the author of more than 120 publications, including the edited volumes *Measuring the Demand for Environmental Quality* and *Environmental Policy with Political and Economic Integration* and articles in the *American Journal of Agricultural Economics, Land Economics,* the *Journal of Environmental Economics and Management,* and *Water Resources Research.*

GARY C. BRYNER is a professor of political science at Brigham Young University and the director of the Public Policy Program. His research interests include policy analysis, environmental and natural resource policy, international development policy, and social welfare policy. His publications include *Bureaucratic Discretion: Law and Policy in Federal Regulatory Agencies* and *Blue Skies, Green Politics: The Clean Air Act of 1990 and Its Implementation.*

THOMAS D. CROCKER is a professor of economics and the director of the School of Environment and Natural Resources at the University of Wyoming. He has held faculty appointments at the Universities of Wisconsin and California and a visiting appointment at Pennsylvania State University. Among his nearly 150 refereed publications are the original proposal for tradable emission rights, the first empirical application of the Coase theorem, and two of the initial treatments in environmental settings of the principal-agent problem. Recently, he has focused on ecosystem valuation methods and the impact of environmental change upon human capital formation.

JAMES CUMMINGS-SAXTON is a principal at Industrial Economics, Inc. in Cambridge, Massachusetts, and a research associate professor at Clark University. Cummings-Saxton previously served as a senior vice president at International Research and Technology Corp., the deputy technical director of the International Energy Program at Argonne National Laboratory, and a supervisor at Bellcomm, Inc. He also served on the Policy Team (Information Cluster) of the Eco-Efficiency Task Force of the President's Council on Sustainable Development. His publications include *Materials Flow in the United States* and *Protection against Depletion of Atmospheric Ozone by Chlorofluorocarbons*.

HALLETT J. HARRIS is the Herbert Fisk Johnson Professor of Environmental Studies and chair of the Department of Environmental Science at the University of Wisconsin at Green Bay. He has published widely in over fifteen different national and international journals. He has worked closely with government agencies to implement ecosystem management practices in the Great Lakes of North America.

KINGSLEY E. HAYNES is a university professor of public policy and the director of the Institute of Public Policy at George Mason University. He has worked in environmental policy and management in Texas, Indiana, Massachusetts, and Quebec and led the Ford Foundation's Office of Resource and Environment in the Middle East, focusing on Nile River management. Most recently his methodological work has related mathematical programming problems and cognitive decision mapping to pollution abatement and environmental justice in New Jersey and Ohio. His recent international work has involved examining trade-offs between regional economic development and environmental management in southern Taiwan for Taiwan's EPA. He has recently published articles in the *Annals of the Association of American Geographers, Environmental Professional,* and *Environmental International.*

JAMES R. KARR is a professor of fisheries, zoology, civil engineering, environmental health, and public affairs at the University of Washington in Seattle. Karr is the author of more than 170 publications, including articles in several dozen journals ranging from *Ecology, Bioscience,* and *Science* to *Environmental Management, Yale Journal of International Law, Human and Ecological Risk Assessment,* and *Ecological Applications*. Karr's research includes studies of tropical forest ecology, water quality and watershed management, conservation biology, ecological health, ecological risk, and environmental policy.

CHANG-GIL KIM is a research associate and a Ph.D. candidate in the Department of Agricultural Economics at Oklahoma State University and a research associate at Korea Rural Economic Institute in Seoul. Formerly a research assistant in the Department of Agricultural Economics at the University of Illinois at Urbana-Champaign, his research interests include nonmarket valuation and waste management. He is currently developing environmental performance indicators in livestock waste management using a data envelopment analysis.

TSCHANGHO JOHN KIM is a professor of urban and regional planning and civil engineering at the University of Illinois at Urbana-Champaign. For twenty years he has been teaching and researching urban and regional systems analysis with a particular emphasis on the development of policy evaluation methods for urban and regional planning, transportation planning, and economic analysis of public plans and policies. His publications include *Integrated Urban Systems Modeling: Theory and Applications, Advanced Transport and Spatial Systems Analysis: Applications to Korea, Expert Systems: Applications to Urban Planning, Expert Systems in Environmental Planning,* and *Spatial Development in Indonesia: Review and Prospects.*

GERRIT J. KNAAP is an associate professor of urban and regional planning at the University of Illinois at Urbana-Champaign and a visiting fellow at the Center for Urban Studies and the Environment at Indiana University. A long-time student of land use and environmental policy, Knaap is the coauthor of *The Regulated Landscape: Lessons on State Land Use Planning from Oregon* and the author of articles published in *Land Economics, Regional Science and Urban Economics,* the *Journal of Policy Analysis and Management,* and *State and Local Government Review.*

MICHAEL E. KRAFT is a professor and the chair of public and environmental affairs and the Herbert Fisk Johnson Professor of Environmental Studies at the University of Wisconsin at Green Bay. Among other works, he is the author of *Environmental Policy and Politics: Toward the Twenty-First Century* and co-editor of *Environmental Policy in the 1990s* and *Public Reactions to Nuclear Waste: Citizens' Views of Repository Siting.*

WILLIAM R. MANGUN is a professor of political science at East Carolina University. He has written on a wide range of environmental policy issues, including air and water pollution control, hazardous waste management, wetlands, wildlife management, natural resource management, and outdoor recreation. His publications include *Managing the Environmental Crisis: Incorporating Competing Values in Natural Resource Administration, American Fish and Wildlife Policy: The Human Dimension, Public Policy Issues in Wildlife Management, The Public Administration of Environmental Policy,* and *Nonconsumptive Use of Wildlife in the United States.* His articles have appeared in such journals as *Environmental Conservation, Environmental Management, Environment Review,* the *Journal of Environmental Systems, Public Administration Review, Policy Studies Review, Policy Studies Journal, Evaluation Review,* and *Leisure Sciences.*

ROGER A. MINEAR is a professor of civil engineering at the University of Illinois at Urbana-Champaign and the director of the Office of Solid Waste Research. He has been an invited scholar to Nankai University in Tianjin in the People's Republic of China and a visiting research scholar at Kyoto University in Japan. He is the author of more than one hundred publications on the nature, origin, transport, and transformation of organic and inorganic compounds in natural waters and wastewaters; the chemistry of aqueous solutions and chemical processes of water and wastewater treatment; and trace and environmental analysis.

DAVID H. MOREAU is a professor of city and regional planning and environmental sciences and engineering at the University of North Carolina at Chapel Hill. His teaching and research have been in the field of water resources planning and management. Since 1993, he has been chair of the North Carolina Environmental Management Commission, the rulemaking body for water and air quality management in that state. He formerly served as chair of the North Carolina Sedimentation Control Commission and director of the Water Resources Research Institute of the University of North Carolina.

MARK A. NANNY is an assistant professor of civil engineering and environmental science at the University of Oklahoma. His general research interests encompass aquatic, sediment, and soil chemistry, specifically nutrient cycling, pollutant-natural organic matter interactions, pollutant chemistry in soils and sediments, and the fate of pollutants during biodegradation. He also is interested in developing and applying experimental methods involving nuclear magnetic resonance spectroscopy to examine the chemical and physical behavior of compounds in the environment. He is the co-editor of *Nuclear Magnetic Resonance Spectroscopy in Environmental Chemistry*, which resulted from an American Chemical Society symposium he co-organized in 1993.

SAMUEL RATICK is an associate professor of geography and the director of graduate studies for the Environmental Science and Policy Program at Clark University. He is the former director of the Center for Technology, Environment, and Development of the George Perkins Marsh Institute and has also served as its acting director. Prior to joining the faculty at Clark, Ratick was on the geography faculty at Boston University and an associate director of the Center for Energy and Environmental Studies. He also served as the legislative assistant for energy and environment to U.S. Senator Daniel Patrick Moynihan and as an environmental scientist with the U.S. Environmental Protection Agency. Ratick has directed research projects addressing the development of analytical methods for environmental analysis, risk assessment, and hazard management funded by, among others, the National Oceanographic and Atmospheric Administration, the U.S. Environmental Protection Agency, the U.S. Army Corps of Engineers, and the National Institute for Global Environmental Change.

ROBERT F. RICH is the director of the Institute of Government and Public Affairs at the University of Illinois at Urbana-Champaign. He is also a professor of law, political science, health resources management, medicine, and community health. Rich has also held faculty positions at Carnegie-Mellon University, Princeton University, the University of Michigan, and the University of Chicago. Rich was a fellow at the Johns Hopkins University Center for the Study of American Government in Washington, D.C. He was also a special guest at the Brookings Institution. He has worked closely with government officials in the areas of strategic planning, organizational management, program evaluation, and management of information.

Contributors

WALTER A. ROSENBAUM is a professor of political science at the University of Florida and an adjunct research professor in the Department of Public Health and Environmental Medicine at Tulane University Medical School. He has published numerous books and articles about environmental politics and policy. Currently, he is a consultant to the Department of Energy on projects related to the environmental restoration of its nuclear weapons production facilities.

DENISE SCHEBERLE is an associate professor of public and environmental affairs at the University of Wisconsin at Green Bay. She has published several articles and book chapters on environmental policy implementation appearing in *Policy Studies Journal*. Her most recent book is *Environmental Federalism: Trust and the Politics of Policy Implementation*.

JASON F. SHOGREN is the Thomas Stroock Distinguished Professor of Natural Resource Conservation and Management and a professor of economics at the University of Wyoming. He has held faculty positions at Appalachian State University and Iowa State University and a visiting appointment at Yale University. Shogren also serves as the senior economist for environmental policy on the Council of Economic Advisors in the Executive Office of the President. Shogren's publications include *Environmental Economics: Theory and Practice* and articles in the *American Economic Review* and the *Journal of Environmental Economics and Management*.

LAWRENCE S. SOLOMON is a senior evaluator at the U.S. General Accounting Office in Washington, D.C. Solomon has directed several studies on environmental issues, including *Waste Minimization: EPA Data Are Severely Flawed, Pollution Prevention: EPA Should Reexamine the Objectives and Sustainability of State Programs*, and *Chemical Accident Safety: EPA's Responsibilities for Preparedness, Response, and Prevention*.

Index

Acid rain, 8, 206–7
Agricultural runoff, 57n3. *See also* Pollution
Air pollution exposure index (APEI), 98
Air pollution index (API), 98–99
Air quality, 91, 263; monitoring of, 135; chemical analytical methods of measuring, 132, 135; evaluation standards of, 88, 211–12, 296; health effects of, 211–12; Maryland Ambient Air Monitoring Program, 96; and motor vehicle emissions, 135; pollution exposure index, 98; pollution index, 98–99; and smokestack emissions, 135; state control programs for, 90–92, 95–96, 97–100. *See also* Clean Air Act; Pollution
Alaska Department of Environmental Conservation (ADEC), 88–89
Ambient Air Monitoring Program (Maryland), 96
American Evaluation Research Association, 30
American Society for Testing and Materials, 58n5
Anodic stripping voltammetry, 134
APEI (Air pollution exposure index), 98
API (Air pollution index), 98–99
Araclors, 131–32
Arizona Department of Environmental Quality (ADEQ), 95–96
Arizona environmental programs, 87, 95–96
Arkansas environmental programs, 149
Atomic adsorption spectroscopy (AA), 134

BCA. *See* Benefit-cost analysis
Benefit-cost analysis (BCA), 11–12, 48, 209–12, 215–25, 338; and Bureau of the Budget, 239; cost estimate models, 251–53; economic valuation, 215–17, 232n12; history of, 238–41, 349; limitations of, 251–53, 353; NAPAP report, 302; and ozone abatement, 211–12; in environmental program evaluation, 10–12, 35–36, 203, 211–12, 231nn1–3; and state programs, 238–53, 353; and VOC abatement technology, 211–12
Benthic index of biotic integrity (B-IBI), 168
Bhopal, India, disaster, 50
Bioaccumulation in food chain, 131, 141, 179
Bioassays, 140–41, 179–80
Biochemical oxygen demand (BOD), 8, 9, 130, 138, 151, 179
Biological assessment approaches, 149
Biological indicators: biomarkers, 141, 179; cumulative impacts on, 149; of fishable and swimmable water, 92–93, 152; impediments to use of, 150–58; measuring of, 8–9; versus chemical indicators, 153–54; of water quality, 150–53, 169, 179. *See also* Indicators
Biological integrity, 148–49, 152–57, 168–71, 180–81, 191; cost-effective approaches to, 158–59; ecoregion concept in, 156; indexes of, 149, 154, 159–68; measuring of, 159; standardization of monitoring, 156–57. *See also* Ecosystem integrity
Biomarkers, 141, 179
BOD (biochemical oxygen demand), 8, 9, 130, 138, 151, 179

Browner, Carol, 75
Bureau of the Budget, 239

CAA. *See* Clean Air Act
California environmental programs, 87, 102
Canadian Council of Ministers of the Environment (CCME), 188
Canadian environmental programs: Canadian Council of Ministers of the Environment (CCME), 188; Ontario IBI application, 165, 167; use of environmental indicators, 188
Carbon tetrachloride, 133
Carson, Rachel, 131
Causality, 48–49
CEA. *See* Cost-effectiveness analysis
CEEPES (Comprehensive Environmental and Economic Planning and Evaluation System), 260
CEQ (Council on Environmental Quality), 296, 309
CFCs (Chlorofluorocarbons), 225
Chemical analytical methods, 130–35; for air, 132, 135; anodic stripping voltammetry, 134; atomic adsorption spectroscopy, 134; detecting low concentrations, 136; electron capture detector, 131, 132; flame ionization detector, 131; future trends in, 134–37; gas chromatography–mass spectroscopy, 130–34, 136; inductively coupled plasma, 134; infrared spectroscopy, 131; impact on environmental programs, 135–37; for inorganic compounds, 134; liquid chromatography, 134, 136; mass spectroscopy, 132–33; measurements in, 137–40; pollution surrogates in, 138; purge and trap techniques, 133–34; remote sensing, 135; standards in, 131–32; ultraviolet-visible spectroscopy, 131; zero contaminant levels, 137
Chemical indicators: chemical-specific toxicological criteria, 156, 157; and risk assessment, 137, 139, 141–42, 179–80; standards, 131–32, 157; and water quality, 178–79. *See also* Indicators
Chemical oxygen demand (COD), 130, 138
Chemical standards, 131–32, 157
Chlorine and chlorinated compounds, 133, 139–40; chlorofluorocarbons (CFCs), 225; chloromethane, 140; DDT, 139–40
Civic environmentalism, 312. *See also* New environmentalism; Pocketbook environmentalism

Clean Air Act (CAA), 62, 68; Amendments (1990), 206–7, 211, 301, 302; economic evaluation of, 203; judging results of, 31; marketable pollution allowances in, 206–7; prevention of significant deterioration (PSD), 91; reporting requirements of, 51, 89; state programs under, 90–92
Clean Water Act (CWA), 9–10, 62, 92, 148, 152–53; Amendments, 1981, 49, 56; Amendments, 1987, 93, 119–20; and biological integrity, 148–71; goals of, 137–38; point versus nonpoint source pollutants in, 93, 119–20, 152, 157; reporting requirements of, 89, 91; and surface waters, 92; waste minimization in, 68. *See also* Water Pollution Control Act; Water quality
Clements, Frederick E., 178
COD (chemical oxygen demand), 130, 138
Colorado environmental programs, 149
Community Right-to-Know Act, 50
Comprehensive Environmental and Economic Planning and Evaluation System (CEEPES), 260
Computable General Equilibrium Analysis (CGE) models, 214–15, 231n10
Congressional Budget and Impoundment Control Act (1974), 336
Congressional Budget Office, 337
Congressional oversight, 321–44; Contract with America agenda, 338–39; coordinating, 335–37; criticisms of, 330–33; definition of, 323; of environmental regulation and programs, 328–30; and environmental statutes, 322; and executive branch, 328–29, 337, 339–41; and federal agencies, 77, 83, 323–25; as form of program evaluation, 357; House committees, 327; improving, 333–42; and political agendas, 83; processes, 325–27; and regulatory review, 337–39; and resistance to economic scrutiny of environmental programs, 203; resources for, 334–35. *See also* Environmental program evaluation; Environmental programs; Environmental regulation
Congressional Research Service (CRS), 58, 337
Connecticut environmental programs, 87
Construction Grant Program, 49, 56
Consumer Product Safety Commission, 55
Contingent valuation method (CVM), 223
Contract with America, 338–39; GOP and environmental deregulation, 305
Cost-effectiveness analysis (CEA): in biological

Index

monitoring, 12; in ecosystem remediation, 192; in environmental program evaluation, 35–36, 212–13, 231n8
Cost-efficiency analysis, 35–36
Council on Competitiveness, 301, 329
Council on Environmental Quality (CEQ), 296, 309
Creek chub (*Semotilus atromaculatus*), 166
CWA. *See* Clean Water Act

Data envelopment analysis (DEA), 272–85; application of, 280–84; compared to TRI ratio method, 283–84; pollution prevention frontier (PPF), 280–84, 355; production functions and efficiency frontiers, 274–77; theoretical and methodological problems in, 272–73, 277–78, 279–80
DDT, 139
DEA. *See* Data envelopment analysis
Delaware River watershed, 9–10
Department of Defense (DOD): site cleanup, 297–98, 310
Department of Energy (DOE): environmental programs, 297–98, 303, 310, 311; nuclear waste repository program, 299; organizational culture, 303, 311
Des Moines Water Works: nitrates release, 263–64
Discharge permits (NPDES), 49, 93, 152
DOD (Department of Defense): site cleanup, 297–98, 310
DOE. *See* Department of Energy
Drinking water standards, 50, 51, 92, 94, 136

Ecological indicators: evolution of, 177–80; in plant ecology, 177–78. *See also* Indicators
Economic benefits and costs. *See* Benefit-cost analysis
Economic models, 208–15; assessment criteria of, 208–9; computable general equilibrium analysis, 214–15; cost-effectiveness analysis (CEA), 212–13; economic valuation, 215–25; input-output analysis (I-O), 213–14, 231n9; Pareto criterion, 208; risk analysis, 225–29, 233n22. *See also* Benefit-cost analysis; Cost-effectiveness analysis; Economic valuation
Economic valuation: benefits and costs of programs, 215–17, 232n12; challenges of environmental programs, 205–6; "command-and-control" versus "incentive-based" programs, 206; computable general equilibrium analysis (CGE) models, 214–15, 231n10; congressional reluctance to accept, 203–4; contingent valuation method (CVM), 223, 232nn18–22; direct methods, 222–24; economic development versus environmental improvement, 304–5; efficiency of, 274, 347; and endogenous risks, 259; estimation of nonmarket values, 206; hedonic price method (HPM), 219–21, 232n16; indirect methods, 219–22, 232n17; limitations of, 207, 355–56; presidential enthusiasm for, 204–6; of program impacts, 215–25; risk analysis, 225–29; survey data use in, 216–19, 223, 225; total economic value, 224–25, 231n11; travel cost method, 221–22; willingness to pay (WTP) and willingness to accept (WTA), 217–19, 233n23
Ecosystem indicators. *See* Environmental indicators
Ecosystem integrity, 8, 9, 156, 176, 180–81, 190, 191; benthic idex of biotic integrity (B-IBI), 168; biological, 12, 148–49, 156–57, 158, 168–71; chemical analytical, 130–40; cost to maintain, 26, 293; ecoregion approach to, 149, 156, 180; and future of biological monitoring, 168–71; monitoring frequency, 46–47; rapid bioassessment protocol III (RBPIII), 168; and social factors, 352; variables in degradation of, 159. *See also* Air quality; Chemical analytical methods; Indicators; Pollution; Water quality
Ecosystem management: goals and objectives, 190–91; Great Lakes, 179–80; Green Bay, Wisconsin, 189–95. *See also* Environmental programs; Environmental Protection Agency; Environmental regulation; General Accounting Office
Efficiency frontier: in production functions, 276–77
Electron capture detector, 131, 132
EMAP. *See* Environmental Monitoring and Assessment Program
Emissions allowance trading, 206–7
Endangered Species Act (ESA) (1993), 8, 299, 304–5
Environmental chemistry, 130. *See also* Chemical analytical methods; Chemical indicators
Environmental indicators, 7–8, 176–96; Environmental Monitoring and Assessment Program (EMAP), 185–87; EPA, 183–84; as measures of environmental progress,

183–84; in program evaluation, 46; search for, 183–85; selection of, 180–82, 189–90, 195, 352; socioeconomic, 8; of water quality, 178–79. *See also* Indicators

Environmental Monitoring and Assessment Program (EMAP): ecological resource categories in, 8, 185; EPA, 8, 185–87; Great Lakes (EMAP-GL), 185–86

Environmental programs, 51, 70, 77–78, 87, 149; abatement, 102; air pollution control, 51, 90–92; in Alaska, 88–89; in Arizona, 87, 95–96; in Arkansas, 149; and benefit-cost analysis (BCA), 238–53, 353; in California, 87, 102; in Colorado, 149; in Connecticut, 87; and environmental policy, 32, 312; evaluation initiatives of, 94–103; expenditures for, 87; in Florida, 87, 149; in Illinois, 91, 165; incentive-based, 229–30, 354–55; in Indiana, 91; innovative, 96–102, 312; in Iowa, 263–64; in Maine, 149; in Maryland, 87, 96; in Massachusetts, 271; in Minnesota, 167; monetary value of, 215–25; nonpoint source, 93; in North Carolina, 241–51; NPDES, 93; in Ohio, 149, 158, 165, 167, 168; operating permit submissions, 92; in Oregon, 87, 97–100, 166, 271; pollution prevention, 47, 50, 271; primacy and EPA oversight, 87–91; in South Carolina, 102; in Tennessee, 91, 167, 168; in Utah, 91; in Vermont, 149; in Washington State, 87, 100–101; water pollution control, 78–79, 92–94; in Wisconsin, 189–95; in West Virginia, 167. *See also* Environmental program evaluation; Environmental Protection Agency; Environmental regulation; General Accounting Office

Environmental program evaluation: approaches to, 28, 33–35, 96–103; bias in, 314–15; congressional oversight of, 321–42; constraints on, 301–3, 314–15; costs of, 26, 50–51; and data collection, 48, 53–54, 55–56, 73–75, 81–82, 89, 309–10, 350; design and concept of, 64–65, 296–303; difficulties and limitations of, 46–51, 309–15; economic analysis, 203–31, 353, 355–56; evaluation research, 26–31; fear of, 49; future prospects for, 56–57, 315–16; GAO approach to, 52–56; goal confusion in, 64–65, 69; history and definition of, 1–4, 23–27, 348–49; and holistic solutions to problems, 183; and inability to determine causality, 48–49; institutional, 300, 307; impact assessments in, 7, 9, 10, 149; implementation of, 5–6, 32, 64–65, 69, 73–75, 78, 86, 299, 307, 351; and industrial self-monitoring, 47, 56; media-based versus integrated or cross-media approaches, 61–62; methodological weaknesses of, 301, 309–11; multidisciplinary efforts in, 358; outcome oriented, 31–32; performance standards in, 50; process oriented, 32, 86, 299, 307; and policy-making, 12–14, 305–9; and professional responsibilities, 169–71; and politics, 66–67, 77, 83, 203–4, 302–5, 349; and "pollution control culture," 75; and regulatory relationships, 53–54, 75; resistance to, 303, 311–13; results of, 12–14, 302–5, 310, 313–14, 356–59; scientific approach to, 23, 24–25, 28–33; state initiatives for, 94–103, 242–44, 353; and survey research, 223; and transfer of problems through time or space, 322. *See also* Economic models; Economic valuation; Environmental programs; Environmental regulation; Evaluation approaches; Risk analysis

Environmental Protection Agency (EPA): congressional oversight of, 324, 328, 330–31, 333–34; and coordinated oversight activities, 335–37; criticism of by GAO and OTA, 302–3, 311; data problems of, 81–82, 89, 309–10; and environmental indicators, 183–87; Environmental Monitoring and Assessment Program (EMAP), 8, 185–87; Green Bay Mass Balance Study, 189; Green Lights Program, 79–80; groundwater program, 78–79; and incentive-based policies, 206–7; lack of staff and funding, 51, 310; organizational resistance of, 311; Office of Air Quality Planning and Standards, 296; Office of Policy, Planning, and Evaluation, 296; outside influence on, 46, 66–67; oversight of state programs, 88–89, 96, 102–3, 124nn1–2; pesticide policy, 260; priorities of, 89; pollution control culture of, 69–70, 77, 82; pollution prevention program, 61–83, 270; and program implementation, 67; regional offices as information sources, 89; and risk assessment and management, 183–87, 258, 260, 294; Science Advisory Board, 338; self-evaluation of, 61–67, 72–83; "Statement of Preferred Options," 68; and standards, 88; Superfund program, 31, 52, 72, 136, 300, 302–3; 33/50 Project, 80. *See also* Economic valuation; Environmental program evaluation; Environmental regulation; Evaluation approaches; Risk analysis

Environmental regulation, 51, 70, 77–78, 87, 149; abatement, 102; of air pollution, 51, 90–92; and benefit-cost analysis (BCA), 238–53, 353; in Bush administration, 301–2, 304, 329; in Canada, 165, 167, 188; in Carter administration, 204, 240; in Clinton administration, 293; in Contract with America, 304; Council on Competitiveness, 1–2, 301, 329; Council on Sustainable Development, 307; and environmental indicators, 183–85; of evaluation initiatives, 94–103; expenditures for, 87; federal departments of, 297–98; in Ford administration, 204; Grace Commission, 4–5, 324; impact on environmental quality, 65, 351; innovative, 96–102, 312; limitations of, 300, 338, 348; National Performance Review Project, 324; of nonpoint source pollution, 93; NPDES, 93; objectives of, 129, 304–5; operating permit submissions, 92; of pollution prevention, 47, 50, 51, 90–92, 271; primacy and EPA oversight, 87–91; and program evaluation, 12–14, 293–94, 302–9, 311–12, 321, 341–42; in Reagan administration, 204–5, 302; by states, 87–89, 91, 95–101, 102, 149, 158, 165–68, 189–95, 241–51, 263–64, 271; Task Force on Regulatory Relief, 328; of water pollution, 78–79, 92–94. *See also* Congressional oversight; Environmental programs; Environmental Protection Agency; General Accounting Office

EPA. *See* Environmental Protection Agency

ESA (Endangered Species Act) (1993), 8, 299, 304–5

Evaluation approaches: alternative, 272–85, 296–99; benefit-cost analysis, 10–12, 35–36, 48, 203–5, 209–12, 215–25, 238–53, 302, 338, 349, 353; cost efficiency analysis, 35–36; cost-effectiveness analysis, 12, 35–36, 192, 212–13; data envelopment analysis (DEA), 272–85; development of performance indicators, 36–37; efficiency methods, 10–12, 351; environmental impact methods, 7–10, 351; "evaluability assessment," 33–34, 38; information management techniques, 37; outcome-based, 87, 296–99; performance monitoring, 34–35; process methods, 5–6, 32, 86, 299, 307, 351; qualitative evaluation, 35; rapid feedback evaluation, 34; research on, 25–31; user satisfaction studies, 37

Fear of evaluation, 49, 51, 61–62
Fecal contamination, 152, 154
Federal Communications Commission, 342n3
Federal Insecticide, Fungicide, and Rodenticide Act (FIFRA), 203
Federal Interagency River Basin Committee, 239
Federal Trade Commission, 342n3
Federal-state partnerships: RCRIS, 52, 53; Superfund program, 52. *See also* Environmental programs; Environmental regulation
Federal Water Quality Control Act (1965), 157, 178, 189
FIFRA (Federal Insecticide, Fungicide, and Rodenticide Act), 203
Fishable and swimmable waters, 92–93, 152
Flame ionization detector, 131
Flood Control Act (1936), 239
Florida environmental programs, 87, 149
Food and Drug Administration, 55

GAO. *See* General Accounting Office
Gas chromatography–mass spectroscopy, 130–34
Gasoline, 206–7
General Accounting Office (GAO), 296, 315; approach to environmental evaluation, 47–48, 52–56, 349–50; Construction Grants Program, 56; ignoring political and institutional factors, 300; oversight function of, 324, 327, 336, 343n5; promotion of evaluation results, 13–14; review of DOE environmental programs, 303; review of EPA data, 48; Superfund program, 302–3; water quality data, 51, 55
Global warming, 8
Government Reform and Oversight Committee (House), 327
Grace Commission, 324
Gramm-Rudman-Hollings Deficit Reduction Act, 343n4
Great Lakes: ecosystem, 8; ecosystem management, 179–81; Environmental Monitoring and Assessment Program (EMAP-GL), 185–86; International Joint Commission (IJC), 187–88; Water Quality Improvement Act (1965), 189
Great Lakes Water Quality Agreement (1978, 1987): search for indicators, 187; and ecosystem integrity, 181

Green Bay: ecosystem management program, 189–95; environmental indicators, 189–95; University of Wisconsin Sea Grant Institute, 189
Green Bay Mass Balance Study, 189
Green Lights Program, 79–80

Hanford Nuclear Reservation, 297
Hazardous and Solid Waste Amendments (1984): goal confusion in, 64–65; Office of Solid Waste, 69; waste minimization in, 68
Hazardous waste: GAO findings, 52–54; number of sites, 47; generators, 53–54
Health effects: of air pollution, 211–12; of dissolved humic material, 134; as environmental indicators, 9; and genetic damage, 141; of low concentration chemicals, 136–37; of reproductive and developmental toxicants, 55; risk evaluation of, 142–43, 228–29, 231n8
Heavy metals, 135. *See also* Toxic chemicals
Hedonic price method (HPM), 219–21
Hilsenhoff index, 154

IBI. *See* Index of biotic integrity
IJC (International Joint Commission), 187–88
Illinois environmental programs, 91, 165
Impact assessment, 7, 9, 10, 149
Index of biotic integrity (IBI): advantages and applications of, 165; geographical adaptability of, 166–68; invertebrate data in, 168; richness and composition metrics in, 160–61, 162–64; stream fish data in, 159–68; use with other environments, 167–68
Indiana environmental programs, 91
Indicators: biological, 8–9, 92–93, 141, 149, 150–58, 179, 352; chemical, 137, 139, 141–42, 156–57, 178–79, 351–52; ecological, 177–80; environmental, 8, 46, 176–95, 351–52; Environmental Monitoring and Assessment Program, 185–87; Green Bay program, 180–95; health effects, 9, 211–12; index of biotic integrity, 159–68, 352; and management objectives, 352; physical, 178–79; physiochemical, 8; selection of, 180–82, 189–90, 195, 352; socioeconomic, 9
Inductively coupled plasma (ICP), 134
Information management techniques, 37–38
Infrared spectroscopy, 131
Inorganic compound analysis, 134
Input-output analysis (I-O), 35–36, 213–14, 231n9

Integrative biological indexes, 149
International Joint Commission (IJC), 187–88
International Trade Commission, 342n3
Invertebrate community index (ICI), 168

Legislation: Clean Air Act (CAA), 31, 51, 62, 68, 89, 90–92, 203, 206–7, 211, 301, 302; Clean Water Act (CWA), 9–10, 49, 56, 62, 68, 89, 91, 92, 93, 119–20, 137–38, 148, 152–53, 157; Community Right-to-Know Act, 50; Endangered Species Act (ESA), 8, 299, 304–5; Federal Insecticide, Fungicide, and Rodenticide Act (FIFRA), 203; Federal Water Quality Control Act, 157, 178, 189; Flood Control Act, 239; Massachusetts Toxic Use Reduction Act, 271; National Environmental Policy Act (NEPA), 7, 179, 299, 300; Nuclear Waste Policy Act, 299; pending, 270; Pollution Prevention Act (PPA), 57n1, 62–63, 65, 70–78, 270, 342n1; Refuse Act, 151; Resource Conservation and Recovery Act (RCRA), 53, 68–69, 72, 73, 300; Risk Assessment and Cost Benefit Act (proposed), 2, 12; Safe Drinking Water Act, 92, 94, 121; Superfund Amendments and Reauthorization Act (SARA), 52, 72; Toxic Substances Control Act, 68, 203; Water Pollution Control Act (WPCA), 148–49, 151–52, 157; Water Quality Act, 153; Water Quality Improvement Act, 178, 189
"Limits of tolerance," 179
Liquid chromatography, 134

Macroeconomic impact of program models, 214
Maine environmental programs, 149
Marketable pollution allowances, 206–7
Market-based incentives, 294
Maryland environmental programs, 87, 96
Massachusetts Toxic Use Reduction Act, 271
Mass spectroscopy, 132–33
Minnesota environmental programs, 167
Motor vehicle emissions: monitoring, 135; pollution allowance trading, 206–7

NAPAP (National Acid Precipitation Assessment Program), 302
National Academy of Sciences: risk assessment and management, 258
National Acid Precipitation Assessment Program (NAPAP), 302
National Energy Strategy (1991), 301–2

Index

National Environmental Policy Act (NEPA) (1969), 7, 179, 299, 300
National Oceanic and Atmospheric Administration (NOAA), 191
National Performance Review project, 324
National Pollutant Discharge Elimination System (NPDES), 49, 93, 152
National Resource Defense Council (NRDC), 271
National Technical Advisory Committee on Water Quality Criteria, 178
NEPA (National Environmental Policy Act) (1969), 7, 179, 299, 300
New environmentalism, 1–2. *See also* Pocketbook environmentalism
Nitrates: released by Des Moines Water Works, 263–64
NOAA (National Oceanic and Atmospheric Administration), 191
Nonmarket valuation techniques, 215–25, 262
Nonpoint pollution sources: agricultural runoff, 57n3; and Clean Water Act Amendments, 93, 152; stream contaminants, 157. *See also* Pollution
North Carolina environmental programs, 241–51
NPDES (National Pollutant Discharge Elimination System), 49, 93, 152
NRDC (National Resource Defense Council), 271
Nuclear Regulatory Commission, 342n3
Nuclear Waste Policy Act (1982), 299

Occupational Safety and Health Administration (OSHA), 55
Office of Management and Budget (OMB): influence on regulatory agencies, 328–29; "quality of life review," 204
Office of Pesticides and Toxics, 76
Office of Policy Evaluation and Enforcement (OPEE), 76
Office of Pollution Prevention (OPP), 71–72, 76; deficient evaluation data of, 81–82; distrust of Congress and EPA, 77; Office of Policy Evaluation and Enforcement (OPEE), 76; "pollution control culture" in, 75; Pollution Prevention Act, 71; startup confusion, 76. *See also* Environmental Protection Agency
Office of Technology Assessment (OTA), 81, 302–3, 337
Ohio environmental programs, 149, 158, 165, 167, 168

O'Leary, Secretary of Energy Hazel, 311
Ontario: IBI application of, 165, 167
Oregon environmental programs, 87, 97–100, 166; Toxic Use Reduction and Hazardous Waste Reduction Act, 271
Oregon Department of Environmental Quality, 97–100
Organizational resistance, 309–14
OSHA (Occupational Safety and Health Administration), 55
Outcome-based evaluation, 87, 296–99
Oversight, 323–25. *See also* Congressional oversight
Oxygen, dissolved, 179
Ozone, 211–12, 231n7

Pareto, Wilfredo, 208
Pareto criterion, 208
PCE (Perchloroethylene), 133
Perchloroethylene (PCE), 133
Performance indicators, 36–37. *See also* Benefit-cost analysis; Cost-effectiveness analysis
Performance standards, 50. *See also* Benefit-cost analysis; Cost-effectiveness analysis
Pesticides, 55–56, 76, 131–32, 260. *See also* Pollution; Toxic chemicals
Physical indicators, 178–79. *See also* Indicators
Pocketbook environmentalism, 1–2, 347, 355, 358
Point pollution sources, 93, 114–18, 151–52, 157. *See also* Pollution
Policy analysis, 25, 301–2, 303
Policy research: information management, 37; management indicators, 38. *See also* Environmental program evaluation
Politics: congressional agendas, 83; congressional distrust of EPA, 77, 83; GAO apoliticism, 67; influence on congressional oversight, 332; influence on program evaluation, 45–46, 66–67, 302–5, 312; oversight as a tool in, 324; in policy initiation and implementation, 65–66, 82; programs off-limits to economic evaluation, 203–4; regulatory, 77; in shaping state environmental programs, 87. *See also* Environmental programs; Environmental regulation
"Pollutant of the month" syndrome, 135
Pollution, 39, 139; control versus prevention, 70–71; costs of control, 45, 47; cross-media strategies, 70; data envelopment analysis (DEA), 272–85; definition of, 63; and economic incentives, 206–7, 229–30, 354–

55; EPA programs, 61–83; Green Lights Program, 79–80; history of prevention of, 68–72; indicators of improvement in, 139–40; by inorganic compounds, 133; market in allowances, 206–7; and measuring surrogates, 138; by metals, 133; methodology of prevention, 270, 271–72; National Pollutant Discharge Elimination System (NPDES), 49, 93, 152; natural sources of, 140; nonpoint sources of, 49, 57n3, 93, 152, 157; by organic compounds, 130–33; point sources of, 93, 114–18, 151–52, 157; "pollutant of the month" syndrome, 135; "pollution control culture" in EPA, 75, 77, 82–83; prevention frontier (PPF), 276–77, 280–84, 355; progress in control, 47–48, 183; and risk reduction, 270; source reduction of, 72; state prevention programs for, 271; 33/50 Project, 80; waste minimization, 63, 68–72, 74–75. *See also* Air quality; Environmental program evaluation; Environmental programs; Environmental Protection Agency; Environmental regulation; Legislation; Pollution Prevention Act; Water quality

Pollution Prevention Act (PPA) (1990), 57n1, 62–63, 71, 270, 342n1; costs to private sector, 77; data-gathering problems, 73–75; goals versus results, 65; institutional confusion, 76–78; objectives and implementation, 70–73

Pollution prevention frontier (PPF), 283–84, 355

Polyaromatic hydrocarbons (PAH), 133

Polychlorinated biphenyl (PCB), 131–32, 133, 136

Polychlorinated dibenzofurans (PCDF), 133, 140

Polychlorinated dioxins (PCDD), 133, 140

Polymerase chain reaction (PCR), 141

PPA. *See* Pollution Prevention Act

PPF (Pollution prevention frontier), 283–84, 355

President's Council on Sustainable Development, 307

President's Council on Competitiveness, 329

Prevention of significant deterioration (PSD), 91

Primacy: state responsibility for law enforcement, 88, 90

Process evaluation, 32, 86, 94–96, 299–300

Production functions: applications in pollution prevention, 274–77

Productivity analysis, 12

PSD (prevention of significant deterioration), 91

Public awareness, 50, 51

Public water supply standards (PWSS), 92, 94

Purge and trap techniques, 133–34

Rapid bioassessment protocol III (RBPIII), 168

RCRA (Resource Conservation and Recovery Act) (1976), 53, 72, 73, 300; Hazardous and Solid Waste Amendments (1984), 68–69

RCRIS (Resource Conservation and Recovery Information System), 52

Reauthorization period, 325–26, 343n3

Recycling: under PPA, 72, 80

Refuse Act (1899), 151

Reilly, William P., 76

Remote sensing, 135

Reproductive and developmental toxicants, 55

Resource Conservation and Recovery Act (RCRA) (1976), 53, 72, 73, 300; Hazardous and Solid Waste Amendments (1984), 68–69

Resource Conservation and Recovery Information System (RCRIS), 52

Resources for the Future Institute, 47, 296

Risk: abatement and transfer, 263–66; assessment, 11, 137, 139, 141–44, 180, 184–85, 225–29, 258–62, 322, 338, 353–54; endogenous, 255–66; exogenous, 255, 257, 261–62; hazard identification, 226; and human health issues, 135–37, 140–42; management, 258–60; reduction, in environmental valuation, 223–24; Risk Assessment and Cost Benefit Act (proposed), 2, 12; and Science Advisory Board, 185; self-protection against, 184–85, 256–59, 261; valuation of, 260–62; Von Neumann-Morgenstern expected utility index (EU), 258; willing to pay versus human capital approach, 228–29; zero contaminant level, 137, 140

Risk Assessment and Cost Benefit Act (proposed), 2, 12

Safe Drinking Water Act (1974), 92, 94, 121

SARA (Superfund Amendments and Reauthorization Act), 52, 72

Science Advisory Board, 185

Scientific method: and environmental policy, 31

Sea Grant Institute, University of Wisconsin, 189, 191

Securities and Exchange Commission, 342n3

Sedimentation and erosion control: in North Carolina, 238, 241–45
Self-protection: against environmental risk, 256–59, 361
Senate Committee on Governmental Affairs, 54–55, 327
Smokestack emissions, 135
South Carolina environmental programs, 102
Standards, 50, 88–89, 239–40; air quality, 88, 211–12, 296; chemical, 131–32; drinking water, 50, 51, 92, 94
Stream-order effect on fish species composition, 163
Sulfur dioxide emissions, 206–7
Superfund Amendments and Reauthorization Act (SARA), 52, 72
Superfund program (EPA), 31, 52, 136, 300, 302
Supreme Court, 1; *Chadha* ruling, 343n4
Surface waters, 92, 153. *See also* Water quality
Sustainable development, 307

Task Force on Regulatory Relief, 328
TCA (Trichloroethane), 133
TCE (Trichloroethylene), 133, 136, 139
Tennessee environmental programs, 91, 167, 168
Tennessee Valley Authority, 167, 168
THM (Trihalomethane), 133
Total organic carbon (TOC), 130, 138
Toxic air emissions: state programs for, 91. *See also* Air quality
Toxic chemicals, 55, 131–33, 136–40, 206–7, 211–12, 263–64; degradation of, 139; natural sources of, 140; Toxic Release Inventory (TRI), 271, 283–84; Toxic Substances Control Act, 68, 203
Toxic Release Inventory (TRI), 271, 283–84
Toxic Substances Control Act, 68, 203
Trading pollution allowances, 206–7
Travel cost method (TCM), 221–22
TRI (Toxic Release Inventory), 271, 283–84
Trichloroethane (TCA), 133
Trichloroethylene (TCE), 133, 136, 139
Trihalomethane (THM), 133

Ultraviolet-visible spectroscopy, 131
U.S. Army Corps of Engineers: and benefit-cost analysis, 239
U.S. Fish and Wildlife Service: habitat evaluation procedures of, 179
U.S. Forest Service: below-cost timber sales, 312–13; habitat evaluation of, 179

U.S. Intergovernmental Task Force on Monitoring Water Quality (ITFMWQ), 181, 184
Utah environmental programs, 91

Vermont environmental programs, 149
Vinyl chloride: as degradation product, 139
VOCs (volatile organic compounds), 133, 211–12, 225
Volatile organic compounds (VOCs), 133, 211–12, 225
Von Neumann-Morgenstern expected utility index (EU), 258

Washington State environmental programs, 87, 100–101
Waste minimization, 63, 68–72, 74–75. *See also* Pollution
Water Pollution Control Act (WPCA), 151; Amendments (1972), 148–49, 151–52, 157
Water quality, 49, 50, 55–56, 78–79, 93–94, 100; biological and chemical indicators of, 169, 178–79; cumulative impact assessment of, 149; drinking water standards, 50, 51, 92, 94; EPA groundwater protection program, 78–79; fishable and swimmable waters, 92–93, 152; laws, 148–49, 150–53, 157, 178, 189; monitoring of, 178–79; nitrates and, 263–64; nonpoint pollution sources, 57n3, 93, 152, 157; and pesticide contamination, 55–56; point pollution sources, 93, 114–18, 152, 157; public water supply standards (PWSS), 92, 94; Safe Drinking Water Act, 92; state programs for, 78–79, 93–94, 238, 242, 245–51; types of water use, 150, 178; versus water rights, 151; wastewater treatment, 49, 56, 136; Wellhead Protection Program, 79. *See also* Clean Water Act; Pollution
Water Quality Act (1987), 153
Water Quality Improvement Act (1965), 178, 189
Water Resources Council (WRC), 239–40
Water Resources Research Institute (WRRI), 247–48
Water Supply Watershed Classification and Protection Act (WSWCPA) (North Carolina), 238, 245–51
Weber, Max, 23–24
Wellhead Protection Program, 79, 80
West Virginia environmental programs, 167
Willingness to pay (WTP): and willingness to

accept (WTA), 217–19, 220, 221, 228–29, 232nn13–14, 233n23
Wisconsin Department of Natural Resources (WDNR), 189–95
WPCA (Water Pollution Control Act), 151; Amendments (1972), 148–49, 151–52, 157

WRC (Water Resources Council), 239–40
WRRI (Water Resources Research Institute), 247–48

"Zero contaminant level," 137, 140